AF389208

Advances in Renewable Energy Research and Applications

Advances in Renewable Energy Research and Applications

Editors

Sharul Sham Dol
Anang Hudaya Muhamad Amin

Basel • Beijing • Wuhan • Barcelona • Belgrade • Novi Sad • Cluj • Manchester

Editors

Sharul Sham Dol
Abu Dhabi University
Abu Dhabi, United Arab
Emirates

Anang Hudaya Muhamad
Amin
Higher Colleges of
Technology
Abu Dhabi, United Arab
Emirates

Editorial Office
MDPI
St. Alban-Anlage 66
4052 Basel, Switzerland

This is a reprint of articles from the Special Issue published online in the open access journal *Energies* (ISSN 1996-1073) (available at: https://www.mdpi.com/journal/energies/special_issues/ Renewable_Energy_Research_Applications).

For citation purposes, cite each article independently as indicated on the article page online and as indicated below:

Lastname, A.A.; Lastname, B.B. Article Title. *Journal Name* **Year**, *Volume Number*, Page Range.

ISBN 978-3-03928-621-8 (Hbk)
ISBN 978-3-03928-622-5 (PDF)
doi.org/10.3390/books978-3-03928-622-5

Cover image courtesy of Sharul Sham Dol

Contents

About the Editors

Sharul Sham Dol

Dr Sharul Sham Dol is a Professor of Mechanical Engineering and Assistant Dean for Research and Innovation, Abu Dhabi University (ADU), United Arab Emirates (UAE). Dr Sharul Dol received his PhD from the University of Calgary, Canada, where he worked on Particle Image Velocimetry (PIV) investigation of pulsatile flow over a backward-facing step, funded by a Talisman Energy Inc. Canada scholarship awarded to outstanding PhD candidates whose research focuses on the advancement of environmentally, socially and economically responsible energy provision, as well as on the challenges and opportunities faced by the oil and gas industry.

Dr Dol's primary research interests are fundamentals of turbulent flows, unsteady flows, vortex dynamics, and instability problems, with applications relating to renewable energies, vehicle aerodynamics, aerospace propulsions, and flow assurances. He has conducted several government- and industry-funded projects related to unsteady turbulent flows, propeller aerodynamics, and turbomachinery. To date, Dr Dol has supervised 17 PhD/Master's students. He is the author of nine books/book chapters on turbulence and renewable energy technologies and has authored or co-authored over 110 scientific papers. Dr Dol's professional qualifications include the title of Chartered Engineer/Chartered Energy Engineer from Energy Institute UK.

Anang Hudaya Muhamad Amin

Dr Anang Hudaya Muhamad Amin is an Assistant Professor in the Computer and Information Science Department, Higher Colleges of Technology, Dubai Men's College. He received a BTech (Hons.) in Information Technology from Universiti Teknologi PETRONAS as well as a Master's and PhD in Network Computing from Monash University. His research interests include artificial intelligence with a specialization in distributed pattern recognition and bio-inspired computational intelligence, wireless sensor networks, cloud computing, and big data analytics.

Preface

This Reprint offers a comprehensive exploration of innovative approaches to engineering sustainability and renewable energy, providing a valuable resource for researchers, academics, and professionals in the field. With a focus on addressing the pressing challenges of sustainability and the need for environmentally friendly technological advancements, this Reprint delves into the core topics shaping the future of renewable energy and power generation.

This Reprint's main goal is to bridge the gap between engineering and innovative ideas by highlighting the interplay between sustainability and technological innovation. It seeks to inspire readers to develop and implement sustainable practices in their respective fields, ultimately contributing to a more environmentally conscious society. Covering a diverse range of topics, this book appeals to researchers, academics, and professionals in engineering, sustainability, environmental studies, and related disciplines. It serves as a valuable reference for those seeking to deepen their understanding of sustainable practices and harness the power of green hydrogen innovations.

Sharul Sham Dol and Anang Hudaya Muhamad Amin
Editors

Article

Investment and Production Strategies of Renewable Energy Power under the Quota and Green Power Certificate System

Min Song [1], **Yu Wang** [2,*] **and Yong Long** [1]

[1] School of Economics and Business Administration, Chongqing University, Chongqing 400030, China; msong@cqu.edu.cn (M.S.); longyongcqu@sina.com (Y.L.)

[2] School of International Business and Management, Sichuan International Studies University, Chongqing 400031, China

* Correspondence: yu.wang@cqu.edu.cn

Abstract: In order to study the impact of a renewable energy quota and green power certificate system on the strategies of energy suppliers, this paper constructs a multi-stage game model of renewable energy power investment and production from the renewable energy interest chain and its stakeholders. Through the calculation and solution of the model, the optimal renewable energy utilization level, pricing and production strategies of renewable energy power of energy suppliers are calculated under the scenarios of direct sale of power and purchase and sale by power grids. The results show that the quota ratio, green certificate price and investment cost are the key influencing factors of energy suppliers' strategies, and changes in the values of the three factors will completely change the renewable energy investment, pricing and production levels of energy suppliers in equilibrium. In addition, the study found that the impact of the renewable energy quota on renewable energy utilization levels of energy suppliers depends on the relative size of investment cost and green power certificate price. At the same time, it was also found that with a change in investment cost, green power certificate price and user preference, the market share and renewable energy utilization level of traditional energy suppliers and new energy suppliers also change.

Keywords: energy suppliers; renewable energy quota; green power certificate; renewable energy power; investment and production strategies

Citation: Song, M.; Wang, Y.; Long, Y. Investment and Production Strategies of Renewable Energy Power under the Quota and Green Power Certificate System. *Energies* **2022**, *15*, 4110. https://doi.org/10.3390/en15114110

Academic Editor: Muhammad Aziz

Received: 27 April 2022
Accepted: 30 May 2022
Published: 2 June 2022

Publisher's Note: MDPI stays neutral with regard to jurisdictional claims in published maps and institutional affiliations.

1. Introduction

Global carbon emissions are 40 billion tons of carbon dioxide every year, 86% of which comes from the use of fossil fuels [1]. Among the total carbon emissions, activities related to energy production and consumption account for the highest proportion of carbon emissions [2]. Therefore, promoting the utilization of renewable energy is the key to achieving green and low-carbon development and the goals of carbon peak and carbon neutrality. Renewable energy power, as an important form of the utilization and development of renewable energy, has attracted extensive attention. In order to promote green and low-carbon development and achieve the goals of carbon peak and carbon neutrality, many countries have formulated corresponding policies and measures to encourage and support the investment and production of renewable energy power [3,4]. However, in the early stage of renewable energy utilization, its development is relatively slow due to its low technological maturity, high cost, unstable returns and other characteristics, leading to serious wind and solar abandonment phenomena. In order to promote the utilization of renewable energy, in the post-subsidy era where subsidies and feed-in tariffs are reduced, governments have formulated and promulgated renewable energy quotas and green power certificate systems to promote the investment and production of renewable energy power. However, how renewable energy quotas and green power certificate systems affect the investment and production of renewable energy power, and how the energy suppliers conduct their investment and production of renewable energy power under renewable energy quotas and green power certificate systems, are facing many problems in practice. Therefore, the investment and production of renewable energy power has become an important topic in both academic and practical circles.

The existing research on renewable energy power is carried out from two angles. One part is carried out from the technical perspectives of power supply [5,6], energy storage [7,8], control [9], relay protection [10], transportation inspection and dispatching [11], and the other part is carried out from the economic management perspectives of policy [12,13], costs and returns [14,15], market mechanisms and configuration [16,17]. However, compared to other aspects of renewable energy power, there is little research on the investment and production of renewable energy power. The existing mechanisms and strategies are mainly about the investment and production of traditional power, and lack the investment and production mechanisms and strategies of renewable energy. Due to the differences in technology, market and interest chain stakeholders, these traditional mechanisms and strategies are not applicable to the investment and production of renewable energy. According to the socio-technical system theory, different technology systems need to be matched to economic management systems. The different technical systems of renewable energy power and thermal power generation lead to the different markets and industrial chains [18]. Therefore, it is necessary to match the mechanisms and strategies to promote the utilization and development of renewable energy. At the same time, the stakeholder theory believes that an organization involves many stakeholders, and each stakeholder has different interest requirements. In order to promote the development of the organization, it is necessary to balance the interests of all parties [19]. Unlike thermal power, the investment and production interest chains of renewable energy power involve many stakeholders, such as the government, traditional energy suppliers, new energy suppliers, large power grids and the users, each of whom has their own interest requirements. If the sum of interest of all stakeholders in the interest chain is not balanced, the quality and efficiency of renewable energy power investment and production will be seriously affected. Therefore, corresponding mechanisms and strategies are needed to meet and balance the interests of all parties in the interest chain as to promote the investment and production of renewable energy. At the same time, compared with the external diseconomy caused by the increase in carbon emissions in traditional thermal power generation, the green and clean production of renewable energy power generation is beneficial for reducing the carbon emissions of the system, which has a high environmental protection value, social value and external economy [20]. However, this kind of external economy has the characteristics of typical public goods and externality, which can lead to market failure, resulting in the power market not being able to allocate the investment and production resources effectively and being Pareto optimal. At the beginning of the utilization of renewable energy, governments formulated measures such as subsidies and feed-in tariffs to guide and optimize the allocation of resources. However, with the changes in technology and the market environment, subsidies and feed-in tariffs have been gradually reduced. In recent years, in the post-subsidy era, governments have formulated and promulgated new measures such as renewable energy quotas and green power certificate systems to guide and support the investment and production of renewable energy power. However, how renewable energy quotas and green power certificate systems affect the investment and production of renewable energy and how the energy suppliers invest and produce renewable energy power under renewable energy quotas and green power certificate systems greatly affect the utilization and development of renewable energy. Corresponding investment and production strategies are needed under renewable energy quotas and green power certificate systems to maximize returns and promote the utilization and development of renewable energy. In order to promote the utilization and development of renewable energy, it is necessary to study the investment and production of renewable energy power under renewable energy quotas and green power certificate systems, analyze relevant influencing factors and design appropriate mechanisms and strategies to promote the utilization and development of renewable energy.

Therefore, in order to study the investment and production of renewable energy power, this paper constructs the investment and production model of renewable energy power under a renewable energy quota and green power certificate system. From the perspectives of the interest chain and stakeholders of renewable energy power, this paper constructs the investment and production model of renewable energy under the scenarios of the direct sale of power in the market and the purchase and sale by large power grids. A multi-stage game model involving the government, traditional energy suppliers, new energy suppliers, large power grids and users is constructed. The optimal investment, pricing and production of renewable energy power under the scenarios of direct power sale and purchase and sale by power grids are solved and calculated; the influences of quota, investment cost and green power certificate price on the investment and production of renewable energy are analyzed; and the relationship and changes between the utilization level of renewable energy and the market share are also analyzed in this paper. On this basis, the paper further analyzes the influences of quota and green power certificate price on the investment, production and returns of traditional energy suppliers and new energy suppliers by using numerical analysis. The analysis results of the model are of great reference significance for the optimization of a renewable energy quota and green power certificate system, and for making the investment and production strategies of renewable energy.

The rest of the paper is organized as follows. Section 2 presents the relevant literature review. Section 3 presents the nomenclature, model description, model assumptions and model establishment. Section 4 analyzes the investment and production strategies of renewable energy power under different scenarios. Section 5 presents the numerical analysis of relevant factors. Section 6 presents the corresponding discussion. Section 7 draws conclusions.

2. Literature Review

In the early stage of research on renewable energy power investment and production, the scholars mainly started from the macro and overall perspective, on the basis of analyzing the value, status quota and efficiency of the utilization of renewable energy and discussing the approaches, modes and policies of the utilization of renewable energy, as to promote the investment and production of renewable energy power. For instance, Zhang et al. [21] analyzed the development background and legislation status of renewable energy, discussed the risks and problems faced by the development of renewable energy and then looked ahead at the prospects of the utilization of renewable energy. Parker [22] studied the economic benefits of renewable energy, and proposed that the development of renewable energy can reduce the energy cost and optimize the energy structure. Banos et al. [23] analyzed the disadvantages of the development of renewable energy and the shortcomings of renewable energy itself, and then put forward the methods to optimize the utilization of renewable energy to promote the development of renewable energy. Wang et al. [24] analyzed the impact of the utilization of renewable energy on climate change mitigation, and then put forward a development path for renewable energy. Chu et al. [25] developed an outlook on the security of renewable energy. Ellabban et al. [26] analyzed the utilization status and policy orientation of renewable energy, and then put forward a development trend and model of renewable energy.

With the development of renewable energy power investment and production, the scholars mainly started from the technical perspective, studying the power generation technology, power generation systems, energy storage systems, power generation predictions, grid-connection, dispatching, relay protection and other technologies of renewable energy, as to discuss the investment and production of renewable energy power. For example, Kunjana et al. [27] studied the impact of renewable energy power generation on the load of the power system, found that renewable energy power has a significant impact on the operation of the power system and then proposed an optimization strategy for the utilization of renewable energy. Mathiesen et al. [28] studied power transmission solutions for renewable energy, and proposed that renewable energy can be used more flexibly through the development of smart energy systems. Weitemeyer et al. [29] studied the energy storage methods of renewable energy, and calculated and optimized the best combination of energy storage, wind power and photovoltaic resources in the region. Andres et al. [30] studied the power generation of renewable energy and believed that it can promote energy diversification and improve the overall economic performance of the power sector. Wang et al. [31] studied the forecasting methods of renewable energy output, aiming at improving the application value of renewable energy through accurate predictions. Dominguez-Navarro et al. [32] studied the optimization methods of storage systems of renewable energy, and found that optimal cost-effectiveness could be obtained by mixing renewable energy and storage systems. Vergara et al. [33] studied the management mechanism of renewable energy systems, and proposed a complete energy management framework for grid-connected hybrid power plants.

In recent years, under the background of carbon peak and carbon neutralization goals and the green and low-carbon transformation of energy as the utilization patterns and technology of renewable energy have developed, scholars now mainly study the investment and production of renewable energy from economic and management perspectives, such as renewable energy digestion and absorption, power transaction, power pricing, investment, medium- and long-term trading and spot trading. For example, Hustveit et al. [34] studied and discussed the operation of a green power certificate system, and believed that a green power certificate system is conducive to supporting the utilization and development of renewable energy. Alizamir et al. [35] studied the impact of the feed-in price on the investment of renewable energy, and found that the feed-in price can ensure the stability of renewable energy investment returns. Sun et al. [36] studied the calculation methods and optimization model of the digestion and absorption of renewable energy according to the output characteristics of the power generation of renewable energy, as to promote the digestion and absorption of renewable energy. Yang et al. [37] studied the impact of government subsidies on the investment of renewable energy, and found that the impact of government subsidies on the investment of renewable energy increases significantly when energy consumption intensity is greater. Zhang et al. [38] studied the renewable energy trading system, and found that the subsidy expenditure of the government can be reduced by subsidizing renewable energy suppliers through market-oriented trading of renewable energy, as to reduce the financial burden of the government. Vlachos et al. [39] studied the pricing methods of renewable energy, and found that market-oriented methods such as green power certificate trading can make up for the cost of renewable energy and reduce the price of renewable energy; thus, renewable energy can be more easily promoted and adopted by the market. Brockway et al. [40] studied the returns on the investment of renewable energy, and found that the returns of renewable energy show a declining trend over time; therefore, appropriate institutions and market-oriented methods are necessary to promote the development of renewable energy. Blaszke et al. [41] used a case study to analyze the local spatial policies and renewable energy development of different municipalities in Poland, and found that the utilization degree and investment level of renewable energy are significantly affected by local spatial policies and development plans.

We follow the trend of research development, starting from the background of the reform of the power market, to study the investment and production of renewable energy power. It is worth mentioning that the existing studies mainly start from the single perspective of renewable energy, and there is a lack of studies from the perspective of investment and production interest chains of renewable energy. However, the investment and production process of renewable energy is the process of benefit sharing and transmission, and the interests in the renewable energy project development are constantly flowing and transmitted between the four interest levels of the government, investment and production, transaction and consumption, and the five stakeholders, i.e., the government, traditional energy suppliers, new energy suppliers, power grids and users; therefore, all parties in the interest chain have significant impacts on the investment and production of renewable energy. For the technical characteristics of renewable energy and the feature of multiple stakeholders, a single perspective is helpful for a deep understanding of corresponding mechanisms, but if the investment and production polices of renewable energy fails to reflect the interest requirements of all the stakeholders and to balance the interests of all parties, the investment and production efficiency and quality of renewable energy can be seriously affected. Therefore, starting from the investment and production interest chains of renewable energy power and the balance between the interests of all parties in the interest chain, we construct the investment and production model under the scenarios of direct sale of power and purchase and sale by power grids to study the investment and production problems of renewable energy, and thus to promote the utilization and development of renewable energy.

3. Model Description and Hypothesis

3.1. Model Description

From the perspective of the interest chain, the investment and production of renewable energy power can be divided into four levels: government, investment and production, transaction and consumption and stakeholders. The five stakeholders are the government, traditional energy suppliers, new energy suppliers, power grids and users. The interests in the investment and production of renewable energy power are constantly flowing and conducting among these four levels and five stakeholders, and there are different influences and correlative relations among various factors. Firstly, at the government level: In the case of the reduction of subsidies and feed-in tariffs, in order to improve the utilization rate of renewable energy, the government has established a corresponding renewable energy quota system to urge energy suppliers in a specific region to fulfill the corresponding renewable energy power production quota, r. At the same time, the government has established a green certificate trading system to enable all parties to trade their excess or unfinished renewable energy power in the market at the price of p_c; thus, the market-oriented way is used to compensate for the environmental protection and social value of renewable energy power and to promote the utilization of renewable energy. Secondly, at the investment and production level: Energy suppliers in the market can be divided into traditional energy suppliers and new energy suppliers. Traditional energy suppliers, t, such as China's five major power generation companies, and national and local energy investment groups, mainly focus on thermal power generation and provide non-renewable energy power. New energy suppliers, n, such as micro-grids, wind power companies, photovoltaic power companies, etc., enter the power market under the reformed power system and through state open market access, mainly producing and providing renewable energy power. Compared with new energy suppliers, traditional energy suppliers have abundant resources, sufficient capital and strong technology in the field of power production, and they are long-established enterprises in the power market. Due to their system and mechanism attributes, their market responses are slow. Compared to their own scale and volume, their new energy businesses have developed relatively slowly. On the contrary, although new energy suppliers are small in scale and volume compared to traditional energy suppliers, their market-oriented operation mechanism is more obvious. They can quickly respond to market demand, engage in the development of new energy projects and actively develop and utilize new energy. In order to compete and gain revenue in the power market, traditional energy suppliers and new energy suppliers decide their own renewable energy technology investments, l_t and l_n, under the government's renewable energy quota, and then decide their own renewable energy prices, p_t and p_n, and production quantities, q_t and q_n. Thirdly, at the market level: traditional energy suppliers and new energy suppliers exchange their own green power certificates according to their production of renewable energy power, and sell or buy green power at the green power certificate price, p_c, according to the completion of their quotas to balance the quota, obtain returns (quota exceeding) or avoid punishment (quota unfinished). Fourthly, at the consumption level: traditional energy suppliers and new energy suppliers provide power of q_t and q_n to users at the prices of p_t and p_c in the case of a direct power sale, and sell power to large power grids at the price of p in the case of the purchase and sale by large power grids according to l_t, l_n, q_t and q_n; then, the power grids distribute and sell power to users to complete consumption. The investment and production process of renewable energy power is shown in Figure 1.

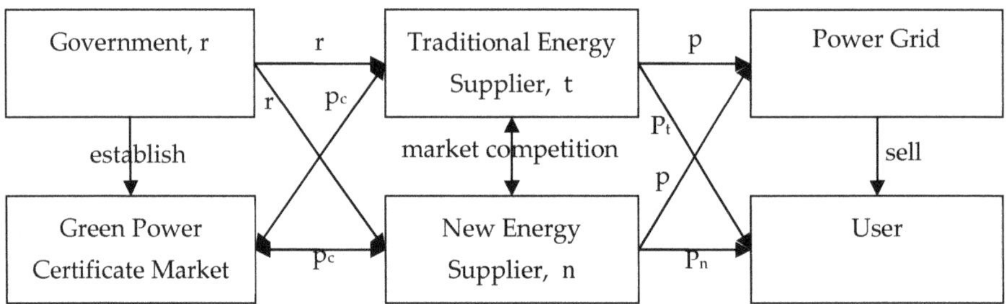

Figure 1. Renewable energy investment and production model.

In the investment and production interest chains of renewable energy, what is the optimal renewable energy investment level, production level and pricing for traditional and new energy suppliers? How do related factors such as renewable energy quotas, investment costs and green power certificate prices affect the investment and production of renewable energy? Meanwhile, what are the problems in the scenarios of direct sale and purchase and sale by power grids? In order to solve these problems, the following model is constructed in this paper.

3.2. Model Hypothesis and Establishment

Hypothesis 1. *In order to improve the utilization of renewable energy, the government has set the renewable energy quota, r; that is, the energy suppliers are required to complete the r proportion of renewable energy power, or to obtain green power certificates of r multiplying the production (1 kW·h of renewable energy power is equal to 1 green power certificate), otherwise the quota will be considered as unfinished. If the energy suppliers cannot meet the renewable energy quota by their own production, they can purchase green power certificates on the green power certificate market to make up their quota, otherwise they will face serious penalties, F.*

Hypothesis 2. *This paper assumes an oligopoly market where there are two kinds of suppliers, traditional energy suppliers, t, and new energy suppliers, n. Faced with a renewable energy quota and grid users' environmental preferences, how do the two kinds of suppliers develop their optimal investment and production strategies, including renewable energy investment, pricing and production? Traditional energy suppliers mainly producing thermal power usually cannot meet the quota, whereas new energy suppliers mainly producing renewable energy power often exceed the quota.*

Hypothesis 3. *Based on previous studies, the technology investment cost, c_i, of renewable energy utilization is set as a one-time expenditure, which can be expressed as:*

$$c_i(l_i) = kl_i^2 \tag{1}$$

where, $l_i (i = t, n)$ is the utilization level of renewable energy of energy suppliers, that is, the level of renewable energy power produced by energy suppliers after technology investment ($l_i \geq 0$); k represents the cost elasticity coefficient of the utilization level of renewable energy ($k > 0$).

Hypothesis 4. *Assume that the green power certificate price is exogenous and $p_c > 0$. The energy supplier's revenue in the green power certificate market is the number of green power certificate transactions multiplied by the green power certificate price. The number of green power certificate transactions of energy suppliers is a supplier's own renewable energy power production* *minus the renewable energy quota that needs to be fulfilled. Therefore, the return that energy suppliers $i(i = t, n)$ can obtain from the green power certificate transaction is:*

$$\pi_{c(i)} = (l_i q_i - r q_i) p_c = (l_i - r) q_i p_c \tag{2}$$

where q_i is the market demand of energy suppliers $i(i = t, n)$. The hidden assumption here is that $(l_i - r) q_i p_c \ll F$. If $F \leq (l_i - r) q_i p_c$, the penalty fee is too low, and the energy suppliers can fulfill the quota requirement only by paying a penalty fee that is relatively small compared to the returns; as a result, the quota and green power certificate system fails to achieve the effect of encouraging the energy suppliers to improve the utilization of renewable energy. Only when $(l_i - r) q_i p_c \ll F$ is the optimal choice for energy suppliers to participate in green power certificate trading to obtain green power certificates; thus, the renewable energy quota and green power certificate system can operate normally.

Hypothesis 5. *According to the definition of demand functions in economics, the demand functions of traditional and new energy suppliers can be composed of an initial market share part, price influence part and user preference influence part, which can be expressed as:*

$$\begin{cases} q_t = M_t - b p_t + s p_n + b_l l_t - s_l l_n \\ q_n = M_n - b p_n + s p_t + b_l l_n - s_l l_t \end{cases} \tag{3}$$

where (p_t, p_n) and (l_t, l_n) represent the pricing of energy suppliers and the utilization levels of renewable energy, respectively; M_t and M_n represent the market influences of energy suppliers at the present stage and $M_t + M_n \leq$ the total market scale; b and s represent the demand elasticity coefficient and substitution elasticity coefficient of price, respectively; and b_l and s_l represent the demand elasticity coefficient and substitution elasticity coefficient of the utilization level of renewable energy, respectively.

Hypothesis 6. *Summing up the above, the revenue functions of traditional and new energy suppliers are composed of sales returns, green power certificate returns and the renewable energy technology investment cost, which can be expressed as:*

$$\begin{cases} \pi_t = p_t q_t - c_t q_t + (l_t - r) q_t p_c - k l_t^2 \\ \pi_n = p_n q_n - c_n q_n + (l_n - r) q_n p_c - k l_n^2 \end{cases} \tag{4}$$

where c_t is the unit power cost.

4. Investment and Production Strategy Analysis of Renewable Energy Power under Different Scenarios

A renewable energy power system usually includes four links: power generation, power transmission and transformation, power distribution and power sale. Power generation is mainly made by energy suppliers, and large power grids are responsible for the purchase, transmission, distribution and sale of power. With the reform of power systems in recent years, the link between distribution and sale has been opened up, and market access has been granted to new suppliers. Energy suppliers and other relevant organizations or institutions are allowed to enter the links of generation, distribution and sale to directly produce and sell power, which leads to the coexistence of the direct power sale of power by energy suppliers and the purchase and sale of power by power grids in the current power market. However, the mode of direct power sale only accounts for a small part of the market, whereas the purchase and sale of power by large power grids accounts for the main part in the power distribution and sale market. Therefore, we analyze the investment and production of renewable energy power in the scenarios of direct power sale and purchase and sale by power grids.

4.1. Investment and Production Strategy Analysis of Renewable Energy Power under the Scenario of Direct Sale of Power

Under the background of the power system reform, the state has opened up market access and allowed energy suppliers to participate in power generation, distribution and sale; as a result, the direct sale of power appears. Generally, the scenario of direct sale of power includes the direct power supply for large enterprises, industrial parks and residential areas, as well as various types of micro-grid projects, energy supply projects of featured buildings, comprehensive energy projects and smart energy projects. In the scenario of direct sale of power, the decision-making process in regard to the investment and production of renewable energy power is shown below in Figure 2. Initially, the government formulates renewable energy quota, r. In the first stage of decision-making, traditional energy suppliers, t, and new energy suppliers, n, will decide their own renewable energy utilization levels, l_t and l_n, according to their own characteristics and planning. In the second stage of decision-making, traditional energy suppliers, t, and new energy suppliers, n, determine their own renewable energy power production, q_t and q_n, and prices, p_t and p_n, according to public information such as the levels of renewable energy utilization, l_t and l_n, in the market. Then, users purchase renewable energy power from energy suppliers at the prices of p_t and p_n, and the power grid also purchases the power of energy suppliers at a specific price p. In the third stage of decision-making, in the green power certificate trading market, traditional energy suppliers, t, and new energy suppliers, n, formulate their market trading strategies according to their own renewable energy power production, p_t and p_n, renewable energy quota, r, and green power certificate price, p_c.

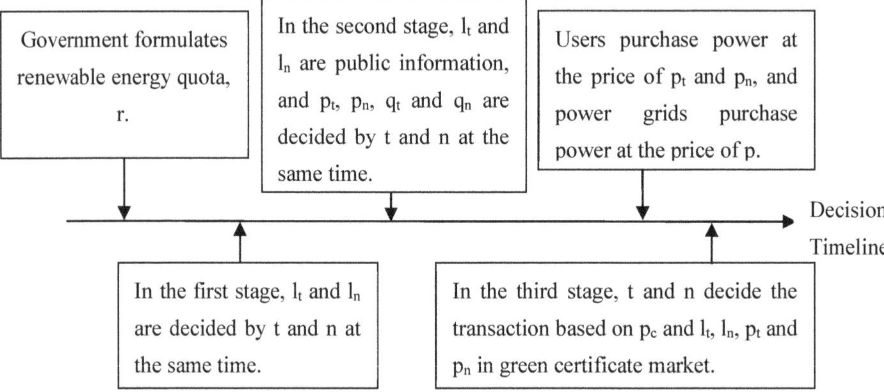

Figure 2. Timeline of renewable energy investment and production decisions.

According to the equilibrium condition, in the second stage, when traditional energy suppliers and new energy suppliers are in equilibrium in a direct power sale market, the first derivative, $\frac{\partial \pi_t}{\partial p_t}$ and $\frac{\partial \pi_n}{\partial p_n}$ are equal to 0. To solve the first derivative, and order $\frac{\partial \pi_t}{\partial p_t} = 0$ and $\frac{\partial \pi_n}{\partial p_n} = 0$, the value of p_t and p_n are obtained by solving the simultaneous equations as follows:

$$\begin{cases} p'_t = \dfrac{2bM_t + sM_n + \left(2bb_l - ss_l - 2b^2p_c\right)l_t - \left(2bs_l - b_l s + bsp_c\right)l_n + 2b^2c_t + bsc_n + (2b+s)brp_c}{4b^2 - s^2} \\ p'_n = \dfrac{2bM_n + sM_t + \left(2bb_l - ss_l - 2b^2p_c\right)l_n - \left(2bs_l - b_l s + bsp_c\right)l_t + 2b^2c_n + bsc_t + (2b+s)brp_c}{4b^2 - s^2} \end{cases} \tag{5}$$

Theorem 1. *In the scenario of direct sale of power, the power price of traditional energy suppliers and new energy suppliers will decrease with the increase in their own utilization level of renewable energy. Meanwhile, the power price of traditional energy suppliers and new energy suppliers will decrease with the increase in the utilization level of renewable energy of the other.*

Proof of Theorem 1.

Taking the derivative of the above equations of p_t and p_n with respect to l_t and l_n, the result is:

$$\frac{\partial p_t'}{\partial l_n} = \frac{\partial p_n'}{\partial l_t} = \frac{-2bs_l + b_l s - bsp_c}{4b^2 - s^2} < 0, \quad \frac{\partial p_t'}{\partial l_t} = \frac{\partial p_n'}{\partial l_n} = \frac{2bb_l - ss_l - 2b^2 p_c}{4b^2 - s^2} < 0 \quad (6)$$

Since the demand elasticity is usually larger than the substitution elasticity, the value of the above equation is less than 0. Thus, Theorem 1 is proved. □

When energy suppliers sell power directly in the market, at the beginning stage of the utilization of renewable energy the power price is usually high due to the immature technology, high risks and cost. It can be seen from Theorem 1 that with the improvement of the utilization level of renewable energy, the development of technology and the balance of the cost-benefit of energy suppliers, the power price will be reduced. At the same time, it can be seen from Theorem 1 that in order to gain market share, when an opponent increases its utilization level of renewable energy and lowers the price, one will be encouraged to also improve its utilization level and lower the price to a certain extent.

Then, substituting (p_t', p_n') into the demand function (3), a new demand function (q_t', q_n') can be obtained. Substituting the new price equation $\begin{cases} P_t'(l_t, l_n) \\ P_n'(l_t, l_n) \end{cases}$ and the new demand function equation $\begin{cases} q_t'(l_t, l_n) \\ q_n'(l_t, l_n) \end{cases}$ into the revenue function (4) of energy suppliers, a new revenue function $\begin{cases} \partial \pi_t(l_t, l_n) \\ \partial \pi_n(l_t, l_n) \end{cases}$ can be obtained. According to the equilibrium conditions, when the first stage is in equilibrium, the first derivative of the revenue function is equal to 0, that is, $\frac{\partial \pi_t}{\partial l_t} = \frac{\partial \pi_n}{\partial l_n} = 0$. Therefore, the renewable energy utilization levels of traditional energy suppliers and new energy suppliers in the first stage of equilibrium can be obtained, and the following theorem can be obtained:

Theorem 2. *In the scenario of direct sale of power, when traditional energy suppliers and new energy suppliers are in equilibrium, the optimal utilization level of renewable energy is (l_t^*, l_n^*).*

Proof of Theorem 2. See Appendix A. □

For the specific equations of l_t^* and l_n^*, see Appendix A Equations (A5) and (A6). l_t^* and l_n^* represent the optimal utilization levels of renewable energy in the scenarios of direct power sale by traditional energy suppliers and new energy suppliers, respectively, in the market, that is, the optimal investment levels. It can also be seen from Theorem 2 and the equations of l_t^* and l_n^* that traditional energy suppliers and new energy suppliers design and plan their own renewable energy investments to carry out market competition according to the characteristics of their own demand and the user demand. In the direct power sale market, traditional energy suppliers, with their abundant resources and strong technology, can usually obtain better projects to promote their investment in renewable energy, but their own system, efficiency and other factors negatively affect the utilization level of renewable energy. Meanwhile, the high degree of marketization and rapid market response of new energy suppliers are beneficial for promoting the improvement of the utilization level of renewable energy; however, due to their own resources and technology, the lack of resources such as for operation and maintenance in the later period may seriously affect user demand and negatively affect their level of renewable energy. According to Theorem 2, the utilization level of renewable energy of the energy suppliers is not the higher the better, but there is an optimal equilibrium point. The optimal utilization level is affected by factors such as market share, investment cost, production cost, green certificate transaction and competition, and energy suppliers should take these factors into consideration when making investment strategies around renewable energy to obtain the optimal investment level and returns.

Then, substituting the equilibrium results (l_t^*, l_n^*) of the first stage into the price equation $\begin{cases} p_t'(l_t, l_n) \\ p_n'(l_t, l_n) \end{cases}$, the demand function $\begin{cases} q_t'(l_t, l_n) \\ q_n'(l_t, l_n) \end{cases}$, the optimal pricing (p_t^*, p_n^*) and the optimal production (q_t^*, q_n^*) of the energy suppliers in the second stage can be obtained. The following theorem can be obtained:

Theorem 3. *In the scenario of direct sale of power, when the competition among traditional energy suppliers and new energy suppliers reaches a balance, the optimal pricing is (p_t^*, p_n^*) and the optimal production is (q_t^*, q_n^*).*

Proof of Theorem 3. See Appendix B. □

For the specific equations of (p_t^*, p_n^*) and (q_t^*, q_n^*), see Appendix B Equations (A7)–(A10). (p_t^*, p_n^*) and (q_t^*, q_n^*) represent the optimal pricing level and the optimal production level, respectively, in the scenario of direct sale of power by traditional energy suppliers and the new energy suppliers in the market. From Theorem 3 and the equations of (p_t^*, p_n^*) and (q_t^*, q_n^*), it can be seen that traditional energy suppliers and new energy suppliers set appropriate prices and production according to their own characteristics and the characteristics of competitors to compete in their own target market and provide power to gain profits. The optimal pricing and production of energy suppliers are affected by factors such as market share, investment cost, production cost, substitution rate and green power certificate price. Energy suppliers should integrate these factors to determine the optimal price and production to carry out competitions. At the same time, it can be seen from Theorem 3 and the above equations that the larger b_l is, the greater the user demand for renewable energy-related technologies. Therefore, the higher the utilization level of renewable energy of the energy suppliers, the more market demand and projects can be obtained, that is, the more renewable energy power production can be obtained. With the needs of low-carbon economic transformation and sustainable development, user demand for renewable energy will be higher and higher. Energy suppliers should improve their own utilization level of renewable energy and improve their own renewable energy power production while meeting the needs of users to obtain more benefits.

4.2. Investment and Production Strategy Analysis of Renewable Energy Power under the Scenario of Purchase and Sale by Power Grids

The traditional operation mode of the power market is that the large power grids purchase power from energy suppliers and then distribute and sell power to users. Although the reform of the power market allows relevant enterprises to enter the distribution and sale link, under the current system, it is mainly the large power grids that carry out the unified purchase, distribution and sale. As state-owned units, large power grids have strong institutional attributes and social responsibilities. In order to implement and respond to the country's green and low-carbon transformation and promote the adjustment of the energy structure, they emphasize the priority of guaranteeing the purchase of renewable energy power when purchasing power, and some regions even emphasize the full purchase of renewable energy power. Different from emphasizing the competition on price in the direct power sale market, the power purchase by power grids emphasizes the high demand for renewable energy power, which has important impacts on the investment and production of the renewable energy of energy suppliers.

In the scenario of purchase and sale by power grids, the power purchased is a single product and there is only a single purchaser; thus, the price is the same, that is, $p_t = p_n = p$. At this time, the demand function becomes:

$$\begin{cases} q_t = M_t - bp + sp + b_l l_t - s_l l_n \\ q_n = M_n - bp + sp + b_l l_n - s_l l_t \end{cases} \tag{7}$$

The revenue function becomes:

$$\begin{cases} \pi_t = pq_t - c_t q_t + (l_t - r)q_t P_c - kl_t^2 \\ \pi_n = pq_n - c_n q_n + (l_n - r)q_n P_c - kl_n^2 \end{cases} \tag{8}$$

Through the above analysis and hypothesis, it can be seen that in the scenario of purchase and sale by power grids, energy suppliers mainly decide the utilization level and production level of renewable energy. When the utilization level of renewable energy in the first stage is in equilibrium, the first-order conditions $\frac{\partial \pi_t}{\partial l_t} = 0$ and $\frac{\partial \pi_n}{\partial l_n} = 0$ are true, and when solving the revenue function with respect to the utilization levels of renewable energy l_t and l_n, the results are as follows:

$$pb_1 - c_t b_1 + (M_t - bp + sp)P_c + 2(b_1 P_c - k)l_t - s_1 P_c l_n - b_1 r P_c = 0 \tag{9}$$

$$pb_1 - c_n b_1 + (M_n - bp + sp)P_c + 2(b_1 P_c - k)l_n - s_1 P_c l_t - b_1 r P_c = 0 \tag{10}$$

Solving the above simultaneous equations can obtain:

Theorem 4. *In the scenario of purchase and sale by power grids, the optimal utilization levels of renewable energy of traditional energy suppliers and new energy suppliers are:*

$$l_t^* = \frac{\left[pb_1 - c_t b_1 + (M_t - bp + sp)P_c - b_1 r P_c\right] * 2(b_1 P_c - k) + \left[pb_1 - c_n b_1 + (M_n - bp + sp)P_c - b_1 r P_c\right] * s_1 P_c}{s_1^2 P_c^2 - 4(b_1 P_c - k)^2} \tag{11}$$

$$l_n^* = \frac{\left[pb_1 - c_t b_1 + (M_t - bp + sp)P_c - b_1 r P_c\right] * s_1 P_c + \left[pb_1 - c_n b_1 + (M_n - bp + sp)P_c - b_1 r P_c\right] * 2(b_1 P_c - k)}{s_1^2 P_c^2 - 4(b_1 P_c - k)^2} \tag{12}$$

The optimal production levels are:

$$\begin{aligned} q_t^* &= M_t - (b - s)p + \frac{\left[pb_1 - c_t b_1 + (M_t - bp + sp)P_c - b_1 r P_c\right] * \left[2(b_1 P_c - k) * b_1 - s_1^2 P_c\right]}{s_1^2 P_c^2 - 4(b_1 P_c - k)^2} \\ &+ \frac{\left[pb_1 - c_n b_1 + (M_n - bp + sp)P_c - b_1 r P_c\right] * (2k - b_1 s_1 P_c)}{s_1^2 P_c^2 - 4(b_1 P_c - k)^2} \end{aligned} \tag{13}$$

$$\begin{aligned} q_n^* &= M_n - (b - s)p + \frac{\left[pb_1 - c_n b_1 + (M_n - bp + sp)P_c - b_1 r P_c\right] * \left[2(b_1 P_c - k) * b_1 - s_1^2 P_c\right]}{s_1^2 P_c^2 - 4(b_1 P_c - k)^2} \\ &+ \frac{\left[pb_1 - c_t b_1 + (M_t - bp + sp)P_c - b_1 r P_c\right] * (2k - b_1 s_1 P_c)}{s_1^2 P_c^2 - 4(b_1 P_c - k)^2} \end{aligned} \tag{14}$$

Proof of Theorem 4.

The optimal utilization levels of renewable energy, l_t and l_n, can be obtained by combining $\frac{\partial \pi_t}{\partial l_t} = 0$ and $\frac{\partial \pi_n}{\partial l_n} = 0$, and q_t and q_n can be obtained by substituting l_t and l_n into the demand function. Theorem 4 is proved. $\square$

The optimal investment levels and production levels of renewable energy of traditional energy suppliers and new energy suppliers are represented by l_t and l_n, and q_t and q_n, respectively, under the scenario of purchase and sale by power grids. It can be seen from Theorem 4 that the optimal investment and production of renewable energy for energy suppliers are affected by the renewable energy quota, power price, cost, green power certificate price and other factors. In particular, the multiple occurrences of green power certificate price in the above equations reflects its significant influence. The possible reason behind this is that the utilization and investment of renewable energy are still at the beginning stage, thus it is difficult to balance the investment risks and returns. However, government subsidies for renewable energy have declined in recent years as the government has encouraged market-oriented approaches to balance the returns. In this case, in order to balance the returns of renewable energy, the green power certificate trading can well supplement the investment cost of renewable energy for energy suppliers, which is beneficial for balancing the benefits of energy suppliers and motivating energy suppliers to invest in renewable energy. Meanwhile, in practice, wind and solar abandonment phenomena are common in some regions, and some regions even have the extreme situation that the complete use of renewable energy leads to the shutdown of all thermal power plants. It can be seen from Theorem 4 that, in the case where the large grids give priority to the purchase of renewable energy power, energy suppliers will invest and produce a large amount of renewable energy in order to maximize profits. At the same time, more investment and production are not always better. There is an optimal investment boundary for the investment and production of renewable energy. Energy suppliers should make the optimal investment level and production level of renewable energy based on the comprehensive consideration of various factors such as the renewable energy quota and green power certificate price.

Proposition 1. *In the scenario of purchase and sale by power grids, when $\frac{(-2b_lp_c+2k-s_lp_c)b_lp_c}{s_l^2p_c^2-4(b_lp_c-k)^2}$ is less than 0, the energy suppliers will reduce the utilization level of renewable energy with the increase in the renewable energy quota; when $\frac{(-2b_lp_c+2k-s_lp_c)b_lp_c}{s_l^2p_c^2-4(b_lp_c-k)^2}$ is greater than 0, the energy suppliers will increase the utilization level of renewable energy with the increase in the renewable energy quota.*

Proof of Proposition 1.
Taking the derivative of renewable energy utilization levels, l_t and l_n, with respect to r, the result is:

$$\frac{\partial l_t}{\partial r} = \frac{\partial l_n}{\partial r} = \frac{-2(b_lp_c-k)b_lp_c - b_lp_c * s_lp_c}{s_l^2p_c^2 - 4(b_lp_c-k)^2} = \frac{(-2b_lp_c+2k-s_lp_c)b_lp_c}{s_l^2p_c^2 - 4(b_lp_c-k)^2} \tag{15}$$

According to the above equation, when $\frac{(-2b_lp_c+2k-s_lp_c)b_lp_c}{s_l^2p_c^2-4(b_lp_c-k)^2}$ is less than 0, quota r is negatively correlated with the utilization levels of renewable energy l_t and l_n; when $\frac{(-2b_lp_c+2k-s_lp_c)b_lp_c}{s_l^2p_c^2-4(b_lp_c-k)^2}$ is greater than 0, quota r is positively correlated with the utilization levels of renewable energy l_t and l_n. Proposition 1 is proved. $\square$

It can be seen from Proposition 1 that at the beginning of the utilization of renewable energy, the investment cost, k, is high, and the green power certificate price, p_c, is low. When $\frac{(-2b_lp_c+2k-s_lp_c)b_lp_c}{s_l^2p_c^2-4(b_lp_c-k)^2}$ is less than zero, the quota is negatively correlated with renewable energy; therefore, in order to promote the utilization of renewable energy, the government and relevant departments should appropriately control the renewable energy quota and green power certificate price to encourage energy suppliers to utilize renewable energy.

With the development of the utilization of renewable energy, the investment cost, k, decreases and the green power certificate price, p_c, increases. When $\frac{(-2b_lp_c+2k-s_lp_c)b_lp_c}{s_l^2p_c^2-4(b_lp_c-k)^2}$ is greater than 0, the quota, r, is positively correlated with the utilization levels of renewable energy, l_t and l_n; therefore, the government and relevant departments can improve the utilization and development of renewable energy by increasing the renewable energy quota and green power certificate price.

Proposition 2. *In the case that other parameters remain unchanged, with the change of the investment cost, green power certificate price and user preference, the relationship between market share difference and the utilization level difference of renewable energy is as follows:*
When $-(2b_lp_c - 2k + s_lp_c) < 0$, $(M_t - M_n)$ is negatively correlated with $(l_t - l_n)$;

When $-(2b_lp_c - 2k + s_lp_c) > 0$, $(M_t - M_n)$ *is positively correlated with* $(l_t - l_n)$;
When $-(2b_lp_c - 2k + s_lp_c) = 0$, $(M_t - M_n)$ *and* $(l_t - l_n)$ *have no direct correlation.*

Proof of Proposition 2.
Subtracting the above $\frac{\partial \pi_t}{\partial l_t} = 0$ and $\frac{\partial \pi_n}{\partial l_n} = 0$, the result is:

$$-c_tb_l + c_nb_l + (M_t - M_n)p_c + (2b_lp_c - 2k + s_lp_c)(l_t - l_n) = 0 \tag{16}$$

When energy suppliers use the same power generation mode, their costs are approximately equal, that is, $c_t \approx c_n$. In order to highlight the analysis of the utilization of renewable energy, the above equation can be simplified to:

$$(M_t - M_n)p_c = -(2b_lp_c - 2k + s_lp_c)(l_t - l_n) \tag{17}$$

According to the above equation, the relation between $(M_t - M_n)$ and $(l_t - l_n)$ changes as the value of $-(2b_lp_c - 2k + s_lp_c)$ changes. Proposition 2 is proved. $\square$

It can be seen from Proposition 2 that there may be an inconsistency between the utilization level of renewable energy of energy suppliers and their market share, The utilization level of energy suppliers is mainly affected by the cost of technology investment and green power certificate price. In practice, although the market share of new energy suppliers is smaller than that of traditional energy suppliers, the utilization level of renewable energy of new energy suppliers is relatively higher than that of traditional energy suppliers. In addition to the renewable energy technology and risks, the possible reason behind this is that the traditional energy suppliers enter the market early, possessing a quality power supply and generation projects, as well as stable returns, and thus lack the incentive to invest in renewable energy, whereas to compare with the relatively saturated thermal power market, new energy companies emerging from the renewable energy power market, as new entrants in the market, have to improve the utilization of renewable energy to gain profits. Given the current cost and returns of renewable energy, it may indeed be uneconomic to make decisions from the perspective of self-interest maximization, but from the perspective of social responsibility, large enterprises should take more responsibility. It is suggested that relevant government departments should adopt corresponding strategies to encourage large enterprises to improve their own utilization level of renewable energy.

On the basis of Proposition 2, Corollary 1 is obtained to elaborate the impacts of relevant factors on the utilization of renewable energy.

Corollary 1. *When* $(M_t - M_n)$ *is negatively correlated with* $(l_t - l_n)$, *the difference in the utilization level among energy suppliers in renewable energy,* $|l_t - l_n|$, *will decrease with the increase in users' preference for renewable energy,* b_l, *and green power certificate price,* p_c, *and will increase with the increase in the renewable energy investment cost coefficient,* k.
When $(M_t - M_n)$ *is positively correlated with* $(l_t - l_n)$, *the difference of the utilization level among energy suppliers in renewable energy,* $|l_t - l_n|$, *will increase with the increase in users' preference for renewable energy,* b_l, *and green power certificate price,* p_c, *and will decrease with the increase in the renewable energy investment cost coefficient,* k.

Proof of Corollary 1.
In order to explore the change of values, both sides of the equation of Proposition 1 are changed into absolute values, where the result is:

$$|M_t - M_n|p_c = |2b_lp_c - 2k + s_lp_c||l_t - l_n| \tag{18}$$

When $(M_t - M_n)$ is negatively correlated with $(l_t - l_n)$, there is the result of $-(2b_lp_c - 2k + s_lp_c) < 0$. Then, $|2b_lp_c - 2k + s_lp_c|$ will increase with the increase in b_l and p_c, whereas $|l_t - l_n|$ will decrease to ensure Equation (18) holds; $|2b_lp_c - 2k + s_lp_c|$

will decrease with the increase in k, whereas $|l_t - l_n|$ will increase. At this time, the effect of price p_c is less than the effect of the cost k and the $|M_t - M_n|$ market share difference changes in the opposite direction.

When $(M_t - M_n)$ is positively correlated with $(l_t - l_n)$, there is the result of $-(2b_1 p_c - 2k + s_1 p_c) > 0$. This means that $|2b_1 p_c - 2k + s_1 p_c|$ will decrease with the increase in b_1 and p_c, whereas $|l_t - l_n|$ will increase to ensure the Equation (18) holds; $|2b_1 p_c - 2k + s_1 p_c|$ will increase with the increase in k, whereas $|l_t - l_n|$ will decrease. At this time, the effect of price p_c is greater than the effect of the cost k and the $|M_t - M_n|$ market share difference changes in the same direction. Corollary 1 is proved. $\square$

According to Corollary 1, in the early stage of the utilization of renewable energy, when $(M_t - M_n)$ is negatively correlated with $(l_t - l_n)$, the impact of green power certificate price p_c is less than the impact of the technology investment cost k; that is, the impact of return acquisition is less than the impact of balancing the cost of development risks. In the early stage of the utilization of renewable energy, if more relevant policies are inclined to the utilization of renewable energy, traditional energy suppliers will also increase their investment in renewable energy, thus expanding the market share gap $(M_t - M_n)$. If green power certificate price p_c is high, the cost of purchasing green power certificates will increase, which will also encourage traditional energy suppliers to increase their investment in renewable energy, expand their market share and reduce their own cost. At this point, if the investment cost of renewable energy is low, traditional energy suppliers have more incentive to invest than new energy suppliers, thus expanding the market share gap $(M_t - M_n)$.

With the development of the utilization of renewable energy, when $(M_t - M_n)$ is positively correlated with $(l_t - l_n)$, the impact of green power certificate price p_c is greater than the impact of the technology investment cost k; that is, the impact of increasing returns is greater than the impact of balancing cost. With the development of renewable energy, if more relevant policies are inclined to the utilization of renewable energy, new energy suppliers will increase their investment in renewable energy, thus narrowing the market share gap $(M_t - M_n)$. If green power certificate price p_c is high, the return of selling green power certificates will increase, which will also encourage new suppliers to increase their investment in renewable energy, expand their market share and increase their own return. At this point, if the investment cost of renewable energy is low, new energy suppliers have more incentive to invest than traditional energy suppliers, thus narrowing the market share $(M_t - M_n)$.

5. Numerical Analysis

Assume that there are five stakeholders in the regional power market: the government, traditional energy suppliers, new energy suppliers, power grids and users. Under the scenario of direct sale of power, the initial market shares of traditional energy suppliers, t, and new energy suppliers, n, are $M_t = 20$ and $M_n = 10$, respectively; under the scenario of purchase and sale by large power grids, their initial market shares are $M_t = 35$ and $M_n = 10$, respectively, and their costs are $c_t = 3.5$ and $c_n = 4$, respectively. In addition, assume that the elasticity coefficient of cost, price demand, price substitution, preference demand and preference substitution of the utilization of renewable energy are $k = 2$, $b = 1$, $s = 2$, $b_1 = 2$ and $s_1 = 0.5$, respectively, that the initial quota, green power certificate price and purchase price of large power grids are $r = 1.5$, $p_c = 1$ and $p = 4$, respectively, that the value of the quota, r, changes between 0.1 and 0.9 and that the value of the green power certificate price changes between 0.1 and 0.5 in order to analyze the impact of the changes of the quota and green power certificate price. Numerical calculations are carried out based on the above formula and MATLAB, and the specific numerical calculation results are shown in the following charts and figures. Next, this paper will further analyze the impact of different factors on the investment and production of renewable energy.

5.1. Impact of Quota Changes on the Investment and Production of Renewable Energy

In order to further analyze the impact of quota changes on the investment and production of renewable energy, Table 1 was determined by using MATLAB based on the above formulas and initial values. As Table 1 shows, the utilization levels of renewable energy, l_t and l_n, increase with the increase in the quota, which shows that the quota has a positive role in promoting the utilization levels of renewable energy of energy suppliers. It can also be seen from Table 2 that although the renewable energy utilization level of traditional energy suppliers l_t has increased, it has not reached a high proportion. The possible reason behind this is that newly-added renewable energy is mainly an incremental part of the power market, and the newly-added market scale is relatively limited compared to the existing traditional power market scale; therefore, the proportion of renewable energy is relatively small. Unless the existing market scale is transformed and replaced by renewable energy, the development of renewable energy will be relatively limited. However, the transformation and replacement of the existing market scale will bring great economic efficiency loss and cost, and thus it is necessary to trade off comprehensively. The power prices, p_t and p_n, decrease with the increase in the quota. The possible reason behind this is that the increase in the quota results from the development of technology, which affects the decline in p_t and p_n. The power quantities, q_t and q_n, increase with the increase in the quota. The possible reason behind this is that energy suppliers will increase the investment in renewable energy to meet the quota, leading to production growth. With the increase in the quota, π_t rises first and then falls. The possible reason behind this is that at the beginning stage of the increase in the quota, traditional energy suppliers can accomplish certain quota requirements; therefore, they can still gain profits despite the fact that the price falls with the increase in production. However, when the quota increases to a certain extent, traditional energy suppliers cannot complete the quota requirements; as a result, the profits decrease under the penalties or the huge cost of buying green power certificates to make up for the quota. It can be seen that the quota has an optimal boundary for traditional energy suppliers. Therefore, in order to achieve the optimal interests of all parties, the government should adjust the quota according to the circumstances. With the increase in the quota, π_n keeps increasing and the speed keeps getting faster. The possible reason behind this is that the output increases with the increase in the quota, and the bigger the quota is, the more benefits new energy suppliers can obtain in the green power certificate market, which further improves their profits. Table 1 also indicates that the impact of the quota changes on the scenario of purchase and sale by large power grids is greater than that on the scenario of direct power sale. The possible reason behind this is that large power grids are generally state-owned units with strong system attributes, and quotas are the means of government regulation; therefore, compared with the market, they will be more positive at promoting the utilization of renewable energy in order to respond to and complete the government's requirements. In addition, it also shows that with the increase in the quota and renewable energy, the gap between the scale of direct power sale $q_t + q_n$ and the scale of purchase and sale by large power grids $q_t + q_n$ is also relatively narrowed, which is also the result of the adjustment of the energy structure and the reform of the power system. As can be seen from the above, the quota has a significant impact on the price, production and revenue of energy suppliers. From the perspective of energy suppliers, in order to obtain benefits and long-term development, energy suppliers should timely adjust their strategies according to the changes of government quota policies. From the perspective of the government, in order to improve the utilization of renewable energy, appropriate quota policies should be formulated according to the market conditions to promote the structural adjustment of the energy system and the green and low-carbon transition.

Table 1. Impact of quota on the investment and production of renewable energy.

The Scenario of Direct Sale of Power

r	l_t^*	l_n^*	p_t^*	p_n^*	q_t^*	q_n^*	π_t	π_n
0.1	0.091	0.562	5.483	5.983	23.235	12.885	52.622	24.422
0.3	0.243	0.653	5.335	5.714	24.496	13.553	53.523	25.083
0.5	0.346	0.744	5.218	5.522	25.153	14.205	54.441	25.770
0.7	0.424	0.835	5.111	5.335	26.285	14.795	53.586	26.445
0.9	0.487	0.926	5.033	5.274	26.515	15.126	52.787	27.265

The Scenario of Purchase and Sale by Large Power Grids

r	l_t^*	l_n^*	$p_t^* = p$	$p_n^* = p$	q_t^*	q_n^*	π_t	π_n
0.1	0.093	0.573	-	-	36.745	12.645	35.745	6.322
0.3	0.242	0.669	-	-	38.473	13.323	36.478	6.915
0.5	0.375	0.765	-	-	40.055	13.815	37.955	7.665
0.7	0.474	0.861	-	-	40.827	14.105	35.202	8.285
0.9	0.543	0.958	-	-	41.265	14.268	34.365	8.891

Table 2. Impact of green power certificate price on the investment and production of renewable energy.

The Scenario of Direct Sale of Power

P_c	l_t^*	l_n^*	p_t^*	p_n^*	q_t^*	q_n^*	π_t	π_n
0.1	0.085	0.533	5.459	5.783	22.915	13.125	43.118	24.266
0.2	0.113	0.609	5.295	5.595	23.838	14.031	44.462	25.167
0.3	0.151	0.685	5.177	5.444	24.585	15.184	45.633	26.017
0.4	0.212	0.761	5.086	5.358	25.072	16.203	44.897	26.978
0.5	0.276	0.839	5.024	5.289	25.328	17.545	43.951	27.863

The Scenario of Purchase and Sale by Large Power Grids

P_c	l_t^*	l_n^*	$p_t^* = p$	$p_n^* = p$	q_t^*	q_n^*	π_t	π_n
0.1	0.093	0.514	-	-	36.785	12.565	21.598	4.164
0.2	0.118	0.553	-	-	38.273	13.161	22.474	4.336
0.3	0.147	0.608	-	-	39.345	13.869	23.332	4.689
0.4	0.192	0.886	-	-	40.172	13.727	22.344	5.213
0.5	0.258	0.778	-	-	40.595	15.855	21.292	5.890

5.2. Impact of Green Power Certificate Price on the Investment and Production of Renewable Energy

In order to further analyze the impact of the green power certificate price on the investment and production of renewable energy, Table 2 was determined by using MATLAB based on the above formulas and the values of parameters. Table 2 shows that the utilization levels of renewable energy, l_t and l_n, increase with the increase in the green power certificate price, which shows that the transaction price of the green power certificate has a positive relationship with the level of renewable energy utilization. The renewable energy utilization level of traditional energy suppliers l_t increases slowly at first and then fast. The possible reason behind this is that when the green power certificate price is lower than the cost, compared with self-production, it is more economical for traditional energy suppliers to buy the green power certificate; when the price is higher than the cost, it is more economical to produce by themselves. The possible reason why l_n increases relatively fast is that the increase in the green power certificate price can bring more interests, thus promoting new energy suppliers to increase the investment and production of renewable energy. With the increase in the green power certificate price, p_t and p_n decrease, but p_n decreases more. The reason is that with the increase in the green power certificate price, new energy suppliers obtain more benefits from the green power certificate market to subsidize the cost and expense of their investment, which is conducive to their price reduction. With the increase in the green power certificate price, q_t and q_n increase, but q_n grows faster. The possible reason is that the benefits brought by the increase in the green power certificate price encourage new energy suppliers to increase their investment and production of renewable energy. With the increase in the green power certificate price, π_t rises first and then falls. The possible reason is that with the increase in the green power certificate price, traditional energy suppliers increase the production of renewable energy at the beginning, but when the green power certificate price reaches a certain level, in order to fulfill the quota requirements, they need to purchase a large number of green power certificates from the market, leading to the decline in benefits. With the increase in the green power certificate price, π_n keeps increasing because of the interests gained from the increasing production of renewable energy and from the green power certificate market. Table 2 also indicates that the impact of the green power certificate price on the scenario of direct power sale is greater than that on the scenario of purchase and sale by large power grids. The possible reason is that the green power certificate system, as a market-oriented system, directly affects the market behaviors of energy suppliers; thus, its impact is greater in the scenario of direct power sale. With the increase in the green power certificate price, the gap between the scale of direct power sale $q_t + q_n$ and the scale of purchase and sale by large power grids $q_t + q_n$ is also relatively narrowed. The reason is that the green power certificate, as an important means of power market reform, has a significant impact on the power market and is conducive to promoting the development of renewable energy utilization. As can be seen from the above, the green power certificate transaction has a significant impact on the renewable energy utilization level, price and production of energy suppliers. From the perspective of energy suppliers, in order to balance the costs and benefits of renewable energy project development, they can actively participate in the transaction of green power certificates. From the perspective of the government, the cost and risk of renewable energy utilization can be compensated in a market-oriented way through the construction and promotion of the green power certificate trading market that promotes the development of renewable energy utilization and reduces the subsidy burden of government renewable energy. By comparing Tables 1 and 2, it can be seen that compared with the green power certificate price, the impact of the quota on the investment and production of renewable energy is greater than that of the green power certificate. The possible reason is that the quota, as a required indicator of the government, has a greater impact on energy suppliers than the market-oriented characteristics of the green power certificate.

5.3. Impact of Technology Cost and User Preference on the Investment and Production of Renewable Energy

In order to further analyze the impact of the technology cost of renewable energy and user preference on the investment and production of renewable energy, Figures 3 and 4 were drawn based on the above formulas and values. As Figure 3 shows, l_t and l_n increase with the decline in the technology cost coefficient k, and when k drops to a certain degree, l_t and l_n increase faster. The possible reason is that at the initial stage, the technological level of renewable energy is low whereas the cost is high. The increase in the technological level and the decrease in the cost can improve the investment of energy suppliers in renewable energy to a certain extent. It can be seen from the above that the technology cost of renewable energy on the supply side directly affects the utilization level of renewable energy. Therefore, when the technological level of renewable energy develops to a certain stage and the technology cost decreases to a certain degree, the energy suppliers investing in renewable energy face less risk and cost, leading energy suppliers to further increase investment in the utilization and production of renewable energy. Therefore, it is necessary to consider the technological level and cost of renewable energy when formulating policies to promote renewable energy, otherwise the policy effect will not be satisfactory and will not be carried out smoothly, and it may bring profit losses to energy suppliers. Figure 4 shows that the utilization levels of renewable energy l_t and l_n increase with the increase in user preference. The l_t and l_n increase at an accelerated rate in the scenario of direct power sale while increasing at a decelerating rate in the scenario of purchase and sale by large power grids. The possible reason is that in the scenario of purchase and sale by large power grids, a unified purchase and sale can cause a large number of energy suppliers to make a centralized decision and cause a rapid response; therefore, rapid changes can be witnessed in the initial stage. Due to the relatively decentralized decision-making of energy suppliers in the scenario of direct power sale, the initial impact is smaller than that of the purchase and sale by large power grids. With the increase in user preferences, the scope and depth of the impact on energy suppliers will be deepened, which will accelerate the improvement of renewable energy utilization levels of energy suppliers in the scenario of direct power sale. It can be seen from the above that the user preference on the demand side has a positive impact on the utilization of renewable energy. Therefore, in the background of the power system reform, in order to improve the utilization level of renewable energy it is necessary not only to improve the technological level of renewable energy utilization and reduce the technology cost, but also to enhance the publicity and guidance according to the circumstances to cultivate the user preference for renewable energy and to promote the development of renewable energy.

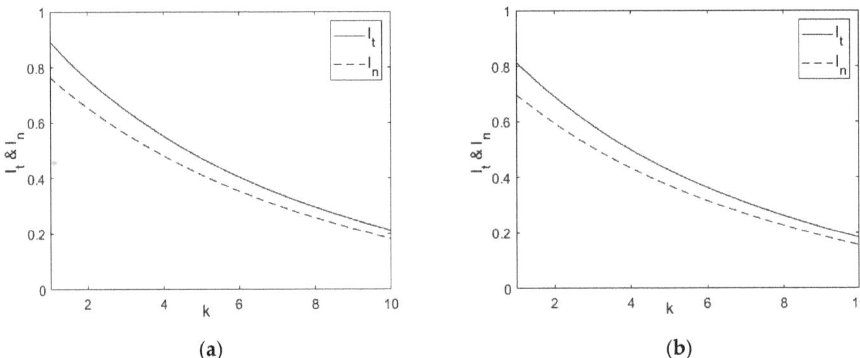

(a) **(b)**

Figure 3. Impact of technology cost elasticity coefficient on the investment and production of renewable energy: (**a**) the scenario of direct sale of power; (**b**) the scenario of purchase and sale by large power grids.

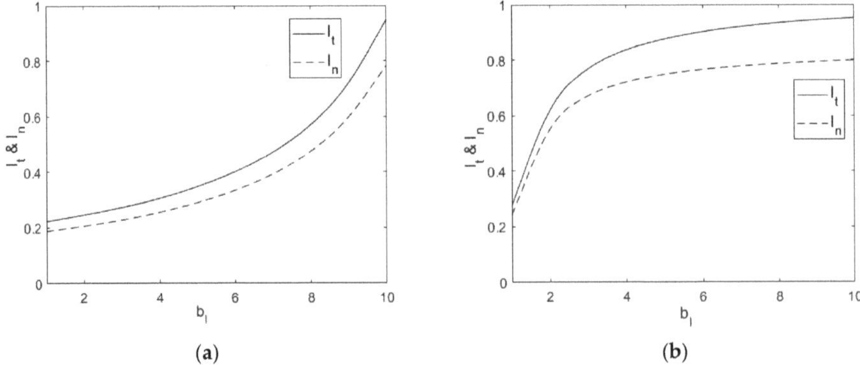

(a) **(b)**

Figure 4. Impact of user preference on the investment and production of renewable energy: (**a**) the scenario of direct sale of power; (**b**) the scenario of purchase and sale by large power grids.

6. Discussion

In order to study how the renewable energy quota and green power certificate system affect the behavior and strategies of energy suppliers, this paper was carried out by constructing a multi-stage game model. Through the calculation of the model, the market competition strategies of traditional energy suppliers and new energy suppliers in equilibrium were obtained. Through the derivation and analysis of the model, the impact of the renewable energy quota, investment cost, green power certificate price and user preference on renewable energy power investment and production of energy suppliers was analyzed. At the same time, numerical analysis was used to analyze the impact of the changes of the renewable energy quota, technology cost, green power certificate price and user preference on the price, production and revenue of energy suppliers.

Previous studies from the perspective of technology mainly focused on the investment and production technology of renewable energy power [27,28]. Previous studies from the perspective of the economy and management mainly focused on the impact of individual policies on renewable energy power investment and production [34–39]. Compared with previous studies focusing on technology and individual policies, this paper started from the renewable energy interest chain and stakeholders. It was conducive to analyze the behavior of stakeholders in the renewable energy interest chain and balance the benefits of all stakeholders. At the same time, the paper better analyzed the impact of the renewable energy quota and green power certificate system on the price, production, revenue and renewable energy utilization level of energy suppliers. It was not only conducive to explore the effects of policies and the behavior of energy suppliers, but also conducive to analyze the changes of the energy structure and energy market share.

The multi-stage game model based on the renewable energy interest chain and stakeholders was conducive to analyze the interaction and relationship among government policy, the green power certificate market and energy suppliers' strategies from an overall perspective. However, because the model involves multiple decision-making stages and multiple stakeholders, it was not conducive to a more detailed analysis of the behavior of a single stakeholder and the impact of a single policy. It can be seen that the in-depth and comprehensive research on individual stakeholders and individual policies is also very valuable.

7. Conclusions

This paper has discussed the investment and production of renewable energy under the quota and green power certificate system, and has drawn some interesting conclusions. Firstly, this paper proposed that the government, traditional energy suppliers, new energy suppliers, power grid companies and users are the main participants in the renewable energy interest chain, and analyzed their respective characteristics and interest requirements. Secondly, based on the demand function, revenue function and cost function of each party, this paper calculated and analyzed the renewable energy utilization level of traditional energy suppliers and new energy suppliers in equilibrium under the scenario of direct power sale and the scenario of purchase and sale by power grids, and calculated and analyzed the optimal pricing, optimal production and income of each party at this time. It was found that the quota, green power certificate price, technology cost coefficient and elasticity coefficients have an important impact on the utilization level, pricing, investment and production of renewable energy. Then, it was analyzed and proved that the utilization level of renewable energy is negatively correlated with pricing; the positive correlation between the quota and the utilization level of renewable energy is affected by relevant conditions; and the changes in the technology cost of renewable energy and green power certificate price affect the change in the relationship between the market share difference and the utilization level difference of renewable energy. It was found that in the initial stage of renewable energy utilization, compared with their market share and scale, traditional energy suppliers have a low utilization level of renewable energy and lack of enthusiasm. Therefore, it is suggested that the government should take corresponding measures to urge large enterprises to take more social responsibilities. In the later stages of renewable energy utilization, the decrease in the cost and the development of the green power certificate market will promote the investment and production of renewable energy to a higher equilibrium level. In the end, we further analyzed the effects of the quota, green power certificate price, technology cost and user preference on the investment and production of renewable energy through numerical analysis. It was found that the renewable energy quota plays a positive role in promoting the utilization of renewable energy. At the same time, the quota of renewable energy has an optimal boundary, which is affected by the scale of the incremental market and the traditional market. The transaction price of the green power certificate also has a positive impact on the utilization of renewable energy. There is also an optimal boundary of green power certificate price, which is affected by the power cost. Therefore, in order to promote the development of renewable energy utilization under the background of power market reform, it is necessary to formulate an appropriate quota and green power certificate trading scheme according to relevant factors. At the same time, it was also found that the reduction in technology cost has a direct impact on the improvement of the utilization level of renewable energy, and the rise of user preference has a positive impact on the utilization of renewable energy. Therefore, in order to promote the development of renewable energy, attention should be paid to improving the technological level of renewable energy and reducing the technology cost of renewable energy, and publicity and guidance should be enhanced to cultivate user preference for renewable energy. These findings provide important references for decisions and actions of all parties in the renewable energy interest chain.

Author Contributions: Conceptualization, M.S. and Y.W.; methodology, Y.L.; software, M.S.; validation, M.S., Y.W. and Y.L.; formal analysis, Y.W.; investigation, M.S.; resources, Y.W.; data curation, M.S.; writing—original draft preparation, Y.W.; writing—review and editing, M.S.; visualization, M.S.; supervision, M.S.; project administration, Y.W.; funding acquisition, Y.L. All authors have read and agreed to the published version of the manuscript.

Funding: This research was funded by The National Social Science Fund of China, grant number 17ZDA065.

Conflicts of Interest: The authors declare no conflict of interest.

Nomenclature

Abbreviations and acronyms

r	Renewable energy quota
F	Penalty fee for not fulfilling the quota
t	Traditional energy supplier
n	New energy supplier

Functions and variables

l_t	Utilization level of renewable energy of traditional energy supplier
l_n	Utilization level of renewable energy of new energy supplier
q_t	Market demand of traditional energy supplier (100 billion kW·h)
q_n	Market demand of new energy supplier (100 billion kW·h)
p_t	Price of traditional energy supplier (0.1 CNY/kW·h)
p_n	Price of new energy supplier (0.1 CNY/kW·h)
p	Purchase price of power grid (0.1 CNY/kW·h)
p_c	Green power certificate price (0.1 CNY/kW·h)
c_t	Cost of traditional energy supplier (0.1 CNY/kW·h)
c_n	Cost of new energy supplier (0.1 CNY/kW·h)
M_t	Market influences of traditional energy supplier (100 billion kW·h)
M_n	Market influences of new energy supplier (100 billion kW·h)
π_t	Revenue of traditional energy supplier (10 billion CNY)
π_n	Revenue of new energy supplier (10 billion CNY)

Parameters

k	Cost elasticity coefficient of the utilization level of renewable energy
b	Demand elasticity coefficient of price
s	Substitution elasticity coefficient of price
b_l	Demand elasticity coefficient of the utilization level of renewable energy
s_l	Substitution elasticity coefficient of the utilization level of renewable energy

Appendix A. Proof of Theorem 2

By substituting Equation (5), (p_t', p_n'), into the demand function (3), the new demand function is gained as follows:

$$\begin{cases} q_t' = \dfrac{2b^2M_t + bsM_n + b\left(2bb_l - ss_l + 2b^2p_c - s^2p_c\right)l_t + b\left(-2bs_l + b_l s - bsp_c\right)l_n - \left(2b^2 - s^2\right)bc_t + b^2scn - (b-s)(2b+s)brp_c}{4b^2 - s^2} \\ q_n' = \dfrac{2b^2M_n + bsM_t + b\left(2bb_l - ss_l + 2b^2p_c - s^2p_c\right)l_n + b\left(-2b_l + b_l s - bsp_c\right)l_t - \left(2b^2 - s^2\right)bc_n + b^2sc_t - (b-s)(2b+s)brp_c}{4b^2 - s^2} \end{cases} \tag{A1}$$

In order to solve the utilization level of renewable energy in the first stage where traditional energy suppliers and new energy suppliers are in equilibrium, substitute $\begin{cases} p_t'(l_t, l_n) \\ p_n'(l_t, l_n) \end{cases}$ and $\begin{cases} q_t'(l_t, l_n) \\ q_n'(l_t, l_n) \end{cases}$ into the revenue function (4) of energy suppliers to solve the optimal investment strategies. It can be obtained from the equilibrium condition that in the first stage of equilibrium, $\frac{\partial \pi_t}{\partial l_t} = \frac{\partial \pi_n}{\partial l_n} = 0$, thus the result can be gained by solving the revenue function:

$$\begin{cases} \dfrac{\partial \pi_t}{\partial l_t} = \dfrac{\partial p_t'}{\partial l_t}q_t' + p_t'\dfrac{\partial q_t'}{\partial l_t} - c_t\dfrac{\partial q_t'}{\partial l_t} + q_t'P_c + (l_t - r)P_c\dfrac{\partial q_t'}{\partial l_t} - 2kl_t = 0 \\ \dfrac{\partial \pi_n}{\partial l_n} = \dfrac{\partial p_n'}{\partial l_n}q_n' + p_n'\dfrac{\partial q_n'}{\partial l_n} - c_n\dfrac{\partial q_n'}{\partial l_n} + q_n'P_c + (l_n - r)P_c\dfrac{\partial q_n'}{\partial l_n} - 2kl_n = 0 \end{cases} \tag{A2}$$

Substituting $\begin{cases} p_t'(l_t, l_n) \\ p_n'(l_t, l_n) \end{cases}$ and $\begin{cases} q_t'(l_t, l_n) \\ q_n'(l_t, l_n) \end{cases}$ into the above equations, we can obtain $\frac{\partial \pi_t}{\partial l_t} = \frac{\partial \pi_n}{\partial l_n} = 0$, which can be respectively expressed as:

$$\begin{aligned} &\frac{2b^2M_t + bsM_n + b\left(2bb_l - ss_l + 2b^2p_c - s^2p_c\right)l_t + b\left(-2bs_l + b_l s - bsp_c\right)l_n - \left(2b^2 - s^2\right)bc_t + b^2scn - (b-s)(2b+s)brp_c}{4b^2 - s^2} \\ &\quad * 2\frac{2bb_l - ss_l + 2b^2p_c - s^2p_c}{4b^2 - s^2} - 2kl_t = 0 \end{aligned} \tag{A3}$$

$$\begin{aligned} &\frac{2b^2M_n + bsM_t + b\left(2bb_l - ss_l + 2b^2p_c - s^2p_c\right)l_n + b\left(-2bs_l + b_l s - bsp_c\right)l_t - \left(2b^2 - s^2\right)bc_n + b^2sc_t - (b-s)(2b+s)brp_c}{4b^2 - s^2} \\ &\quad * 2\frac{2bb_l - ss_l + 2b^2p_c - s^2p_c}{4b^2 - s^2} - 2kl_n = 0 \end{aligned} \tag{A4}$$

Solving the above equations, we can obtain:

$$\begin{aligned} l_t^* = &-\frac{\left[\left(2bb_l - ss_l + 2b^2p_c - s^2p_c\right)^2 * b - \left(4b^2 - s^2\right)k\right] * \left(2bb_l - ss_l + 2b^2p_c - s^2p_c\right) * \left[2b^2M_t + bsM_n - \left(2b^2 - s^2\right)bc_t + b^2scn - (b-s)(2b+s)brp_c\right]}{\left[\left(2bb_l - ss_l + 2b^2p_c - s^2p_c\right)^2 * b - \left(4b^2 - s^2\right)k\right]^2 - \left(2bb_l - ss_l + 2b^2p_c - s^2p_c\right)^2 * b^2 * \left(-2bs_l + b_l s - bsp_c\right)^2} \\ &+\frac{\left(2bb_l - ss_l + 2b^2p_c - s^2p_c\right)^2 * b\left(-2bs_l + b_l s - bsp_c\right) * \left[2b^2M_n + bsM_t - \left(2b^2 - s^2\right)bc_n + b^2sc_t - (b-s)(2b+s)brp_c\right]}{\left[\left(2bb_l - ss_l + 2b^2p_c - s^2p_c\right)^2 * b - \left(4b^2 - s^2\right)k\right]^2 - \left(2bb_l - ss_l + 2b^2p_c - s^2p_c\right)^2 * b^2 * \left(-2bs_l + b_l s - bsp_c\right)^2} \end{aligned} \tag{A5}$$

$$\begin{aligned} l_n^* = &-\frac{\left[\left(2bb_l - ss_l + 2b^2p_c - s^2p_c\right)^2 * b - \left(4b^2 - s^2\right)k\right] * \left(2bb_l - ss_l + 2b^2p_c - s^2p_c\right) * \left[2b^2M_n + bsM_t - \left(2b^2 - s^2\right)bc_n + b^2sc_t - (b-s)(2b+s)brp_c\right]}{\left[\left(2bb_l - ss_l + 2b^2p_c - s^2p_c\right)^2 * b - \left(4b^2 - s^2\right)k\right]^2 - \left(2bb_l - ss_l + 2b^2p_c - s^2p_c\right)^2 * b^2 * \left(-2bs_l + b_l s - bsp_c\right)^2} \\ &+\frac{\left(2bb_l - ss_l + 2b^2p_c - s^2p_c\right)^2 * b\left(-2bs_l + b_l s - bsp_c\right) * \left[2b^2M_t + bsM_n - \left(2b^2 - s^2\right)bc_t + b^2scn - (b-s)(2b+s)brp_c\right]}{\left[\left(2bb_l - ss_l + 2b^2p_c - s^2p_c\right)^2 * b - \left(4b^2 - s^2\right)k\right]^2 - \left(2bb_l - ss_l + 2b^2p_c - s^2p_c\right)^2 * b^2 * \left(-2bs_l + b_l s - bsp_c\right)^2} \end{aligned} \tag{A6}$$

The optimal renewable energy utilization levels l_t^* and l_n^* of traditional energy suppliers and new energy suppliers are obtained. Theorem 2 is proved.

Appendix B. Proof of Theorem 3

By substituting the optimal utilization level of renewable energy (l_t^*, l_n^*) of the energy suppliers in the first stage into the price Equation (5) and demand function (3), the equilibrium pricing (p_t^*, p_n^*) and production (q_t^*, q_n^*) of the energy suppliers in the second stage can be obtained as follows:

The optimal pricing is:

$$
\begin{aligned}
p_t^* =\ & \frac{2bM_t+sM_n+2b^2c_t+bsc_n+(2b+s)brp_c}{4b^2-s^2} + \frac{\left(2bb_l-ss_l+2b^2p_c-s^2p_c\right)*\left[2b^2M_t+bsM_n-\left(2b^2-s^2\right)bc_t+b^2sc_n-(b-s)(2b+s)brp_c\right]}{\left[\left(2bb_l-ss_l+2b^2p_c-s^2p_c\right)^2*b-\left(4b^2-s^2\right)k\right]^2-\left(2bb_l-ss_l+2b^2p_c-s^2p_c\right)^2*b^2*\left(-2bs_l+b_ls-bsp_c\right)^2}* \\
& \frac{\left(2bb_l-ss_l-2b^2p_c\right)*\left[-\left(2bb_l-ss_l+2b^2p_c-s^2p_c\right)^2*b+\left(4b^2-s^2\right)k\right]+\left(2bb_l-ss_l+2b^2p_c-s^2p_c\right)b*\left(-2bs_l+b_ls-bsp_c\right)^2}{4b^2-s^2} + \\
& \frac{\left(2bb_l-ss_l+2b^2p_c-s^2p_c\right)*\left(-2bs_l+b_ls-bsp_c\right)*\left[2b^2M_n+bsM_t-\left(2b^2-s^2\right)bc_n+b^2sc_t-(b-s)(2b+s)brp_c\right]*\left[k-\left(2bb_l-ss_l+2b^2p_c-s^2p_c\right)bp_c\right]}{\left[\left(2bb_l-ss_l+2b^2p_c-s^2p_c\right)^2*b-\left(4b^2-s^2\right)k\right]^2-\left(2bb_l-ss_l+2b^2p_c-s^2p_c\right)^2*b^2*\left(-2bs_l+b_ls-bsp_c\right)^2}.
\end{aligned}
\tag{A7}
$$

$$
\begin{aligned}
p_n^* =\ & \frac{2bM_n+sM_t+2b^2c_n+bsc_t+(2b+s)brp_c}{4b^2-s^2} + \frac{\left(2bb_l-ss_l+2b^2p_c-s^2p_c\right)*\left[2b^2M_n+bsM_t-\left(2b^2-s^2\right)bc_n+b^2sc_t-(b-s)(2b+s)brp_c\right]}{\left[\left(2bb_l-ss_l+2b^2p_c-s^2p_c\right)^2*b-\left(4b^2-s^2\right)k\right]^2-\left(2bb_l-ss_l+2b^2p_c-s^2p_c\right)^2*b^2*\left(-2bs_l+b_ls-bsp_c\right)^2}* \\
& \frac{\left(2bb_l-ss_l-2b^2p_c\right)*\left[-\left(2bb_l-ss_l+2b^2p_c-s^2p_c\right)^2*b+\left(4b^2-s^2\right)k\right]+\left(2bb_l-ss_l+2b^2p_c-s^2p_c\right)b*\left(-2bs_l+b_ls-bsp_c\right)^2}{4b^2-s^2} + \\
& \frac{\left(2bb_l-ss_l+2b^2p_c-s^2p_c\right)*\left(-2bs_l+b_ls-bsp_c\right)*\left[2b^2M_t+bsM_n-\left(2b^2-s^2\right)bc_t+b^2sc_n-(b-s)(2b+s)brp_c\right]*\left[k-\left(2bb_l-ss_l+2b^2p_c-s^2p_c\right)*bp_c\right]}{\left[\left(2bb_l-ss_l+2b^2p_c-s^2p_c\right)^2*b-\left(4b^2-s^2\right)k\right]^2-\left(2bb_l-ss_l+2b^2p_c-s^2p_c\right)^2*b^2*\left(-2bs_l+b_ls-bsp_c\right)^2}.
\end{aligned}
\tag{A8}
$$

The optimal production is:

$$
\begin{aligned}
q_t^* =\ & \frac{2b^2M_t+bsM_n-\left(2b^2-s^2\right)bc_t+b^2sc_n-(b-s)(2b+s)brp_c}{4b^2-s^2} + \\
& \frac{\left(2bb_l-ss_l+2b^2p_c-s^2p_c\right)^2*b*\left[2b^2M_t+bsM_n-\left(2b^2-s^2\right)bc_t+b^2sc_n-(b-s)(2b+s)brp_c\right]}{\left[\left(2bb_l-ss_l+2b^2p_c-s^2p_c\right)^2*b-\left(4b^2-s^2\right)k\right]^2-\left(2bb_l-ss_l+2b^2p_c-s^2p_c\right)^2*b^2*\left(-2bs_l+b_ls-bsp_c\right)^2}* \\
& \frac{-\left(2bb_l-ss_l+2b^2p_c-s^2p_c\right)^2b+\left(4b^2-s^2\right)k+\left(-2bs_l+b_ls-bsp_c\right)^2b}{4b^2-s^2} + \\
& \frac{\left(2bb_l-ss_l+2b^2p_c-s^2p_c\right)b*\left(-2bs_l+b_ls-bsp_c\right)*\left[2b^2M_n+bsM_t-\left(2b^2-s^2\right)bc_n+b^2sc_t-(b-s)(2b+s)brp_c\right]*k}{\left[\left(2bb_l-ss_l+2b^2p_c-s^2p_c\right)^2*b-\left(4b^2-s^2\right)k\right]^2-\left(2bb_l-ss_l+2b^2p_c-s^2p_c\right)^2*b^2*\left(-2bs_l+b_ls-bsp_c\right)^2}.
\end{aligned}
\tag{A9}
$$

$$
\begin{aligned}
q_n^* =\ & \frac{2b^2M_n+bsM_t-\left(2b^2-s^2\right)bc_n+b^2sc_t-(b-s)(2b+s)brp_c}{4b^2-s^2} + \\
& \frac{\left(2bb_l-ss_l+2b^2p_c-s^2p_c\right)^2b*\left[2b^2M_n+bsM_t-\left(2b^2-s^2\right)bc_n+b^2sc_t-(b-s)(2b+s)brp_c\right]}{\left[\left(2bb_l-ss_l+2b^2p_c-s^2p_c\right)^2*b-\left(4b^2-s^2\right)k\right]^2-\left(2bb_l-ss_l+2b^2p_c-s^2p_c\right)^2*b^2*\left(-2bs_l+b_ls-bsp_c\right)^2}* \\
& \frac{-\left(2bb_l-ss_l+2b^2p_c-s^2p_c\right)^2b+\left(4b^2-s^2\right)k+\left(-2bs_l+b_ls-bsp_c\right)^2b}{4b^2-s^2} + \\
& \frac{\left(2bb_l-ss_l+2b^2p_c-s^2p_c\right)*b\left(-2bs_l+b_ls-bsp_c\right)*\left[2b^2M_t+bsM_n-\left(2b^2-s^2\right)bc_t+b^2sc_n-(b-s)(2b+s)brp_c\right]*k}{\left[\left(2bb_l-ss_l+2b^2p_c-s^2p_c\right)^2*b-\left(4b^2-s^2\right)k\right]^2-\left(2bb_l-ss_l+2b^2p_c-s^2p_c\right)^2*b^2*\left(-2bs_l+b_ls-bsp_c\right)^2}.
\end{aligned}
\tag{A10}
$$

The optimal pricing (p_t^*, p_n^*) and optimal production (q_t^*, q_n^*) of traditional energy suppliers and new energy suppliers are obtained. Theorem 3 is proved.

References

1. Ding, Z.L. Research on China's "Carbon Neutral" Framework Route. Available online: https://wenhui.whb.cn/third/baidu/202105/30/407093.html (accessed on 30 May 2021).
2. National Development and Reform Commission of China; National Energy Board of China. 14th Five-Year Modern Energy System Planning. Available online: http://zfxxgk.nea.gov.cn/2022-01/29/c_1310524241.htm (accessed on 22 March 2022).
3. National Development and Reform Commission of China; National Energy Board of China. Opinions on Improving the Institutional Mechanism and Policy Measures for Energy Green and Low-Carbon Transformation. Available online: http://zfxxgk.nea.gov.cn/2022-01/30/c_1310464313.htm (accessed on 10 February 2022).
4. European Parliament and Council. The Promotion of the Use of Energy from Renewable Sources. Available online: https://eur-lex.europa.eu/eli/dir/2018/2001/oj (accessed on 11 December 2018).
5. Sgouridis, S.; Carbajales, M.; Csala, D. Comparative net energy analysis of renewable electricity and carbon capture and storage. *Nat. Energy* **2019**, *4*, 456–465. [CrossRef]
6. Hansen, K.; Breyer, C.; Lund, H. Status and perspectives on 100% renewable energy systems. *Energy* **2019**, *175*, 471–480. [CrossRef]

7. Nagaraju, G.; Sekhar, S.C.; Ramulu, B. An Integrated Approach Toward Renewable Energy Storage Using Rechargeable Hybrid Supercapacitors. *Small* **2019**, *15*, 1805418. [CrossRef] [PubMed]
8. Talluri, G.; Lozito, G.M.; Grasso, F.; Iturrino Garcia, C.; Luchetta, A. Optimal Battery Energy Storage System Scheduling within Renewable Energy Communities. *Energies* **2021**, *14*, 8480. [CrossRef]
9. Mehrasa, M.; Pouresmaeil, E.; Sepehr, A. Control technique for the operation of grid-tied converters with high penetration of renewable energy resources. *Electr. Power Syst. Res.* **2019**, *166*, 18–28. [CrossRef]
10. Tang, Z.G.; Yang, Y.H.; Blaabjerg, F. An Inter linking Converter for Renewable Energy Integration into Hybrid Grids. *IEEE. T. Power. Electr.* **2021**, *36*, 2499–2504. [CrossRef]
11. Li, L.L.; Wen, S.Y.; Tseng, M.L. Renewable energy prediction: A novel short-term prediction model of photovoltaic output power. *J. Clean. Prod.* **2019**, *228*, 359–375. [CrossRef]
12. Lund, P.D. Effects of energy policies on industry expansion in renewable energy. *Renew. Energy* **2008**, *34*, 53–64. [CrossRef]
13. Nowak, M.J.; James, V.U.; Golubchikov, O. The Role of Spatial Policy Tools in Renewable Energy Investment. *Energies* **2022**, *15*, 2393. [CrossRef]
14. De, F.R.A.; Vogel, E.P.; Korzenowski, A.L.; Oliveira, R.L.A. Stochastic model to aid decision making on investments in renewable energy generation: Portfolio diffusion and investor risk aversion. *Renew. Energy* **2020**, *162*, 1161–1176.
15. Milone, D.; Curto, D.; Franzitta, V.; Guercio, A.; Cirrincione, M.; Mohammadi, A. An Economic Approach to Size of a Renewable Energy Mix in Small Islands. *Energies* **2022**, *15*, 2005. [CrossRef]
16. Elkadeem, M.R.; Wang, S.R.; Sharshir, S.W. Feasibility analysis and techno-economic design of grid-isolated hybrid renewable energy system for electrification of agriculture and irrigation area: A case study in Dongola, Sudan. *Energy Convers. Manag.* **2019**, *196*, 1453–1478. [CrossRef]
17. Lund, P.D.; Lindgren, J.; Mikkola, J. Review of energy system flexibility measures to enable high levels of variable renewable electricity. *Renew. Sustain. Energy Rev.* **2015**, *45*, 785–807. [CrossRef]
18. Long, Y.; Wang, Y.; Pan, C. Auction Mechanism of Micro-Grid Project Transfer. *Sustainability* **2017**, *9*, 1895. [CrossRef]
19. Long, Y.; Wang, Y.; Pan, C. Incentive Mechanism of Micro-grid Project Development. *Sustainability* **2018**, *10*, 163. [CrossRef]
20. Long, Y.; Pan, C.; Wang, Y. Research on a Microgrid Subsidy Strategy Based on Operational Efficiency of the Industry Chain. *Sustainability* **2018**, *10*, 1519. [CrossRef]
21. Zhang, X.L.; Wang, R.S.; Huo, M.L. A study of the role played by renewable energies in China's sustainable energy supply. *Energy* **2009**, *35*, 4392–4399. [CrossRef]
22. Parker, T. Sustainable energy: Cutting science's electricity bill. *Nature* **2011**, *480*, 315–316. [CrossRef]
23. Banos, R.; Manzano-Agugliaro, F.; Montoya, F.G. Optimization methods applied to renewable and sustainable energy: A review. *Renew. Sustain. Energy Rev.* **2011**, *15*, 1753–1766. [CrossRef]
24. Wang, J.H.; Conejo, A.J.; Wang, C.S.; Yan, J.Y. Smart grids, renewable energy integration, and climate change mitigation—Future electric energy systems. *Appl. Energy* **2012**, *96*, 1–3. [CrossRef]
25. Chu, S.; Majumdar, A. Opportunities and challenges for a sustainable energy future. *Nature* **2012**, *488*, 294–303. [CrossRef]
26. Ellabban, O.; Abu-Rub, H.; Blaabjerg, F. Renewable energy resources: Current status, future prospects and their enabling technology. *Renew. Sustain. Energy Rev.* **2014**, *39*, 748–764. [CrossRef]
27. Kunjana, C.; Somboon, N. Impact assessment of renewable generation on electricity demand characteristics. *Renew. Sustain. Energy Rev.* **2014**, *39*, 995–1004.
28. Mathiesen, B.V.; Lund, H.; Connolly, D. Smart Energy Systems for coherent 100% renewable energy and transport solutions. *Appl. Energy* **2015**, *145*, 139–154. [CrossRef]
29. Weitemeyer, S.; Kleinhans, D.; Vogt, T. Integration of Renewable Energy Sources in future power systems: The role of storage. *Renew. Energy* **2015**, *75*, 14–20. [CrossRef]
30. Andrés, I.; Rodrigo, M.; Alejandro, B.; Hugh, R. CVaR constrained planning of renewable generation with consideration of system inertial response, reserve services and demand participation. *Energy Econ.* **2016**, *59*, 104–117.
31. Wang, H.Z.; Lei, Z.X.; Zhang, X. A review of deep learning for renewable energy forecasting. *Energy Convers. Manag.* **2019**, *198*, 111799. [CrossRef]
32. Dominguez-Navarro, J.A.; Dufo-Lopez, R.; Yusta-Loyo, J.M. Design of an electric vehicle fast-charging station with integration of renewable energy and storage systems. *Int. J. Elec. Power* **2019**, *105*, 46–58. [CrossRef]
33. Vergara-Dietrich, J.D.; Morato, M.M.; Mendes, P.R.C. Advanced chance-constrained predictive control for the efficient energy management of renewable power systems. *J. Process. Contr.* **2019**, *74*, 120–132. [CrossRef]
34. Hustveit, M.; Forgner, J.S.; Fleten, S.E. Tradable green certificates for renewable support: The role of expectations and uncertainty. *Energy* **2017**, *141*, 1717–1727. [CrossRef]
35. Alizamir, S.; de Véricourt, F.; Sun, P. Efficient feed-in-tariff policies for renewable energy technologies. *Rairo-Oper. Res.* **2016**, *64*, 52–66. [CrossRef]
36. Sun, J.; Li, M.; Zhang, Z.; Xu, T.; He, J.; Wang, H.; Li, G. Renewable Energy Transmission by HVDC Across the Continent: System Challenges and Opportunities. *CSEE. J. Power Energy* **2017**, *3*, 353–364. [CrossRef]
37. Yang, X.L.; He, L.Y.; Xia, Y.F.; Chen, Y.F. Effect of government subsidies on renewable energy investments: The threshold effect. *Energy Policy* **2019**, *132*, 156–166. [CrossRef]

38. Zhang, Q.; Wang, G.; Li, Y. Substitution effect of renewable portfolio standards and renewable energy certificate trading for feed-in tariff. *Appl. Energy* **2018**, *227*, 426–435. [CrossRef]
39. Vlachos, A.G.; Biskas, P.N. Embedding renewable energy pricing policies in day-ahead electricity market clearing. *Electr. Power Syst. Res.* **2014**, *116*, 311–321. [CrossRef]
40. Brockway, P.E.; Owen, A.; Brand-Correa, L.I. Estimation of global final-stage energy-return-on-investment for fossil fuels with comparison to renewable energy sources. *Nat. Energy* **2019**, *4*, 612–621. [CrossRef]
41. Blaszke, M.; Nowak, M.; Śleszyński, P.; Mickiewicz, B. Investments in Renewable Energy Sources in the Concepts of Local Spatial Policy: The Case of Poland. *Energies* **2021**, *14*, 7902. [CrossRef]

 *energies*

Article

Transition to Low-Carbon Hydrogen Energy System in the UAE: Sector Efficiency and Hydrogen Energy Production Efficiency Analysis

Mustapha D. Ibrahim *, Fatima A. S. Binofai and Maha O. A. Mohamad

Industrial Engineering Technology, Higher Colleges of Technology, Sharjah 7947, United Arab Emirates
* Correspondence: mibrahim1@hct.ac.ae

Abstract: To provide an effective energy transition, hydrogen is required to decarbonize the hard-to-abate industries. As a case study, this paper provides a holistic view of the hydrogen energy transition in the United Arab Emirates (UAE). By utilizing the directional distance function undesirable data envelopment analysis model, the energy, economic, and environmental efficiency of UAE sectors are estimated from 2001 to 2020 to prioritize hydrogen sector coupling. Green hydrogen production efficiency is analyzed from 2020 to 2050. The UAE should prioritize the industry and transportation sectors, with average efficiency scores of 0.7 and 0.74. The decomposition of efficiency into pure technical efficiency and scale efficiency suggests policies and strategies should target upscaling the UAE's low-carbon hydrogen production capacity to expedite short-term and overall production efficiency. The findings of this study can guide strategies and policies for the UAE's low-carbon hydrogen transition. A framework is developed based on the findings of the study.

Keywords: hydrogen; sector coupling; efficiency; data envelopment analysis; UAE

Citation: Ibrahim, M.D.; Binofai, F.A.S.; Mohamad, M.O.A. Transition to Low-Carbon Hydrogen Energy System in the UAE: Sector Efficiency and Hydrogen Energy Production Efficiency Analysis. *Energies* **2022**, *15*, 6663. https://doi.org/10.3390/en15186663

Academic Editors: Nuno Carlos Leitão and Peter V. Schaeffer

Received: 8 August 2022
Accepted: 8 September 2022
Published: 13 September 2022

Publisher's Note: MDPI stays neutral with regard to jurisdictional claims in published maps and institutional affiliations.

1. Introduction

The United Arab Emirates (UAE) has set clear climate goals. This includes a significant reduction in CO_2 emissions by 2030 and net-zero emissions by 2050 [1]. The UAE net-zero 2050 strategic initiative aligns with the Paris Agreement that calls for long-term strategies to reduce greenhouse gas (GHG) emissions and limit the global temperature to 1.5C compared to pre-industrial levels [2]. This goal means a significant transition for the UAE's energy, economic, and environmental system. A large share of the country's energy comes from natural gas and oil to produce electricity or industrial processes that require a vast amount of heat. The share of gas in power generation is 96.7%, while the share of low carbon sources is 2.7% [3]. For some sectors, they can easily switch to green electricity. However, for high-temperature heat applications, this may not be feasible. For example, at high temperatures, the efficiency of a solar-driven desalination unit drops [4]. These challenging sectors can be decarbonized by switching to low-carbon hydrogen [5,6].

Hydrogen is key to having a complete set of alternatives to decarbonize the current and future energy consumption [7] and a crucial component of supporting longer-term climate neutrality and strategic independence for several large countries [8]. Compared to the available energy sources, hydrogen has attracted significant interest as a contributor to the sustainable development of industries worldwide due to its sustainable and reliable characteristics [9]. Hydrogen is a growth enabler for the multisectoral transition to a low-carbon economy based on clean energy sources. Hydrogen could contribute to reaching Goal 7 of the United Nation's Sustainable Development Goals (SDGs), i.e., affordable and clean energy [10]. Hydrogen will allow the decarbonization of areas where electrification is not a solution, either for technical or economic reasons [11]—namely, heavy industry, heavy transportation, air transportation, sea transportation, and many industrial processes. Additionally, hydrogen could play a role in long-term electricity storage. The electricity storage potential has significant economic and technical advantages in improving the efficiency of renewable energy (RE) sources [12]. Hydrogen is also a valuable chemical feedstock in petrochemical and metallurgical processes, food, and microelectronics [13]. Despite the momentum behind hydrogen, there is some skepticism regarding adopting hydrogen energy compared to alternative sources. Midilli et al. [13] highlight two major concerns regarding hydrogen: First is the quick volatility of hydrogen when combined with air at low temperatures, which raises safety concerns. Secondly, hydrogen storage in liquid form requires low temperature, making it difficult to store and transport. In addition, the commercial attractiveness of hydrogen compared to alternative energy sources is also a concern due to the cost associated with production, storage, and transportation. Hydrogen's environmental and ecological impact is linked to potential leakage during transport [7]. Few studies have identified hydrogen as having indirect greenhouse gas potential; however, they are significantly lacking compared to fossil fuel-based energy sources [8]. These concerns regarding hydrogen are the key motivation of this study to identify the appropriate sectors for hydrogen implementation.

The renewed focus on hydrogen is a result of the convergence of multiple factors: improvement in the cost and efficiency of electrolyzers, the increased competitive cost of renewable energy technologies, increased global regulatory efforts on climate change, and the broader application of hydrogen [9]. As technology advances, it is expected that the cost of hydrogen will continue to drop. National governments are making a significant effort to transition into a sustainable energy system. This transition has an economic and environmental impact on multiple sectors of the country. The hydrogen economy is envisioned as a burgeoning phenomenon that would ensure economic and environmental sustainability [10]. This paper provides a two-phase efficiency analysis of sectoral hydrogen transition based on energy, economic, and environmental factors, and low-carbon hydrogen production efficiency. Subsequently, a framework for the uptake of hydrogen energy transition in the United Arab Emirates (UAE) is presented.

The Hydrogen Roadmap of Europe stated that accomplishing the European Union's energy transition will require upscaling hydrogen. Without establishing the projects, the EU would not fulfil the decarbonization objective. The fuel offers a diverse, sustainable, and scalable energy vector to achieve the transition objectives [11]. Concrete model calculations demonstrate that 'blue' hydrogen can temporarily supplement the needs-based supply of consumers via existing gas storage. A concomitant expansion of renewable electricity generation is necessary to raise the percentage of green hydrogen. To ensure the competitiveness of climate-neutral hydrogen on the energy market, standard and suitable framework conditions are required to build a hydrogen industry in line with the market [12]. Noussan et al. [9] discussed the market and geopolitical perspectives of hydrogen generation via green or blue pathways, transportation, storage, and final use in various sectors. In addition, they covered the critical aspects of implementing an energy system based on hydrogen technologies. Moreover, this hydrogen transition study towards a low-carbon society covered three analytical perspectives: an energy model, an economic model, and a socio-technical case study. The results showed that, in Norway, it is required to have access to renewable power and hydrogen to decarbonize transport and industrial sectors and that hydrogen will drive the momentum and maintain a high level of economic activity [13]. Considering the spatiotemporal fluctuations in energy demand and supply, the researchers developed a comprehensive methodology for co-optimizing infrastructure investments across the power and H_2 supply chains. The conclusion showed that the deployment of carbon capture and storage (CCS) for power generation is less cost-effective than its usage for low-carbon H_2 production due to the grid flexibility allowed by sector coupling [14].

Hydrogen Potential of the UAE

The UAE is a major player in the energy market and is positioned to lead the hydrogen energy transition in the region and globally. The UAE plans to decarbonize its system and increase the generation of low-carbon hydrogen by having half of its installed power capacity come from nuclear and renewable sources by the year 2050 [15]. The UAE plans to diversify from fossil fuels, and export renewable energy and clean hydrogen to foreign consumers [16].

To maximize the contribution of hydrogen to the attainment of net-zero emissions, consumption should increase to 212 million metric tons (mt) by 2030 [17]. The UAE declared its goal of capturing 25% of the global low-carbon hydrogen market by 2030 [18]. The UAE's national oil facility, Abu Dhabi National Oil Company (ADNOC), already produces more than 300 kt of hydrogen annually. Efforts are made to expand its reach and increase production to 500 kt annually [19], particularly for the decarbonization of high-polluting heavy sectors. The increased output of low-carbon hydrogen, which may be used in the domestic economy or transformed into exportable goods such as ammonia, may be achieved with relatively modest additional investment [20]. Policies are streamlined with international organizations to accelerate the growth of hydrogen in the country. For example, an agreement was established to jointly build a waste-to-hydrogen project in the UAE between the British waste-to-energy company Chinook Sciences and the Emirati waste management company Bee'ah [21]. The UAE could be the major exporter of green hydrogen to the South Asia market, which is close enough to be connected by a pipeline [22]. Therefore, large-scale solar application is a viable investment for the UAE to become the leading green hydrogen exporter in the future [22,23].

Based on electrolyzer cost assumptions, solar predictions, and learning rates, Gandhi et al. [24] highlighted the economic viability of green hydrogen production in UAE industries between 2032 and 2038 with production costs at USD 0.95/kg and USD 1.35/kg. The UAE is constantly rated among the cheapest producers of renewable energy. They recorded the lowest solar energy cost in Dubai in 2015 at USD 5.6/kWh and in 2016 at USD 2.99/kWh. The most recent record was in 2020 in Abu Dhabi for the 2 GW Al Dhafra project with an offer of just USD 1.35 kWh [25]. According to IEA, the Middle East has the lowest hydrogen production cost from natural gas for blue hydrogen production at USD 0.43 kg/h [26].

The UAE has several advantages in leading the hydrogen market. For blue hydrogen, advantages include reliable and affordable hydrocarbons, large-scale ammonia and hydrogen production facilities, and sizable, well-characterized underground formations for carbon dioxide storage [27]. The outstanding sun generation conditions and the low levelized cost of solar power generation are favorable for green hydrogen [28]. The existing infrastructure, such as port facilities, LNG export and import terminals (which could be modified to support hydrogen trade), gas pipelines, salt domes for hydrogen storage, a central geographic location between critical markets, and a steady, business-friendly, and innovative approach are all benefits for hydrogen projects. UAE also benefits from an advanced, extensive network of petrochemical and refining facilities, including the Ruwais ammonia plant, and future possibilities for manufacturing methanol and synthetic fuels [29].

The economic dimension of the hydrogen transition is imperative. The UAE is considered a reliable and sustainable investment environment with a solid track record of public–private collaborations with domestic and foreign companies across the energy and industrial sectors. The establishment of the Abu Dhabi Hydrogen Alliance by ADNOC, Mubadala, ADQ, and the Ministry of Energy and Infrastructure is a critical first step towards a coordinated effort to achieve the UAE's low-carbon economy. However, other strategic initiatives, such as creating carbon accounting guidelines and protocols for hydrogen production, usage, and trade are required [22].

The energy transition at the national or global level is a complex system to unpack, given the existing energy infrastructure. A transition would require many infrastructural upgrades or changes. Estimating the low-carbon hydrogen production efficiency and prioritizing the sector coupling to obtain maximum energy, environmental, and economic impact is imperative to the energy transition, hence the motivation of this study. Many studies have focused primarily on hydrogen energy transition in transportation without socio-economic and environmental basis. Few studies have looked into the UAE's H$_2$ economy potential. Kazim [30] performed a conceptual and techno-economic analysis of the UAE's potential to utilize hydrogen energy to satisfy its energy needs. Eveloy and Gebreegziabher [31] studied the UAE's excess electricity and power-to-gas potential. More recently, Gandhi et al. [24] quantified the potential of green hydrogen in the UAE for domestic use and energy exports. This contributes to the literature by presenting an efficient hydrogen sector coupling framework for the UAE by first performing an energy, economic, and environmental analysis of the UAE sectors to justify the notion of hydrogen sector coupling and identifying the sector that should be prioritized in the hydrogen energy transition. The so-called data envelopment analysis technique is employed to evaluate efficiency and create an empirical basis for the UAE's hydrogen energy transition sector (DEA). DEA is a robust performance analysis technique developed by Charnes et al. [32] based on the work of Farrel [33]. DEA has been applied in renewable energy [34] and hydrogen-related subjects [35]. For example, Chi et al. [36] evaluated the efficiency of hydrogen-listed enterprises in China using DEA and the Malmquist index. Huang and Liu [37] presented an analysis of China's potential for producing hydrogen using solar and wind energy. This paper takes a holistic view of the UAE's hydrogen energy transition. To the best of our knowledge, this study is the first to analyze the UAE's sector efficiencies and production efficiency in achieving its hydrogen transition target. Furthermore, the hydrogen energy transition framework introduced presents an empirical rationale for future hydrogen policy implementation strategies. Figure 1 illustrate a diagrammatic overview of the study. The remainder of the article is organized as follows: Section 2 details low-carbon hydrogen. Sector coupling is discussed in Section 3 with the analysis framework in Section 4. Section 5 presents the methodology of the study. The results and discussions are provided in Section 6 with the proposed UAE hydrogen transition framework. Finally, the summary and conclusions are presented in Section 7.

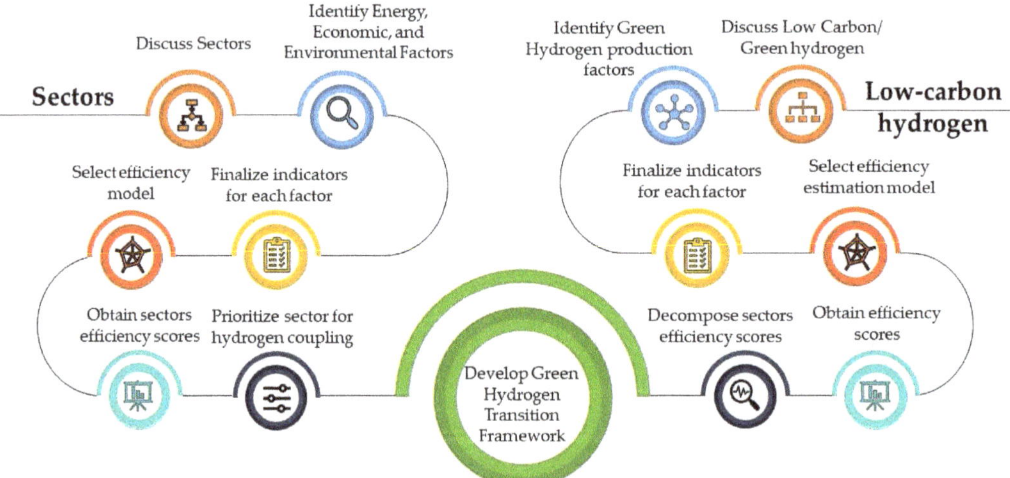

Figure 1. Research overview.

2. Low-Carbon Hydrogen Energy

The economic development pathways for many countries are increasingly influenced by the urgent need to tackle climate change. However, the success of environmental protection measures, such as GHG emission reduction, is being used to gauge economic growth [38]. Some countries are implementing efforts to assume leadership roles in the movement to create low-carbon economies as the international community works towards an effective global system to price the use of carbon [39]. The UAE has undertaken an ambitious national policy for energy saving and emission mitigation in response to international challenges to reduce its CO_2 emissions. The nation intends to lower its carbon intensity while expanding its industrial activities.

To minimize GHG emissions and decelerate climate change, the global energy system must fully transition to a decarbonized system [40]. Energy sustainable development is the only viable path [39]. The role of alternative fuels and technologies intended to support the energy transition are imperative. Studies have shown that hydrogen fits into several integrated energy system models as an alternative fuel in the low-carbon economy [13].

High costs and the availability of infrastructure have prevented hydrogen from having a significant impact on energy systems [38]. However, hydrogen is vital in the UAE's low-carbon pathway due to its versatility as an energy carrier and its potential to reduce emissions in hard-to-abate industries [41]. Hydrogen can be produced in several ways. The primary production method is steam methane reforming, but for hydrogen to be a low-carbon energy carrier, the existing generating methods need to be altered using renewable electrolysis [42]. Utilizing renewable electrolysis can help enhance the penetration of renewable energy [43].

Hydrogen is categorized according to colors based on the energy source used [24]. Hydrogen produced from fossil fuels such as petroleum, coal, and natural gas is denoted as 'black', 'brown', and 'gray' hydrogen, respectively. Low-carbon hydrogen includes blue hydrogen (hydrogen from fossil fuels with CO_2 emissions reduced by the use of carbon capture use and storage (CCU/S), green hydrogen (hydrogen from renewable electricity), and aqua hydrogen (hydrogen from fossil fuels via new technology) [44]. Natural gas reformation is currently used on a wide scale to produce hydrogen [45]. It is a well-developed industrial method that creates hydrogen using high-temperature steam. In this process, the steam reacts with a hydrocarbon fuel to produce hydrogen. Today, the steam reforming of natural gas produces 95% of global hydrogen [46]. Electrolysis is a potentially clean hydrogen generation process, although its environmental impact is well documented. As opposed to hydrogen from natural gas reforming, only low-carbon electrical sources such as hydro, solar, or wind power that enable a significant reduction in GHG emissions are utilized [47]. To produce green hydrogen, the energy source must be renewable energy. There are commercially available alkaline and polymer electrolyte membrane (PEM) electrolyzers. Both are secure and dependable [48].

To meet the enormous demand for low-carbon hydrogen, national and regional efforts are made to boost the electrolyzer capacity. Green hydrogen production is still limited due to renewable energy constraints due to geographical factors and electrolyzer capacity [49,50]. Wind power is a vital energy source for renewable energy [51] and green hydrogen production. Technological advancements such as wind power predictability will enhance renewable energy [52] to support green hydrogen production. To boost low-carbon production, Huang and Liu [53] recommend an integrated scheme of fossil-based hydrogen with CCSU/S and renewable energy hydrogen. Utilizing CCS as an option for low-carbon hydrogen production comes with challenges such as leakage. Selecting the appropriate underground natural gas storage could enhance energy transition and sustainable development goals [54]. To facilitate the permanent storage of CO_2, AlRassas et al. [55] developed an integrated static and dynamic modeling framework to tackle the challenge of CO_2 storage capacity.

The cost of hydrogen production is important in attaining a low-carbon hydrogen economy [13]. The Middle East has the lowest hydrogen production cost using natural gas, with and without carbon capture [3]. Various technical and economic considerations, with gas prices and capital expenditures, affect the cost of producing hydrogen from natural gas [49].

3. Sector Coupling

The sector coupling (SC) concept has been recognized since the turn of the 20th century. There is a broad spectrum of definitions of SC and sectors in the literature. Initial studies defined the sectors as power, heat, hydrogen, and natural gas [50]. Recent studies stated that sectors are not limited to energy sources but also cover the industry, energy economics, residential, commercial, and mobility fields. Figure 2 shows the sector coupling system for direct or indirect electrification; for indirect electrification, power-to-heat (P2H) techniques using combined heat and power (CHP) or electric boilers can be used (EBs). Another option is power-to-gas (P2G), which converts power into hydrogen through electrolysis which can then be processed to produce methanol or methane (CH_4) using a source of CO_2 (H_2) [51].

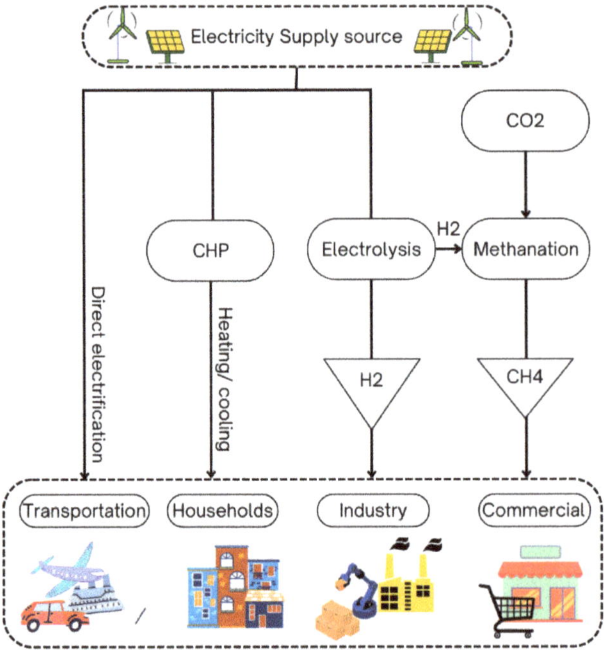

Figure 2. Sector coupling.

There are several definitions of SC, and the researchers described it based on the nature of the study [52]. Energy networks have previously been viewed independently, and only optimization interventions for a single network have been used. Instead, sector coupling enables the integration of all existing networks into a comprehensive energy system, typically referred to as a multi-carrier energy system (MCES). MCES is made up of various energy carriers such as hydrogen, heat, electricity or gas [53]. Economically, MCES lowers the integrated system's operational costs by 1.3% and potentially reduces the impact of wind power output unpredictability on operating expenses by 20% [54]. These benefits add to the growing support for SC in many countries.

SC primarily refers to electrifying end-use industries such as transportation and heating to increase the proportion of renewable energy in these sectors and offer balancing services to the power sector [55]. Supply-side coupling has been integrated into SC more recently through technologies such as power-to-gas that aim to incorporate the gas and electric power sectors [52,56]. The cost-effective decarbonization of the energy system depends heavily on electrifying end-use sectors, but efficient integration of the energy supply sectors can ensure flexibility and efficiency to meet the net-zero national and global target [57]. SC has been extended to include system adaptability with technologies such as electrolyzers for hydrogen production, which has increased demand [58]. This enables a flexible system integration with transportation, electrical, and thermal energy grids to optimize the whole operations as one system economically [59]. End-use SC and cross-vector integration are the two categories under SC [52]. End-use sector coupling strengthens the relationship between electricity supply and end-use while electrifying energy demand. Cross-vector coupling is the coordinated use of various energy infrastructures and vectors, particularly electricity, heat, and gas at the supply side, for example, when surplus electricity is converted to hydrogen, or at the demand side when residual heat from power plants or industrial processes is used for district heating. Sector coupling, according to numerous studies, can reduce the overall costs of the energy transition [57].

Direct electrification is problematic for end-use SC, such as heavy-duty transportation, where refilling time and the volume of energy density are essential factors in fuel selection. Studies show hydrogen energy is essential in the decarbonization of hard-to-electrify end-users [14]. Integrating renewable energy at a higher rate is demanding since passenger and freight transport vehicles require affordable replacements for all fossil fuels used in combustion engines. Additionally, alternatives to high-temperature industrial process heat are crucial. To effectively handle these difficulties and meet the 2050 decarbonization target, SC must expand rapidly, and renewable energy can play an increasingly important role in the hard-to-abate sectors, either directly or through the creation of hydrogen or synthetic fuels [60].

4. Analysis Framework

To adopt a holistic view of hydrogen energy transition, a two-phase efficiency analysis is applied. Figure 3 illustrates the conceptual analysis framework of the paper. First, the sector efficiency of the UAE's system is analyzed to prioritize the less efficient sector. We assume that early transition in the less efficient sectors will have more impact on the transition target rather than in the relatively efficient sector. Subsequently, the low-carbon hydrogen production efficiency is estimated for future planning and action.

The first phase measures energy, economic, and environmental efficiency for each sector. The sectors selected are transportation, industry, and electricity and heat producers. The sectors were considered according to IEA Energy Technology Perspectives' (ETPs) 2017 end-use sectors (industry, transportation, and buildings), which are integrated and linked robustly. The industrial sector covers mining and quarrying, construction, and manufacturing. Buildings cover residential, non-residential, commercial, and public service sectors that use cooling, water heating, lighting, and appliances that require electricity [61]. In our paper, this sector is referred to as electricity and heating deployed in the UAE. The third sector is transportation, which includes all primary motorized forms of transportation such as road, rail, shipping, and air services that offer passenger and cargo services [61].

The variables used to model the efficiency of sectors were applied as the dimensions of each sector. The final energy consumption by sectors represents the energy dimension, while the sector contribution to gross domestic product (GDP) represents the economic dimension [62,63]. The CO_2 emissions of each sector represent the environmental dimensions of the analysis [63]. The energy dimension is the input of the DEA analysis, while the environmental and economic dimensions are the output variables in the analysis. Total energy consumption has been used as an input variable in previous DEA studies [64]. Similarly, GDP [65] and CO_2 emissions [64] have been used to represent the economic and environmental outputs of DEA efficiency analysis. Data for sectoral total final energy consumption and CO_2 emissions were sourced from the IEA's World energy balance [66]. Sector contribution to GDP was sourced from the UAE's Federal Competitiveness and Statistics Center [67].

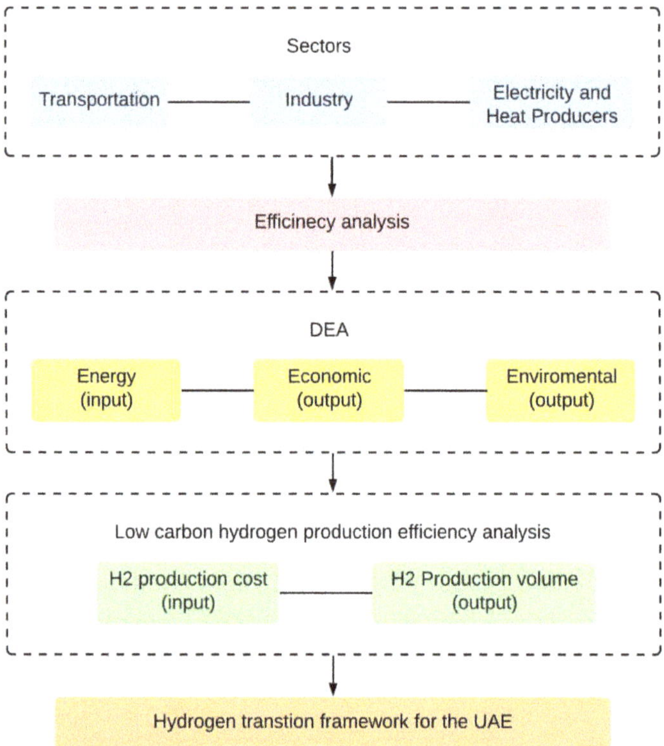

Figure 3. Analysis framework.

The second phase of the analysis looks into the efficiency of the UAEs' low-carbon hydrogen production potential. The results will contribute to developing a hydrogen energy transition strategy for the decarbonizing sectors identified in phase one. Two critical factors are considered to estimate the low-carbon hydrogen production efficiency of the UAE: low-carbon hydrogen production cost and the amount of hydrogen produced annually. A major concern for hydrogen decarbonisation is cost efficiency [68]. To estimate future low-carbon hydrogen production efficiency, we rely on the data from the recently published study of Gandhi et al. [24].

The study of Gandhi et al. [24] applied economics-based aggregate estimates of future demand for green H_2 in the UAE based on existing and planned uses. Our study considered the business-as-usual (BAU) data regarding the adoption across sectors. The BAU method presents a more realistic and less assumptive scenario. This practical perspective assumes the current situation is maintained. Based on the cost and adoption results of the scenarios, the UAE's green H_2 demand and production scenarios were visualized using growth trajectories across all sectors. We take into account capacity utilization, typical hydrogen requirements, and anticipated market growth rates to estimate annual hydrogen demand per industry. This estimate represents the entire green H_2 demand from the industries examined. The absolute demand is evaluated based on the current projections for domestic production rates and those in the future according to Equation (1).

$$ID_{H2,n} = IC_{UAE} \times U_{plant} \times CF_{H2} \tag{1}$$

ID_{H2} represents the industrial H_2 demand in a year in units of tonnes, while IC_{UAE} indicates the domestic industrial installed capacity in units of tonnes. Ultimately, CF_{H2} is the conversion factor from unity output to hydrogen tonnes.

5. Data Envelopment Analysis Efficiency Estimation

Data envelopment analysis is a data-driven, non-parametric technique that effectively evaluates the efficiency of systems known as decision-making units (DMUs). DMUs are a set of homogenous entities that carry out an input–output transformation process. DMUs are characterized by an input–output vector that comprises the amount of different inputs consumed and different outputs produced. The basis of DEA methodology is the derivation of the production possibility set (PPS) that contains all points of the input–output vectors that are considered feasible based on the observed DMUs.

DEA evaluates the efficiency of a set of $j = 1, 2, \ldots, n$ observed DMUs. These observations transform a vector of $i = 1, 2, \ldots, m$ inputs $x \in \Re_{++}^m$ Xer into a vector of $i = 1, 2, \ldots, s$ outputs $y \in \Re_{++}^s$ using the PPS of Equation (2), following constant return to scale, and Equation (3), following variable return to scale.

$$P = \{(x,y)|x \geq X\lambda, y \leq Y\lambda, \lambda \geq 0\} \tag{2}$$

$$P = \{(x,y)|x \geq X\lambda, y \leq Y\lambda, e\lambda = 1, \lambda \geq 0\} \tag{3}$$

where $X = (x)_j \Re_{++}^{s \times n}, Y = (Y)_j \Re_{++}^{m \times n}$ and $\lambda = (\lambda_1, \ldots, \lambda_n)^T$ is a semipositive vector.

An efficient DMU is derived with maximum outputs with constant inputs or minimum inputs with constant outputs. Apart from inputs and desirable outputs, undesirable outputs are also considered inefficient evaluation, especially in energy utilization systems. The undesirable outputs are related to the environment, such as CO_2 emissions [69], SO_2 emissions, and industrial waste [70]. Therefore, to take into account the undesirable outputs produced, DEA evaluates efficiency by finding the maximum desirable output and minimum undesirable outputs utilizing the same inputs. To estimate efficiency, this paper adopts the undesirable output direction distance function model (DDF). Compared to the traditional DEA model. The DDF model introduced by Chambers et al. [71] is a non-radial, non-oriented DEA model. The Chambers et al. [71] model estimates efficiency given technology T and a non-zero direction vector $(g^x, g^y) \in (\Re_+^m \times \Re_+^s)$, and DDF $D_T(X_0, Y_0, g^x, g^y)$ is defined as:

$$
\begin{aligned}
D_T(X_0, Y_0, g^x, g^y) &= Max\ \beta \\
&s.t. \\
\hat{X} &= \sum_{j=1}^{n} \lambda_j X_j \leq X_0 - \beta g^x \\
\hat{Y} &= \sum_{j=1}^{n} \lambda_j Y_j \geq Y_0 + \beta g^y \\
(\lambda_1, \lambda_2, &\ldots, \lambda_n) \in \Lambda^T\ \beta\ free
\end{aligned} \tag{4}
$$

where Λ^T can be constant return to scale or variable return to scale. The first and second constraints of Equation (1) show that $D_T(X_0, Y_0, g^x, g^y)$ reduces the inputs and increases the outputs simultaneously using the direction (g^x, g^y) as a projection direction. Equation (4) corresponds to technical efficiency (TE) from the constant return to scale (CRS) assumption. Pure technical efficiency (PTE) with respect to variable return to scale (VRS) is calculated by adding the VRS constant, $\sum_{j=1}^{n} \lambda_j = 1$. Consequently, scale efficiency (SE) can be calculated as: $SE = TE - PTE$.

To simultaneously increase desirable output and decrease undesirable output in efficiency estimation, Chung et al. [72] introduce an undesirable output DDF model to ensure asymmetry between both types of production. Treating both sets of outputs differently requires the redefinition of the PPS as $P = \left\{ \left(x, y^d, y^u\right) \middle| x \geq X\lambda, y^d \leq Y\lambda,\ y^u = Y\lambda,\ \lambda \geq 0 \right\}$

where the outputs are separated into desirable and undesirable outputs as: $y = \left(y^d, y^u\right)$ with $y^d \in R^q_{++}$ and $y^u \in R^r_{++}$, respectively. Correspondingly, the directional efficiency measure of a DMU $\left(x_0, y_0^d, y_0^u\right)$ is projected along the direction output vector $g_y = \left(y^d, y^u\right) \neq 0_{m+s}$. To prevent inconsistency in the original approach, Álvarez et al. [73] present a robust undesirable output DDF model as:

$$
\begin{aligned}
&\max_{\beta, \lambda} \; \beta \\
&subject\ to \\
&\quad X\lambda \leq x_0 \\
&\quad Y^d\lambda \geq y_0^d + \beta y_0^d \\
&\quad Y^u\lambda \leq y_0^u + \beta y_0^u \\
&\quad \max\{y_i^u\} \geq y_0^u + \beta y_0^d \\
&\quad \lambda \geq 0.
\end{aligned}
\tag{5}
$$

The solution of the linear program presents the efficiency score. If the optimal solution $\beta^* = 0$ with $\lambda_0 = 1$, $\lambda_j = 0$ $(j \neq 0)$, then the unit is directionally efficient. Otherwise, $\beta^* > 0$ implies the unit is inefficient, and efficiency score is presented as: $1 - \beta^*$.

6. Results and Discussions

Table 1 present the descriptive statistics of the variables used in the first phase of the analysis. The result shows that the industry sector has the highest variation among all variables, which indicates that the values are spread out over a broader range. This variation may be attributed to the variability in industry activities compared to other sectors. On the other hand, the lowest standard deviation is the electricity and heat producers' sector. The electricity and heat producers' sector has a higher consistency and predictability in energy consumption and other factors than other sectors. In addition, the industry sector consistently has the highest mean among all the variables.

Table 1. Descriptive statistics for sector efficiency analysis.

Sectors		Total Final Consumption (TJ)	Sector GDP Contribution (Billion USD)	CO_2 Emissions by Sector (mt CO_2)
Industry	Mean	1,021,346.5	147.45	57.65
	Std. Dev	300,881.02	28.28	17.22
	Min.	606,997	90.24	34
	Max.	1,459,132	181.81	83
Transport	Mean	407,791.64	19.67	29.2
	Std. Dev	100,384.21	2.60	7.08
	Min.	226,644	14.88	16
	Max.	577,048	26.03	41
Electricity and heat producers	Mean	101,964.04	7.41	58.75
	Std. Dev	35,234.752	2.89	12.16
	Min.	39,107	3.21	32
	Max.	148,246.85	11.36	72

In estimating the sectors' energy, economic, and environmental efficiency, model (5) is applied using Matlab R2020a. The results are presented in Table 2. Total final energy consumption is considered as the input, sector contribution to GDP is the desirable output, and sector CO_2 emission is an undesirable output. Figure 4 illustrates the average efficiency scores in different time intervals. The electricity and heat producers' sector was the most efficient over time, with an efficiency score between 0.89 and 0.99. The average efficiency of electricity and heat producers improved over time, especially over the last 3–5 years.

Table 2. Sector efficiency scores.

Year	Industry	Transport	Electricity and Heat Producers
2001	0.66	0.99	0.99
2002	0.69	0.96	1.00
2003	0.77	0.94	0.77
2004	0.87	1.00	0.77
2005	0.92	0.94	0.78
2006	1.00	0.95	0.77
2007	0.88	0.97	0.79
2008	0.82	0.94	0.66
2009	0.67	0.97	0.89
2010	0.66	0.72	0.84
2011	0.69	0.74	0.90
2012	0.65	0.75	0.89
2013	0.64	0.70	0.90
2014	0.67	0.74	0.92
2015	0.64	0.82	0.97
2016	0.64	0.73	1.00
2017	0.71	0.70	0.98
2018	0.85	0.74	0.98
2019	0.74	0.79	1.00
2020	0.70	0.69	0.98

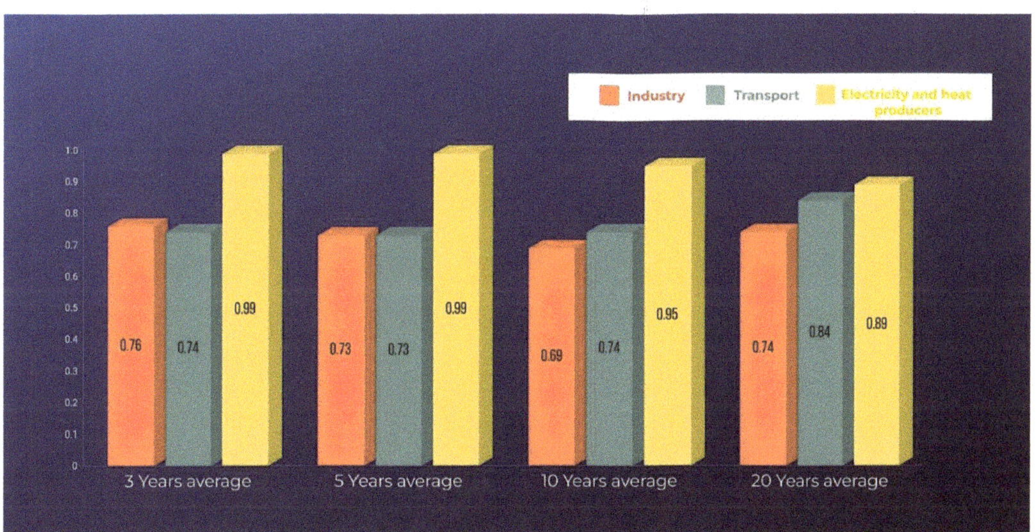

Figure 4. Average energy, economic, and environmental efficiency of UAE sectors.

Conversely, the industry and transport sector observed a relative decline in average efficiency. Average efficiency indicated the transport sector to be the least efficient in the last 3–5 years. Table 2 presents the detailed annual efficiency scores. Electricity and heat producers had three periods—2002, 2016, and 2019—identified as efficient. The transport (2004) and industry (2006) sectors had only one period identified as efficient. The efficiency of the transport sector in 2020 had the least efficiency scores of all periods in the transport sector. The industry sector showed a slight improvement over time. However, the last 3–5 years were relatively inefficient.

The results of the efficiency scores of the industry, transport, and electricity and heat producer sectors, therefore, suggest the transportation and industry sectors be prioritized in the hydrogen energy transition, given the relatively inefficient performance of the sectors. Coupling hydrogen energy with the transportation and industry sector could be more impactful, and the net-zero target could be efficiently attained.

The second phase of the analysis looks into the green hydrogen production potential of the UAE. Annual green hydrogen production cost is the input, while the output is the amount of green hydrogen to be produced. To estimate green hydrogen production efficiency, Equation (4) is applied. Figure 5 presents the estimated annual technical efficiency (TE) score of green hydrogen production. TE is decomposed into PTE and SE for further analysis. Figures 6 and 7 illustrate PTE and SE scores, respectively. TE is consistent with overall efficiency, which includes operational efficiency and scale efficiency. PTE corresponds to the operational aspect of efficiency. The relationship between TE and PTE is the scale efficiency, which refers to the utilization of capacity with respect to the input.

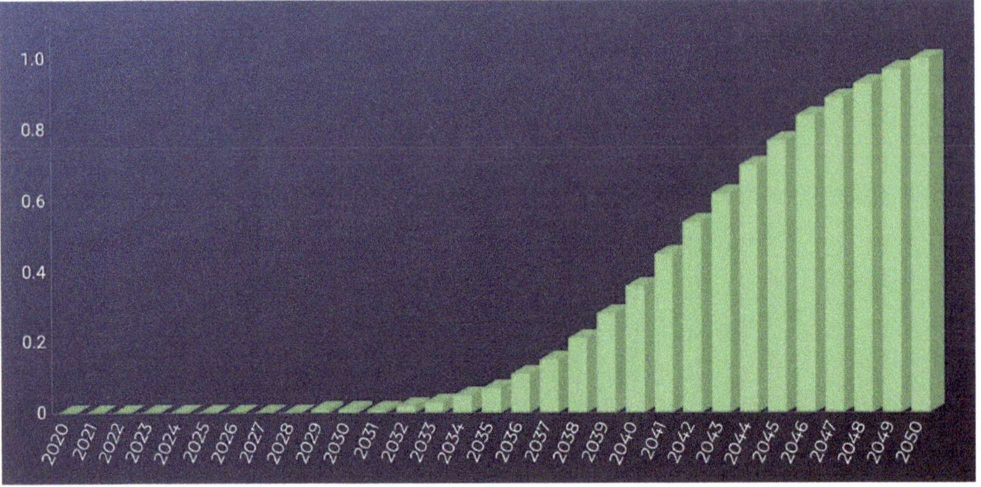

Figure 5. Technical efficiency (TE) of green hydrogen production in the UAE.

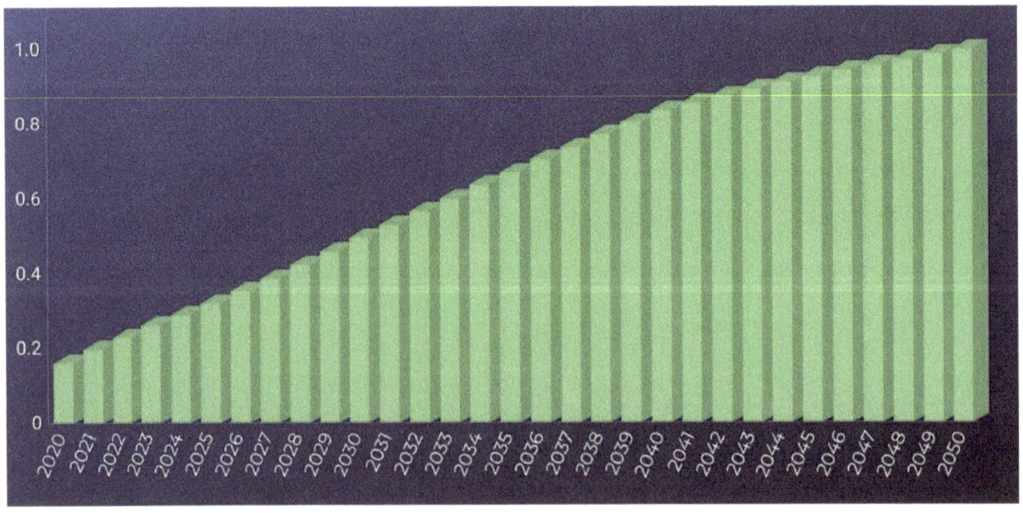

Figure 6. Pure technical efficiency (PTE) of green hydrogen production in the UAE.

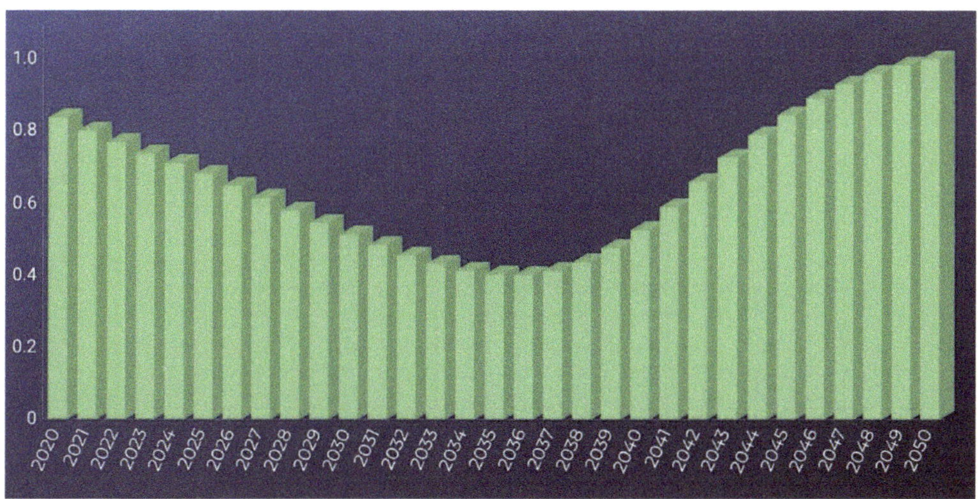

Figure 7. Scale efficiency of green hydrogen production in the UAE.

Results of TE indicate a steady growth in green hydrogen production efficiency in the UAE. For the next 10 years, the UAE will continue to experience an inefficient green hydrogen production system. This period can be considered a growth and capacity-building stage. Across the UAE, various projects are developed to improve production. In addition, the UAE is building operational competency for green hydrogen production, as observed in PTE efficiency. In 2030, the UAE will experience a boost in operational efficiency with a PTE score of 0.5. The result shows that the UAE is focused on building operational competency for the next ten years. There is a decline in scale efficiency, as observed in Figure 7. The decline is relative to the cost of hydrogen production. This decline in scale efficiency signifies that the amount of hydrogen produced does not complement the hydrogen production cost. The decomposition of the green hydrogen production efficiency results highlights the need to scale infrastructure for the storage and transportation of green hydrogen from both supply and demand sides. Since low-carbon hydrogen requires significant infrastructure for effective implementation, investment is needed to meet the short-term target. The results show a shortfall of technological advances to meet the growing technical efficiency of green hydrogen in the short term. Since the results point to short-term inefficiency in the UAE, the international hydrogen market is a viable alternative for the UAE to export low-carbon hydrogen to countries such as Japan, contributing to the global hydrogen energy transition by maintaining its role in the energy market. Therefore, the scale needs to be increased in these periods. This decline is projected to continue until 2036, when there will be a significant improvement in scale efficiency. This improvement in scale efficiency corresponds to the growth in the overall efficiency of hydrogen production in the UAE. Therefore, scale efficiency is imperative to the overall efficiency of the UAE's green hydrogen production system. Policies and strategies should, therefore, target upscaling the UAE's green hydrogen production capacity.

The low-carbon hydrogen transition is a global effort towards decarbonizing our energy systems. To our knowledge, no study has explored sector efficiency and green hydrogen production efficiency in the UAE. However, the findings of this study are consistent with a Chinese provincial study by Huang and Liu [42]. Their analysis points to a promising green hydrogen production for Chinese provinces using wind and solar energy up to 2030. Huang and Liu [37] recommend an integrated scheme to promote low production costs to boost low-carbon hydrogen production. Our study adds to the literature by further decomposing efficiency into pure technical and scale efficiencies, presenting an empirical improvement strategy combined with the sector efficiency analysis for an efficient hydrogen economy. Furthermore, the study provides a long-run efficiency estimation, offering technical and managerial insight toward attaining the 2050 target.

Hydrogen Transition Framework for the UAE

Developing a hydrogen-integrated energy system requires a broad and integrated set of changes, including infrastructure and technology; policies and incentives; regulations and codes; behaviors and habits; and investment patterns. Such broad changes are considered a "technological transition" from one sociotechnical system to another [74]. The dynamics of sociotechnical change are imperative in developing a framework for the hydrogen energy transition. The energy, economic, and environmental efficiency results indicate that the industry and transportation sectors should be the priority for the UAE's hydrogen energy transition. Following the increase in low-carbon hydrogen production efficiency, the UAE shows competency in achieving its hydrogen energy transition through sector coupling. Based on global hydrogen transition strategies of leading hydrogen economies such as the United Kingdom [75] and Germany [76], a UAE hydrogen energy transition framework is proposed in Figure 8.

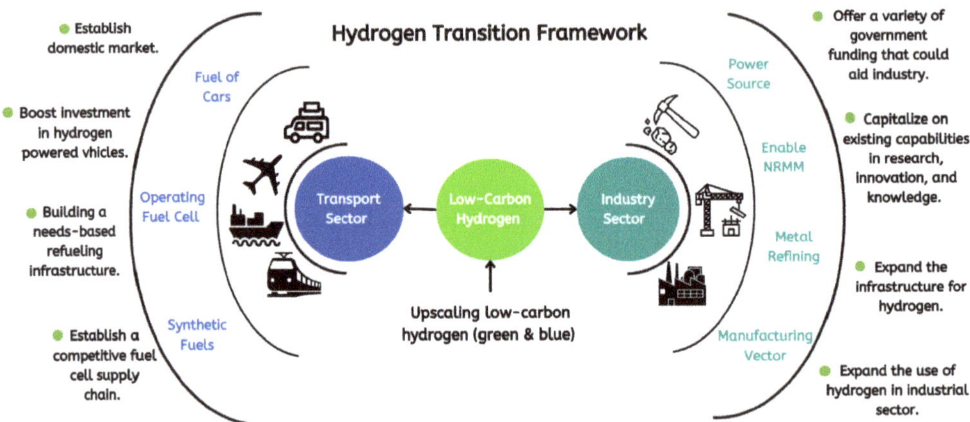

Figure 8. Hydrogen energy transition framework for the UAE's transportation and industry sectors.

The sector efficiency analysis in phase one suggests the transportation sector be prioritized before the industry sector due to its lower average efficiency score. However, the infrastructural constraints of the transport sector necessitate the industry to come first. This strategy could improve the transition's short-term inefficiency and achieve decarbonization by 2030. The low-carbon hydrogen strategy targets the decarbonization of transport and industry sectors through upscaling the production of blue and green hydrogen. The UAE transportation sector comprises cars, buses, metro, trams, airplanes, and marine transportation. The industry sector includes the manufacturing, mining, and construction operations [3], which are considered busy sub-sectors in the UAE.

Establishing a robust and sustainable domestic market for hydrogen production and domestic use is the first step that needs to be taken to accelerate the adoption of hydrogen technology. The following measures should be implemented to expedite the adoption of hydrogen technology between the public and private sectors to realize its full potential. The transition strategy should serve as a basis for private sector investment in hydrogen generation that is both commercially and environmentally feasible.

In the industry sector, the government has to offer a variety of funding that can help the industry make the transition to low-carbon technologies. Capital expenditure (CAPEX) support mechanisms and operating expenditure (OPEX) mechanisms need to be in place to support and boost the transition. The UAE has initiatives that encourage the use of hydrogen in production. However, this needs to be scaled to meet the net-zero target. Effective collaboration in a public–private partnership is key to a successful transition. To fully realize hydrogen's potential to aid in reducing carbon emissions, we must invest in the research and innovation of the hydrogen value chain [75]. Hydrogen might be used to reduce a significant amount of emissions brought on by using non-road mobile machinery (NRMM) such as diggers and excavators. Robust and scalable hydrogen infrastructure must be built with significant investments in CAPEX and OPEX to utilize it as a solution in various industry options [75]. Industrial users are anticipated to generate the majority of the new demand for hydrogen through industrial fuel switching, and the industrial sector is likely to pave the way for large-scale low-carbon hydrogen consumption. This is achieved by accelerating the industrial use of current hydrogen to decarbonize current processes. In order to expand into other sub-sectors of the industry and the larger energy system, the hydrogen economy must thrive. A rise in the number of locations having access to low-carbon hydrogen ongoing technological advancements to broaden the variety of processes that can use hydrogen, and a change in associated expenses, such as the price of carbon, would all contribute to the transition [76]. There is significant market demand for low-carbon hydrogen. International collaborations for hydrogen transport could enhance the hydrogen energy transition.

In the transport sector, hydrogen is an energy source that can be deployed to create synthetic fuels or used in fuel cells to power hydrogen-powered vehicles. Synthetic fuels are necessary for decarbonizing both sea and air transportation. For some mobility requirements, fuel cells and battery-powered drives may be an option in aviation and coastal and inland transportation. Despite the opportunities in this field, technological advancements are required to push transport development. Therefore, increasing financial investment in hydrogen-powered transportation (light and heavy trucks, buses, railroads, and cars) will boost the advances in the sector [76]. Building a needs-based refueling infrastructure is necessary for vehicles, particularly heavy-duty road haulers, vehicles used in public transportation, and local passenger rail services. The support for developing a competitive fuel cell supply chain includes the establishment of an industrial basis for large-scale fuel cell stack production for vehicle applications. Investigation of the feasibility of establishing a hydrogen technology and innovation hub will facilitate the emergence of hydrogen energy in the transport sector [76]. More importantly, financial investment is insufficient for private companies to collaborate and achieve the sector's transition [77]. The IEA recommends developing funding schemes, such as a capital expenditure tax decrease, to overcome the high cost of new technologies. Moreover, establishing emission restrictions, carbon pricing, and mandates for renewable energy adoption in the sector will encourage the higher adoption of hydrogen demand [78]. Approaches such as tax incentives for switching to hydrogen will be beneficial. Moreover, reducing investment risks will significantly enable infrastructure expansion and market penetration for hydrogen [79].

7. Summary and Conclusions

This paper takes a holistic view of hydrogen energy transition in the UAE. Using an undesirable output DDF DEA model, the energy, economic, and environmental efficiency of UAE sectors are analyzed in the first phase for the period of 2001–2020 to identify the sectors that should be prioritized in hydrogen energy sector coupling. The second phase of the analysis evaluates the production efficiency of low-carbon hydrogen, i.e., green hydrogen, in the UAE for the period of 2020–2050, using the business-as-usual (BAU) case of green hydrogen production cost and green hydrogen production volume in the UAE. The sector efficiency analysis result shows the transport and industry sectors are the most inefficient due to their declining efficiency averages. Therefore, the two sectors should be prioritized in the decarbonization strategy.

The green hydrogen production efficiency results show great potential in the long run. The relative underperformance of short-term hydrogen production efficiency is attributed to scale inefficiency. Therefore, to improve short-term efficiency and for the UAE to attain its

2030 hydrogen vision of capturing 25% of the global low-carbon hydrogen market, there is a need to scale low-carbon hydrogen production to meet the short-term target. The current trajectory shows that the UAE is on track to attain its 2050 green hydrogen energy vision.

The paper comes with some limitations. The low-carbon hydrogen production efficiency focuses primarily on green hydrogen production, and the model does not consider carbon emission in the production process. Process emissions during electrolyzer/PV manufacturing, construction, and transportation are, of course, present. However, the green hydrogen production process itself does not involve any emissions. Other low-carbon H_2 options, such as blue hydrogen, do involve GHG emissions. Future research should consider blue hydrogen production and GHG emissions in the model.

Author Contributions: M.D.I., F.A.S.B. and M.O.A.M. developed the conceptualization, investigation and data curation; M.D.I. was involved in project administration, validation, and the supervision of the work, processed the data, performed the analysis, drafted the manuscript, and performed formal analysis; F.A.S.B. designed the figures and visualized the results; F.A.S.B. and M.O.A.M. aided in interpreting the results and worked on the manuscript. All authors discussed the results and commented on the manuscript and reviewed and edited the paper. All authors have read and agreed to the published version of the manuscript.

Funding: The Authors thank Higher Colleges of Technology and Applied research committee for SURF_2022 grant [Fund no. 243009].

Data Availability Statement: Data for the study can be found in the repository: doi. Ibrahim, Mustapha D (2022), "Transition to Hydrogen energy_UAE", Mendeley Data, V1, doi: 10.17632/sccn7k4k3y.1.

Conflicts of Interest: The authors declare no conflict of interest.

References

1. UAE. The UAE's Response to Climate Change. Available online: https://u.ae/en/information-and-services/environment-and-energy/climate-change/theuaesresponsetoclimatechange#:~{}:text=Clean%20fossil%20fuels-,UAE%20Net%20Zero%202050,MENA)%20nation%20to%20do%20so (accessed on 29 June 2022).
2. UNFCCC. The Paris Agreement. Available online: https://unfccc.int/process-and-meetings/the-paris-agreement/the-paris-agreement (accessed on 29 June 2016).
3. IEA. Energy Transition Indicators. Available online: https://www.iea.org/countries/united-arab-emirates (accessed on 29 June 2019).
4. Ghaffour, N.; Reddy, V.; Abu-Arabi, M. Technology development and application of solar energy in desalination: MEDRC contribution. *Renew. Sustain. Energy Rev.* **2011**, *15*, 4410–4415. [CrossRef]
5. Hanley, E.S.; Deane, J.; Gallachóir, B.Ó. The role of hydrogen in low carbon energy futures–A review of existing perspectives. *Renew. Sustain. Energy Rev.* **2018**, *82*, 3027–3045. [CrossRef]
6. Rissman, J.; Bataille, C.; Masanet, E.; Aden, N.; Morrow, W.R.; Zhou, N.; Elliott, N.; Dell, R.; Heeren, N.; Huckestein, B.; et al. Technologies and policies to decarbonize global industry: Review and assessment of mitigation drivers through 2070. *Appl. Energy* **2020**, *266*, 114848.
7. Sarı, A.; Sulukan, E.; Özkan, D.; Uyar, T.S. Environmental impact assessment of hydrogen-based auxiliary power system onboard. *Int. J. Hydrogen Energy* **2021**, *46*, 29680–29693. [CrossRef]
8. European Comission. Environmental Impacts of Hydrogen-based Energy Systems. October 2006. Available online: https://ec.europa.eu/environment/integration/research/newsalert/pdf/39na1_en.pdf (accessed on 27 July 2021).
9. Noussan, M.; Raimondi, P.P.; Scita, R.; Hafner, M. The Role of Green and Blue Hydrogen in the Energy Transition—A Technological and Geopolitical Perspective. *Sustainability* **2020**, *13*, 298. [CrossRef]
10. YiDou. Chapter 10—Opportunities and Future Challenges in Hydrogen Economy for Sustainable Development. *Hydrogen Econ.* **2017**, 277–305. [CrossRef]
11. FCH. *Hydrogen Roadmap Europe: A Sustainable Pathway for the European Energy Transition*; Fuel Cells and Hydrogen: Luxembourg, 2019.
12. Adam, P.; Heunemann, F.; von dem Bussche, C.; Engelshove, S.; Thiemann, T. Hydrogen infrastructure—The pillar of energy transition. *Siemens Energy Conf. Eur. Comm.* **2020**. Available online: https://www.gascade.de/fileadmin/downloads/wasserstoff/whitepaper-h2-infrastructure.pdf (accessed on 24 July 2022).
13. Espegren, K.; Damman, S.; Pisciella, P.; Graabak, I.; Tomasgard, A. The role of hydrogen in the transition from a petroleum economy to a low-carbon society. *Int. J. Hydrogen Energy* **2021**, *46*, 23125–23138. [CrossRef]
14. He, G.; Mallapragada, D.S.; Bose, A.; Heuberger-Austin, C.F.; Gençer, E. Sector coupling via hydrogen to lower the cost of energy system decarbonization. *Energy Environ. Sci.* **2021**, *14*, 4635–4646. [CrossRef]

15. UAE. *UAE Energy Strategy 2050*; United Arab Emirates Government: Abu Dhabi, United Arab Emirates, 2021.
16. Lin, M.T. *UAE Wants to Transform from a Petrostate to Renewable and Hydrogen Powerhouse*; 2022; Available online: https://www.bloomberg.com/news/articles/2021-01-19/uae-can-be-major-low-cost-blue-hydrogen-producer-adnoc-ceo-says (accessed on 24 July 2022).
17. Adams, A. Net Zero with Hydrogen is on the Horizon, We Need the World on Board to Get There. Available online: https://hydrogencouncil.com/en/net-zero-with-hydrogen-is-on-the-horizon-we-need-the-world-on-board-to-get-there/ (accessed on 27 July 2021).
18. UAE. UAE Targets 25% of Hydrogen Market by 2030. Available online: https://www.energyconnects.com/news/renewables/2021/november/uae-targets-25-of-hydrogen-market-by-2030/ (accessed on 27 July 2021).
19. ADNOC. Hydrogen; ADNOC. 2022. Available online: https://adnoc.ae/en/news-and-media/press-releases/2022/adnoc-expands-strategic-partnerships-across-the-hydrogen-value-chain-with-leading-german-companies (accessed on 24 July 2022).
20. ADNOC. ADNOC to Build World-Scale Blue Ammonia Project. Available online: https://www.adnoc.ae/news-and-media/press-releases/2021/adnoc-to-build-world-scale-blue-ammonia-project (accessed on 27 July 2021).
21. Hebert, J. Bee'ah and Chinook to work on UAE hydrogen first. *Proj. Financ.* **2021**.
22. Friedmann, J.; Mills, R. *The Uae's Role in the Global Hydrogen Economy*; Qamar Energy: Dubai, United Arab Emirates; Available online: https://www.qamarenergy.com/sites/default/files/The%20UAE%27s%20Role%20in%20the%20Global%20Hydrogen%20Economy.pdf (accessed on 24 July 2021).
23. Joubi, A.; Akimoto, Y.; Okajima, K. A Production and Delivery Model of Hydrogen from Solar Thermal Energy in the United Arab Emirates. *Energies* **2022**, *15*, 4000. [CrossRef]
24. Joubi, A.; Akimoto, Y.; Okajima, K. Catching the hydrogen train: Economics-driven green hydrogen adoption potential in the United Arab Emirates. *Int. J. Hydrogen Energy* **2022**, *47*, 22285–22301.
25. DubaiFuture. *Hydrogen: From Hype to Reality*; Dubai Future Foundation: Duabi, United Arab Emirates, 2021.
26. IEA. The Future of Hydrogen. Available online: https://www.iea.org/reports/the-future-of-hydrogen (accessed on 27 July 2018).
27. Halff, A.; Mills, R. UAE to Play Leading Role in Emerging Global Hydrogen Market. Available online: https://www.ceoforlifeawards.com/blog/uae-to-play-leading-role-in-emerging-global-hydrogen-market/ (accessed on 27 July 2018).
28. Calnan, S.; Bagacki, R.; Bao, F.; Dorbandt, I.; Kemppainen, E.; Schary, C.; Schlatmann, R.; Leonardi, M.; Lombardo, S.A.; Milazzo, R.G.; et al. Development of various photovoltaic-driven water electrolysis technologies for green solar hydrogen generation. *Solar RRL* **2022**, *6*, 2100479. [CrossRef]
29. Downs, D.E.; Halff, A.; Mills, R. *Having It Both Ways*; Center for Global Energ Policy, Columbia University CGEP: New York, NY, USA, 2021.
30. Kazim, A. Strategy for a sustainable development in the UAE through hydrogen energy. *Renew. Energy* **2010**, *35*, 2257–2269. [CrossRef]
31. Eveloy, V.; Gebreegziabher, T. Excess electricity and power-to-gas storage potential in the future renewable-based power generation sector in the United Arab Emirates. *Energy* **2019**, *166*, 426–450. [CrossRef]
32. Charnes, A.; Cooper, W.W.; Rhodes, E. Measuring the efficiency of decision making units. *Eur. J. Oper. Res.* **1978**, *2*, 429–444. [CrossRef]
33. Farrell, M.J. The measurement of productive efficiency. *J. R. Stat. Soc. Ser. A* **1957**, *120*, 253–281. [CrossRef]
34. Kara, S.; Ibrahim, M.; Daneshvar, S. Dual efficiency and productivity analysis of renewable energy alternatives of OECD countries. *Sustainability* **2021**, *13*, 7401. [CrossRef]
35. Mei, M.; Chen, Z. Evaluation and selection of sustainable hydrogen production technology with hybrid uncertain sustainability indicators based on rough-fuzzy BWM-DEA. *Renew. Energy* **2021**, *165*, 716–730. [CrossRef]
36. Chi, Y.; Xiao, M.; Pang, Y.; Yang, M.; Zheng, Y. Financing Efficiency Evaluation and Influencing Factors of Hydrogen Energy Listed Enterprises in China. *Energies* **2022**, *15*, 281. [CrossRef]
37. Huang, Y.-S.; Liu, S.-J. Chinese green hydrogen production potential development: A provincial case study. *IEEE Access* **2020**, *8*, 171968–171976. [CrossRef]
38. Ishaq, H.; Dincer, I. Comparative assessment of renewable energy-based hydrogen production methods. *Renew. Sustain. Energy Rev.* **2021**, *135*, 110192. [CrossRef]
39. Nader, S. Paths to a low-carbon economy—The Masdar example. *Energy Procedia* **2009**, *1*, 3951–3958. [CrossRef]
40. Gielen, D.; Boshell, F.; Saygin, D.; Bazilian, M.D.; Wagner, N.; Gorini, R. The role of renewable energy in the global energy transformation. *Energy Strategy Rev.* **2019**, *24*, 38–50. [CrossRef]
41. Niaz, S.; Manzoor, T.; Pandith, A.H. Hydrogen storage: Materials, methods and perspectives. *Renew. Sustain. Energy Rev.* **2015**, *50*, 458–469. [CrossRef]
42. Fan, J.-L.; Zhang, Y.-J.; Wang, B. The impact of urbanization on residential energy consumption in China: An aggregated and disaggregated analysis. *Renew. Sustain. Energy Rev.* **2017**, *75*, 220–233. [CrossRef]
43. Guilbert, D.; Vitale, G. Hydrogen as a Clean and Sustainable Energy Vector for Global Transition from Fossil-Based to Zero-Carbon. *Clean Technol.* **2021**, *3*, 51. [CrossRef]
44. Yu, M.; Wang, K.; Vredenburg, H. Insights into low-carbon hydrogen production methods: Green, blue and aqua hydrogen. *Int. J. Hydrogen Energy* **2021**, *46*, 21261–21273. [CrossRef]

45. Khan, M.H.A.; Daiyan, R.; Neal, P.; Haque, N.; MacGill, I.; Amal, R. A framework for assessing economics of blue hydrogen production from steam methane reforming using carbon capture storage & utilisation. *Int. J. Hydrogen Energy* **2021**, *46*, 22685–22706.
46. Boretti, A.; Banik, B.K. Advances in hydrogen production from natural gas reforming. *Adv. Energy Sustain. Res.* **2021**, *2*, 2100097. [CrossRef]
47. Antonini, C.; Treyer, K.; Streb, A.; Van Der Spek, M.W.; Bauer, C.; Mazzotti, M. Hydrogen production from natural gas and biomethane with carbon capture and storage–A techno-environmental analysis. *Sustain. Energy Fuels* **2020**, *4*, 2967–2986. [CrossRef]
48. Ursúa, A.; Sanchis, P. Static–dynamic modelling of the electrical behaviour of a commercial advanced alkaline water electrolyser. *Int. J. Hydrogen Energy* **2012**, *37*, 18598–18614. [CrossRef]
49. International Energy Agency. *The Future of Hydrogen*; International Energy Agency: Paris, France, 2019.
50. Lu, Y.; Pesch, T.; Benigni, A. Simulation of Coupled Power and Gas Systems with Hydrogen-Enriched Natural Gas. *Energies* **2021**, *14*, 7680. [CrossRef]
51. Clegg, S.; Mancarella, P. Integrated modeling and assessment of the operational impact of power-to-gas (P2G) on electrical and gas transmission networks. *IEEE Trans. Sustain. Energy* **2015**, *6*, 1234–1244. [CrossRef]
52. Ramsebner, J.; Haas, R.; Ajanovic, A.; Wietschel, M. The sector coupling concept: A critical review. *Wiley Interdiscip. Rev. Energy Environ.* **2021**, *10*, e396. [CrossRef]
53. Rehman, O.A.; Palomba, V.; Frazzica, A.; Cabeza, L.F. Enabling Technologies for Sector Coupling: A Review on the Role of Heat Pumps and Thermal Energy Storage. *Energies* **2021**, *14*, 8195. [CrossRef]
54. Mirzaei, M.A.; Nazari-Heris, M.; Zare, K.; Mohammadi-Ivatloo, B.; Marzband, M.; Asadi, S.; Anvari-Moghaddam, A. Evaluating the impact of multi-carrier energy storage systems in optimal operation of integrated electricity, gas and district heating networks. *Appl. Therm. Eng.* **2020**, *176*, 115413. [CrossRef]
55. Wei, M.; McMillan, C.A. Electrification of industry: Potential, challenges and outlook. *Curr. Sustain. Renew. Energy Rep.* **2019**, *6*, 140–148. [CrossRef]
56. Child, M.; Koskinen, O.; Linnanen, L.; Breyer, C. Sustainability guardrails for energy scenarios of the global energy transition. *Renew. Sustain. Energy Rev.* **2018**, *91*, 321–334. [CrossRef]
57. Nuffel, L.V.; Dedecca, J.G.; Smit, T.; Rademaekers, K. *Sector Coupling: How Can It Be Enhanced in the EU to Foster Grid Stability and Decarbonise?* Policy Department for Economic, Scientific and Quality of Life Policies, European Parliament: Brussels, Belgium, 2018.
58. Rabiee, A.; Keane, A.; Soroudi, A. Green hydrogen: A new flexibility source for security constrained scheduling of power systems with renewable energies. *Int. J. Hydrogen Energy* **2021**, *46*, 19270–19284. [CrossRef]
59. IRENA. *Sector Coupling in Facilitating Integration of Variable Renewable Energy in Cities*; International Renewable Energy Agency (IRENA): Abu Dhabi, United Arab Emirates, 2021.
60. Gils, H.C.; Simon, S.; Soria, R. 100% Renewable Energy Supply for Brazil—The Role of Sector Coupling and Regional Development. *Energies* **2017**, *10*, 1859. [CrossRef]
61. IEA. *ETP Model 2017*; IEA: Paris, France, 2020.
62. FCSC. *Data Related to Gross Domestic Product, Consumer Price Index, Trade, Investment and Other Related Datasets*; FCSC: London, UK, 2020.
63. IEA. *United Arab Emirates*; IEA: Paris, France, 2019.
64. Fong, W.; Sun, Y.; Chen, Y. Examining the Relationship between Energy Consumption and Unfavorable CO2 Emissions on Sustainable Development by Going through Various Violated Factors and Stochastic Disturbance–Based on a Three-Stage SBM-DEA Model. *Energies* **2022**, *15*, 569. [CrossRef]
65. Xu, T.; You, J.; Li, H.; Shao, L. Energy efficiency evaluation based on data envelopment analysis: A literature review. *Energies* **2020**, *13*, 3548. [CrossRef]
66. IEA. IEA World Energy Balances. Available online: https://www.iea.org/data-and-statistics/data-product/world-energy-statistics-and-balances (accessed on 15 July 2022).
67. FCSC. Distribution of Gross Domestic Product at Constant (2010) Prices By Economic Activities. Available online: https://fcsc.gov.ae/en-us (accessed on 13 July 2022).
68. Brauner, G. Efficiency Through Sector Coupling. In *System Efficiency by Renewable Electricity*; Springer: Berlin/Heidelberg, Germany, 2022; pp. 209–224.
69. Xie, B.-C.; Tan, X.-Y.; Zhang, S.; Wang, H. Decomposing CO2 emission changes in thermal power sector: A modified production-theoretical approach. *J. Environ. Manag.* **2021**, *281*, 111887. [CrossRef] [PubMed]
70. Ma, D.; Xiong, H.; Zhang, F.; Gao, L.; Zhao, N.; Yang, G.; Yang, Q. China's industrial green total-factor energy efficiency and its influencing factors: A spatial econometric analysis. *Environ. Sci. Pollut. Res.* **2022**, *29*, 18559–18577. [CrossRef]
71. Chambers, R.G.; Chung, Y.; Färe, R. Benefit and distance functions. *J. Econ. Theory* **1996**, *70*, 407–419. [CrossRef]
72. Chung, Y.H.; Färe, R.; Grosskopf, S. Productivity and undesirable outputs: A directional distance function approach. *J. Environ. Manag.* **1997**, *51*, 229–240. [CrossRef]
73. Álvarez, I.C.; Barbero, J.; Zofío, J.L. A data envelopment analysis toolbox for MATLAB. *J. Stat. Softw.* **2020**, *95*, 1–49. [CrossRef]

74. NREL. H2A: Hydrogen Analysis Production Models. Available online: https://www.nrel.gov/hydrogen/h2a-production-models.html (accessed on 30 June 2022).
75. UK. *UK Hydrogen Strategy*; Secretary of State for Business, Energy & Industrial Strategy: London, UK, 2021.
76. The German Government. *The National Hydrogen Strategy*; Division, P.R., Ed.; Federal Ministry for Economic Affairs and Energy: Berlin, Germany, 2020; Available online: https://www.bmwk.de/Redaktion/EN/Publikationen/Energie/the-national-hydrogen-strategy.pdf?__blob=publicationFile&v=6 (accessed on 24 July 2022).
77. UNECE. Draft Roadmap for Production and Use of Hydrogen in Ukraine. Available online: https://unece.org/sites/default/files/2021-03/Hydrogen%20Roadmap%20Draft%20Report_ENG%20March%202021.pdf (accessed on 8 August 2021).
78. De Blasio, N.; Hua, C.; Nuñez-Jimenez, A. Sustainable Mobility: Renewable Hydrogen in the Transport Sector. *Environ. Nat. Resour. Program Policy Briefs* **2021**. Available online: https://www.belfercenter.org/sites/default/files/2021-06/HydrogenPB3.pdf (accessed on 24 July 2022).
79. EC. Powering a Climate-Neutral Economy: Commission Sets out Plans for the Energy System of the Future and Clean Hydrogen. Available online: https://ec.europa.eu/commission/presscorner/detail/en/ip_20_1259 (accessed on 8 August 2020).

Article

Numerical and Experimental Performance Evaluation of a Photovoltaic Thermal Integrated Membrane Desalination System

Sajid Ali *, Fahad Al-Amri and Farooq Saeed

Mechanical and Energy Engineering Department, Imam Abdulrahman Bin Faisal University, Dammam 31441, Saudi Arabia
* Correspondence: sakzada@iau.edu.sa

Abstract: Membrane desalination (MD) is preferred over other desalination techniques since it requires a lower temperature gradient. Its performance can be further enhanced by preheating the intake of saline water. In this context, a novel solar-assisted air gap membrane desalination (AGMD) system was hypothesized. The motivation was derived from the fact that the use of solar energy to provide power and a pre-heating source for the intake of saline water can offer a sustainable alternative that can further enhance the acceptance of MD systems. Since solar panels suffer from a loss of efficiency as they heat up during operation, a solar-assisted air gap membrane desalination (AGMD) system can help to improve the overall system performance by (1) providing the necessary pumping power to operate the system and (2) improving solar panel performance by exchanging heat using water that is (3) used to pre-heat the saline water necessary for increased performance of the AGMD system. To verify the hypothesis, a solar-assisted AGMD system for freshwater production was theoretically designed, fabricated locally, and then tested experimentally. The effect of the process operating parameters and the ambient conditions on the overall performance of the proposed solar-assisted AGMD desalination unit is presented in detail, both theoretically and experimentally. The results indicated a direct correlation between the permeate flux, saline hot feed temperature, and hot feed flow rate. In addition, an inverse relationship between the cold feed temperature, cold feed flow rate, and the air gap thickness of the module was also observed and reported, thus, validating the hypothesis that a solar-assisted air gap membrane desalination (AGMD) system can help to boost performance.

Keywords: membrane desalination; renewable energy; solar-assisted desalination; permeate flux; air gap membrane desalination

Citation: Ali, S.; Al-Amri, F.; Saeed, F. Numerical and Experimental Performance Evaluation of a Photovoltaic Thermal Integrated Membrane Desalination System. *Energies* **2022**, *15*, 7417. https://doi.org/10.3390/en15197417

Academic Editors: Sharul Sham Dol and Anang Hudaya Muhamad Amin

Received: 9 March 2022
Accepted: 21 April 2022
Published: 10 October 2022

Publisher's Note: MDPI stays neutral with regard to jurisdictional claims in published maps and institutional affiliations.

1. Introduction

The continuing increase in the global human population has led to a continuous decline in access to fresh water [1]. The problem is further compounded by climate change due to the uneven redistribution of freshwater resources around the globe [2]. Well over a billion people around the world do not have access to safe drinking water [3]. It is estimated that by 2025, more than 2.8 billion people worldwide will face water shortages [4]. The urgency of increasing access to fresh water and providing alternatives is a major concern for humanity, in general, and the scientific community. Distillation has the potential to address this concern and it has been acknowledged and duly recognized as a practical approach to overcoming the problem of freshwater scarcity [5,6]. Distillation techniques employing various methods, such as reverse osmosis, multi-effects, multistage flashes, and membranes, are some of the possible ways to desalinate seawater [7,8].

In various regions of the world, large-scale reverse osmosis (RO) has been used to provide drinkable water from seawater [9]. The reverse osmosis and multistage flash techniques share the credit for producing the highest volume of fresh water in the world

(43.5%) [10,11]. Reverse osmosis, which is a pressure-driven membrane separation process, is highly effective for desalinating seawater along the coastal areas, such as those around the Arabian peninsula [12]. Although this method has a lower maintenance cost, the installation and operating cost, along with susceptibility to fouling, which is an undesirable phenomenon, make this process unsuitable for small-scale desalination [13–15]. In the search for distillation techniques that are independent of conventional fuel resources, membrane distillation (MD) stands out since it has the ability to directly utilize solar thermal energy [16,17]. Membrane distillation relies on the difference in temperature between both sides of the hydrophobic membrane. The membrane used in this process prevents the liquid phase from being present but allows for the transfer of the vapor phase [18,19]. The temperature difference between both sides of the membrane results in a vapor pressure difference, which is the main driving force behind the transfer of vapors from the hot saline feed side of the membrane to the cold feed side [20]. The MD process has several distinguishable attributes, making it a better choice over other desalination techniques [21], such as lower resistance to mass transfer, lower operating hydrostatic pressure, and lower operating temperature [22,23]. In the MD process, low-grade thermal energy can be used as a heat source [24]. Four widely used MD configurations are (i) air gap membrane distillation (AGMD), (ii) vacuum membrane distillation (VMD), (iii) direct contact membrane distillation (DCMD), and (iv) sweep gas membrane distillation (SGMD) [25,26].

In comparison to these processes, AGMD has relatively low energy consumption. In addition, compared to DCMD and VMD, AGMD can operate at lower operating temperatures and lower operating hydrostatic pressures [27,28]. AGMD's ability to attain extremely high salt rejection factors adds to its appeal. Membrane fouling is also less of an issue than with AGMD as compared to other pressure-driven desalination techniques, such as RO. There is no need for substantial pre-treatment, which is required in the RO procedure [29]. Desalination necessitates a large amount of energy, which comes at a considerable expense. It is reported in [30] that the desalination technique necessitates a 9-to-22-fold increase in energy expenditure over typical surface water treatment methods. Huyen et al. [31] provided a thorough overview of the energy requirements of different desalination processes. The comparison is made in terms of total equivalent electric consumption (kWh/m^3), which is the sum of electric and thermal energy usage (kWh/m^3). Multi-stage flash (MSF), thermal-vapor compression (TVC), electrodialysis (ED), multi-effect distillation (MED), and reverse osmosis (RO) have electrical energy consumption needs of $5\ kWh/m^3$, $1.8\ kWh/m^3$, $3.2\ kWh/m^3$, $2.2\ kWh/m^3$, and $7\ kWh/m^3$, respectively [32,33].

In AGMD, a condensing surface is placed between the membrane and the cold flow channel. Compared to other MD configurations, the presence of the air gap significantly reduces the conduction losses, thus increasing the thermal efficiency [34,35]. Additionally, the air gap layer allows for direct collection of the permeate [36]. An optimized AGMD model was developed for seawater desalination in [37]. An innovative AGMD module with improved efficiency was presented by Tian et al. [38]. The proposed AGMD design significantly improved productivity. The primary performance index was estimated using the regression model. The thermal efficiency of the AGMD module was evaluated for a particular flow rate in [39]. The effect of the material and type of membrane used in an AGMD module on its performance was experimentally studied in [40]. It was reported in [41] that a larger membrane pore size results in higher permeate flux. It was shown that replacing the flat condensing plate with a channeled plate improved the process efficiency by 50% [42]. For household applications in UAE, a novel AGMD system was designed and developed to co-produce hot and fresh water [43]. The efficiency of a solar panel is directly linked to its surface temperature. Usually, up to 80% of the solar radiation is transformed into heat [44]. It has been shown that for a single degree increase in temperature, the effectiveness of a PV solar cell is reduced by 0.45% [45]. The electrical efficiency of the solar panels requires constant cooling to extract the excessive heat from its surface [46]. Thus, in the proposed design, by extracting thermal energy from the panel, it will act as a passive cooling agent to reduce the panel surface temperature. The proposed system was tested experimentally. Operation and control-related issues of an AGMD system were discussed by Change et al. [47,48].

A thorough review of the literature revealed that there have been only a limited number of studies on the performance analysis of solar-assisted AGMD desalination systems. In the quest for further improvements, the current paper presents a novel design of a solar-assisted AGMD desalination system for freshwater production. The salient feature of this new design is that the saline water is pre-heated as it is passed through a liquid container integrated at the back of a solar PV panel. It is postulated that the pre-heated saline water will result in improved performance of an AGMD desalination system. In this regard, the current paper presents the design of the proposed system along with a detailed theoretical thermodynamic and heat and mass transfer analysis accompanied by experiment observations and measurements to help validate the improved performance of the proposed system. The effect of various process operating parameters is also presented in detail.

In the sections that follow, the first details of the design of the proposed model, including the in-house designed and built AGMD module, are presented. Next, the theoretical model of the complete process is described, along with its numerical implementation. Then, the experimental setup is described, along with different instrumentations for different measurements. The numerical and experimental results are then used to present the effect of various process operating parameters on the AGMD system performance. Finally, the study ends with important conclusions and some recommendations for future work.

2. Model Development

As shown in Figure 1, the saline solution is heated by passing it through the hot water at the back of the solar panel. The main components of the experimental setup are the air gap membrane module, water chiller, and solar panel integrated with a passive water-cooling system. The rear of the panel is filled with water, as the surface temperature of the solar panel rises, the temperature of the water at the rear of the panel also increases. A pipe (main pipe) carrying the saline solution, supplied from a tank via a pump, is passed through the heated water and given as an input to the AGMD module through a connecting pipe, Figure 2. Desalinated water is collected in a container placed below the AGMD module.

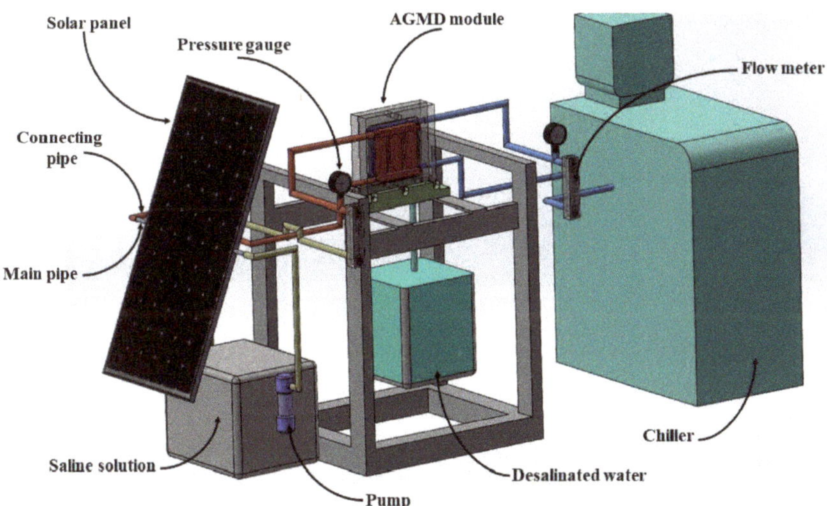

Figure 1. Proposed system.

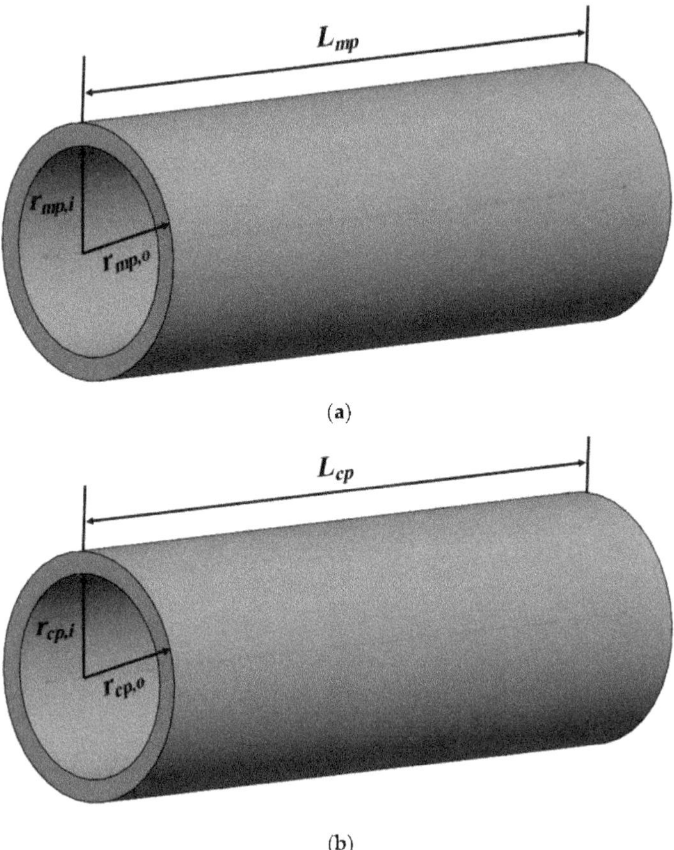

Figure 2. (**a**) Main pipe (mp) passing through the hot water and (**b**) connecting pipe (cp) that connects the main pipe with the AGMD module.

The temperature of the hot water surrounding the main pipe is "T_w" and temperature of the ambient air surrounding the connecting pipe is "T_a". The mean inlet temperatures for main and connecting pipes are "$T_{mp,mi}$" and "$T_{cp,mi}$", respectively. Similarly, the mean outlet temperatures for the main and connecting pipes are "$T_{mp,mo}$" and "$T_{cp,mo}$", respectively, Figure 3. The temperature differences for the main and connecting pipes are defined as

$$\Delta T_{mp,o} = T_w - T_{mp,mo} \tag{1}$$

$$\Delta T_{mp,i} = T_w - T_{mp,mi} \tag{2}$$

$$\Delta T_{cp,o} = T_w - T_{cp,mo} \tag{3}$$

$$\Delta T_{cp,i} = T_w - T_{cp,mi} \tag{4}$$

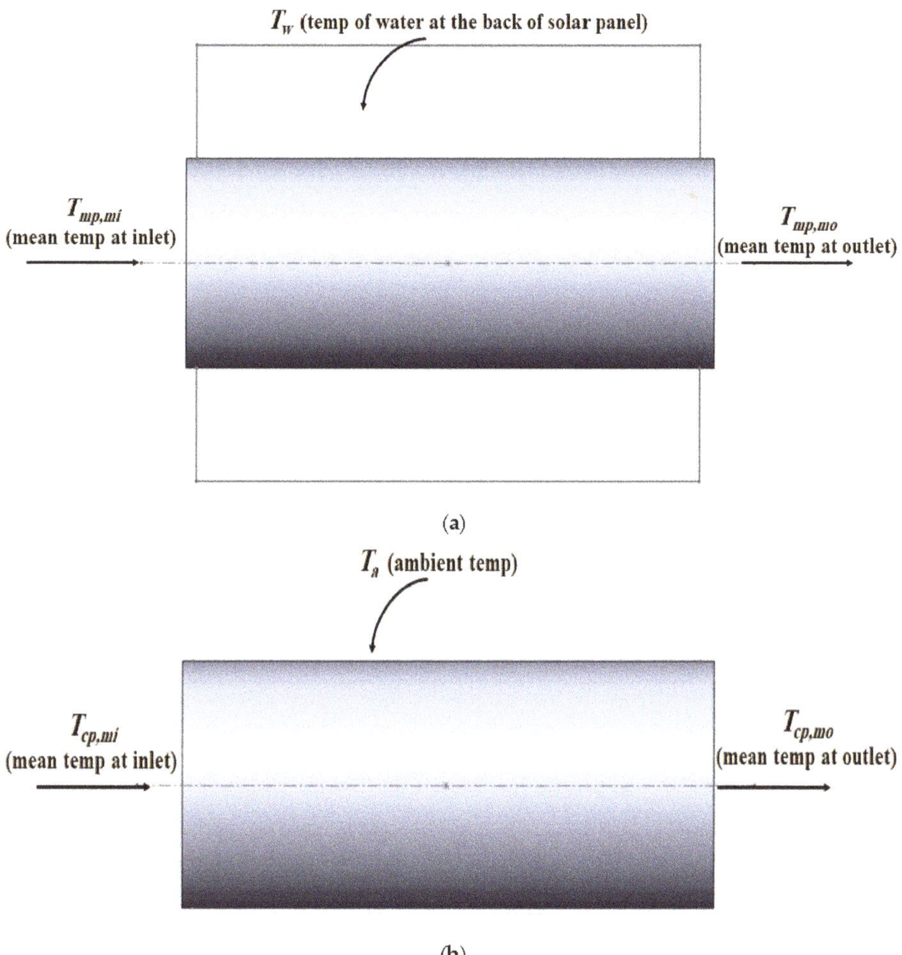

Figure 3. Inlet, outlet, and surrounding temperatures at the (**a**) main pipe (surrounded by hot water) and (**b**) connecting pipe (surrounded by ambient air).

The log-mean temperature differences for the main and connecting pipes are

$$\Delta T_{mp,lm} = \frac{\Delta T_{mp,o} - \Delta T_{mp,i}}{\ln\left(\frac{\Delta T_{mp,o}}{\Delta T_{mp,i}}\right)} \tag{5}$$

$$\Delta T_{cp,lm} = \frac{\Delta T_{cp,o} - \Delta T_{cp,i}}{\ln\left(\frac{\Delta T_{cp,o}}{\Delta T_{cp,i}}\right)} \tag{6}$$

The rates of heat transfer in the main and connecting pipes are given as

$$q_{mp} = \bar{U}_{mp} A_{mp,s} \Delta T_{mp,lm} \tag{7}$$

$$q_{cp} = \bar{U}_{cp} A_{cp,s} \Delta T_{cp,lm} \tag{8}$$

since we can write

$$\overline{U}_{mp}A_{mp,s} = \frac{1}{R_{mp}} \tag{9}$$

$$\overline{U}_{cp}A_{cp,s} = \frac{1}{R_{cp}} \tag{10}$$

where the corresponding thermal resistances are

$$R_{mp} = \frac{1}{h_{mp,1}2\pi r_{mp,i}L_{mp}} + \frac{\ln\left(\frac{r_{mp,o}}{r_{mp,i}}\right)}{2\pi k_{mp}L_{mp}} + \frac{1}{h_{mp,2}2\pi r_{mp,o}L_{mp}} \tag{11}$$

$$R_{cp} = \frac{1}{h_{cp,1}2\pi r_{cp,i}L_{cp}} + \frac{\ln\left(\frac{r_{cp,o}}{r_{cp,i}}\right)}{2\pi k_{cp}L_{cp}} + \frac{1}{h_{cp,2}2\pi r_{cp,o}L_{cp}} \tag{12}$$

The surface heat transfer coefficients for the main pipe in the above equations can be found from the associated Nusselt number correlations:

$$h_{mp,1} = \frac{K_{mp}}{D_{mp,i}}\overline{Nu}_{mp,1}$$

$$h_{mp,2} = \frac{K_{mp}}{D_{mp,0}}\overline{Nu}_{mp,2}$$

Here,

$$\overline{Nu}_{mp,1} = 0.023 Re_{mp}^{\frac{4}{5}} Pr^n$$

$$\overline{Nu}_{mp,2} = 0.36$$

Similarly, for the connecting pipes, the associated heat transfer coefficients and corresponding Nusselt number correlations are

$$h_{cp,1} = \frac{K_{cp}}{D_{cp,i}}\overline{Nu}_{cp,1}$$

$$h_{cp,2} = \frac{K_{cp}}{D_{cp,0}}\overline{Nu}_{cp,2}$$

$$\overline{Nu}_{cp,1} = 0.023 Re_{cp}^{\frac{4}{5}} Pr^n$$

$$\overline{Nu}_{cp,2} = 0.40$$

The rates of heat transfer in the main and connecting pipes in terms of the flow rates and heat capacities are

$$q_{mp} = \dot{m}C_P\left(T_{mp,mi} - T_{mp,mo}\right) \tag{13}$$

$$q_{cp} = \dot{m}C_P\left(T_{cp,mi} - T_{cp,mo}\right) \tag{14}$$

Substituting Equations (5) and (6) into Equations (7) and (8) and equating the resultant with Equations (13) and (14), respectively, and then rearranging the resulting equations gives

$$T_{mp,mo} = T_w - \left(T_w - T_{mp,mi}\right)\exp\left(\frac{-\overline{U}_{mp}A_{mp,s}}{\dot{m}C_P}\right) \tag{15}$$

$$T_{cp,mo} = T_a - \left(T_a - T_{cp,mi}\right) \exp\left(\frac{-\bar{U}_{cp}A_{cp,s}}{\dot{m}C_P}\right) \tag{16}$$

Thus, for given main and connecting pipe mean inlet temperatures, pipe geometry, and flow rates, the Nusselt number correlations were used to determine the surface heat transfer coefficients and the thermal resistance. Then, the overall heat transfer coefficients (U) were determined, and finally, the mean outlet temperatures were calculated. Since the fluid properties were evaluated at an initial guess temperature, an iterative procedure was adopted to ensure that the fluid properties corresponded to the film temperature based on the mean inlet and converged outlet temperatures.

Different sections of the proposed AGMD module are shown in Figure 4, which are the (i) hot flow chamber, (ii) membrane layer, (iii) air gap, (iv) cooling plate, and (v) cold flow chamber.

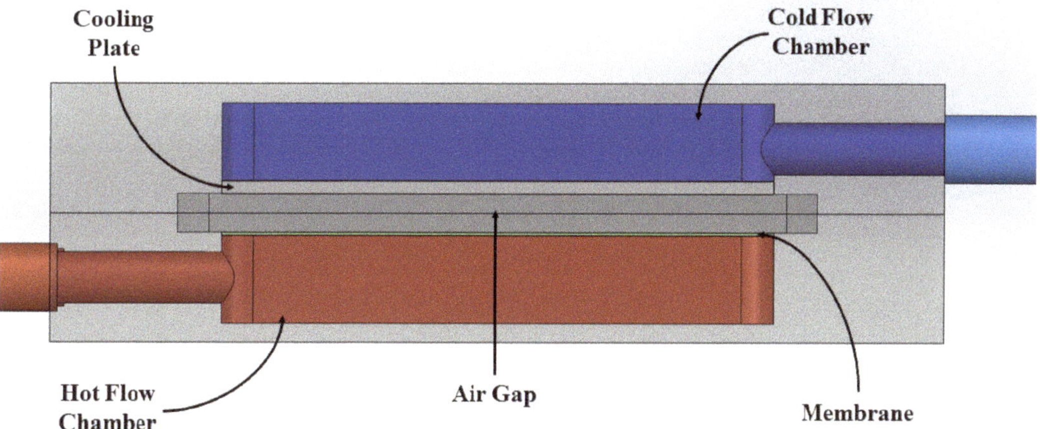

Figure 4. Proposed AGMD module.

The mass transferred along the material of the membrane relies on the difference in vapor pressure between the membrane sides. The correlation is described between mass transferred (J_{wvp}) and vapor pressure difference (ΔP) along the membrane and it can be written as follows [49]:

$$J_{wvp} = M_c\left(P_{hm} - P_{cm}\right) \tag{17}$$

where Mc is the mass transfer coefficient, P_{hm} is the vapor pressure on the feeding side of the membrane, and P_{cm} is the vapor pressure on the condensing side of the membrane. Substituting the overall mass transfer coefficient (Mc) into Equation (17), a simplified equation of the water vapor permeate flux is given as

$$J_{wvp} = \frac{\varepsilon P_{total} D_{water-air}}{RT_m \frac{\left(P_{afm} - P_{acm}\right)}{\ln\left(\frac{P_{afm}}{P_{acm}}\right)}}\left(P_{hm} - P_{cm}\right) \tag{18}$$

where $D_{water-air}$ is the diffusion of the mass in the water vapor and the air, ε is the membrane porosity, and Tm is the mean temperature. Antoine equations [49] are applied to obtain the values of P_{hm} and P_{cm}. In the case of salt dissolved in the hot fluid, Raoult's law is used to express the vapor pressure, where P_{hm} is replaced with P_{hhm} to include the salt concentrate effect in the hot feed as follows:

$$P_{hhm} = (1 - CM_{NaCl})P_{hm} \tag{19}$$

where

$$CM_{NaCl} = \frac{\text{NaCl concentration } \left(\frac{g}{L}\right)}{58.44 \left(\frac{g}{mol}\right)} \tag{20}$$

Heat transfer analysis over the membrane is performed to calculate T_{hm} and T_{cm}. Through the hot saline solution, the heat transferred is presented as

$$Q_h = \left(h_h + J_{wvp}C_f\right)\left(T_{cp,mo} - T_{hm}\right) \tag{21}$$

where h_h is the heat transfer coefficient at the hot feed side, C_f is the coefficient of specific heat, and T_{hm} is the temperature at the hot side of the membrane.

Applying the methodology adopted in [49], the heat transfer from the membrane to the cold solution passing through the air gap and cooling plate is given as (i) heat transferred from the hot side of the membrane to the condensation liquid interface Q_1, (ii) heat transferred via convection through the condensate layer Q_2, (iii) heat transfer via conduction into the cooling plate Q_3, and (iv) heat transferred via convection between the coolant solution and the cold plate Q_4, which are expressed as

$$Q_1 = h_1\left(T_{hm} - T_{cp}\right) + J_{wvp}Q_{vp} \tag{22}$$

$$Q_2 = h_2\left(T_{cp} - T_p\right) \tag{23}$$

$$Q_3 = k_c\left(T_p - T_{cc}\right) \tag{24}$$

$$Q_4 = h_3\left(T_{cc} - T_c\right) \tag{25}$$

Here, h_i is the coefficient of heat transfer, T_{cp} is the temperature of the condensate side of the cooling plate, T_p is the temperature of the cooling plate, T_{cc} is the temperature of the cold fluid side of the plate, T_c is the temperature of the cold fluid, k_c is the thermal conductivity per unit thickness of the plate, and Q_{vp} is the enthalpy of vaporization. The overall heat moved from the condensate layer to the cold solution can be written as

$$Q_c = h_T\left(T_{cp} - T_c\right) \tag{26}$$

where h_T is the heat transfer coefficient from membrane to the cold solution and is given as

$$h_T = \left(\frac{1}{h_2} + \frac{1}{k_c} + \frac{1}{h_3}\right)^{-1} \tag{27}$$

Therefore, the temperature of the hot side of the membrane and cold side of the cooling plate is obtained in the following form:

$$T_{hm} = T_{cp,mo} - \frac{H_T}{h_h}\left(\left(T_{cp,mo} - T_c\right) + \frac{J_{wvp}Q_{vp}}{h_1}\right) \tag{28}$$

and

$$T_{cp} = T_c + \frac{H_T}{h_h}\left(\left(T_{cp,mo} - T_c\right) + \frac{J_{wvp}Q_{vp}}{h_1}\right) \tag{29}$$

with

$$H_T = \left(\frac{1}{h_h} + \frac{1}{h_1} + \frac{1}{h_T}\right)^{-1} \tag{30}$$

The solution of Equation (18), along with Equations (28) and (29), will yield the permeate flux through the membrane under given setup conditions.

3. Experimental Setup

The designed and developed experimental setup used in the study is shown in Figure 5. The key components of the experimental setup were the (i) air gap membrane module, (ii) water chiller, and (iii) solar PV panel integrated with a passive water-cooling system.

Figure 5. Experimental setup.

A commercial-scale PV panel with a 330 W peak output and size of 1990 × 995 × 35 mm, as shown in Figure 5, was used in the experimental setup. The rear of the PV panel was enclosed in a tin sheet to form a closed container to hold the water (57 L) such that the water was in direct contact with the backside of the solar panel to absorb the heat from the solar panel. As the surface temperature of the solar panel rose, it heated the water at the back of the panel.

A pipe carrying the saline solution was passed through the heated water. The temperature of the saline solution increased as a result of heat transfer between water at the back of the panel and the pipe. The AGMD unit was manufactured from acrylic with the dimensions of 100 mm × 100 mm to accommodate a membrane with dimensions of 60 mm × 60 mm to desalinate the saline water with a high enough availability of resources. Each chamber was machined from the inside to allow for a thickness of 15 mm for water. A Thermo Scientific Arctic A40/AC200 (Chiller) with a built-in pump (Figure 6a) was used to pump the cold water through the coolant chamber of the AGMD module through the condensation plate that was located between the air gap and the coolant channel, and the coolant water went back to the chiller. The inlet of the AGMD module was connected to the pipe's exit containing the saline solution and passed through the back of the solar panel. The AGMD module consisted of two identical flow chambers, namely, hot and cold flow. A membrane was placed adjacent to the hot flow chamber. At the same time, an aluminum plate (condensation surface) was placed adjacent to the cold flow chamber. Rubber spacers were used within the module components to create the air gap. The pressure difference between the sections was the main driving force causing a permeate flux to flow from the hot solution to the condensation section. Other components of the setup included (i) two flow meters (Omega FL 50000; Figure 6b) to measure and control the flow of hot saline solution and cold-water solution, (ii) two Baumer pressure gauges (Figure 6c) at the inlet of the cold and hot flow chambers, and (iii) thermocouples to measure temperatures at both sides of the AGMD module.

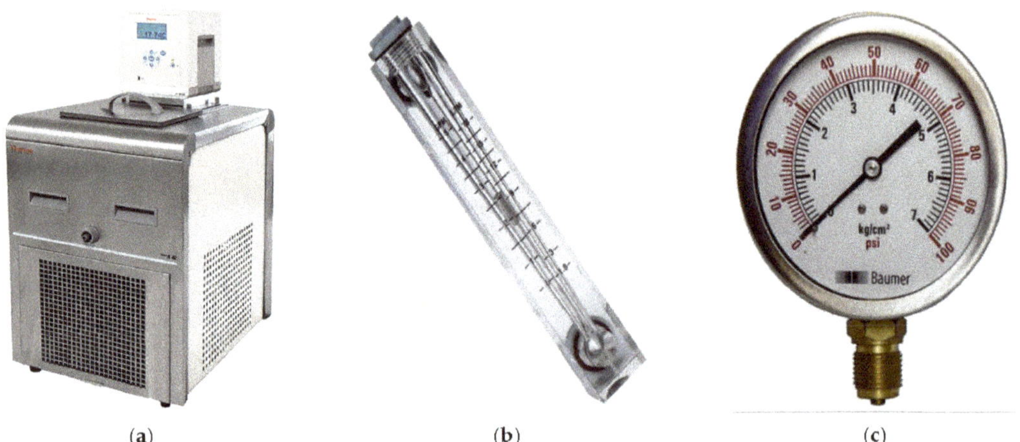

(a)   (b)   (c)

Figure 6. (a) Chiller with a built-in pump, (b) flow meters (Omega FL 50,000), and (c) Baumer pressure gauges.

As per the product's technical specification, k-type thermocouples have a precision of ± 2.5 percent. The calibration of these inserted thermocouples was further tested by utilizing a standard calibration process for more accurate results. These thermocouples provide a linear relationship across a wide temperature range of up to 250 °C. The permeate flux was measured using a beaker (10 cm diameter) with a calibrated scale with an accuracy of ± 1 mm. This translated into a measurement accuracy of the permeate flux to within ± 0.785 kg/m^2/h. For the flow measurement, the Omega FL 50000A was used, and for pressure measurement, the Baumer pressure gauge was used, with the accuracies of $\pm 5\%$ and $\pm 1.6\%$, respectively. The FL 50000A has a robust, one-piece acrylic shell, metal associated components, panel-mountable backside suction, and discharge ports with high accuracy.

4. Results and Discussions

The results of the numerical and experimental investigations of the process parameters are presented in this section. One of the critical process parameters was the temperature of the hot feed, which depends on the PV panel surface temperature and the ambient temperature. The ambient temperature and solar panel surface temperature were recorded, as shown in Figures 7 and 8.

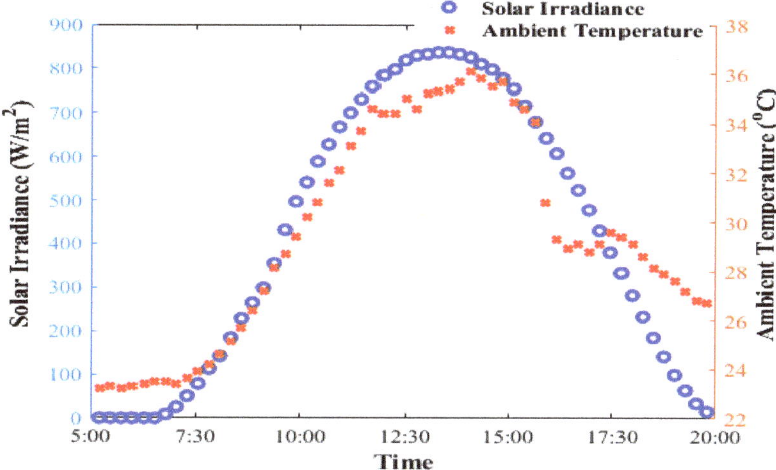

Figure 7. Measured solar irradiance and ambient temperature.

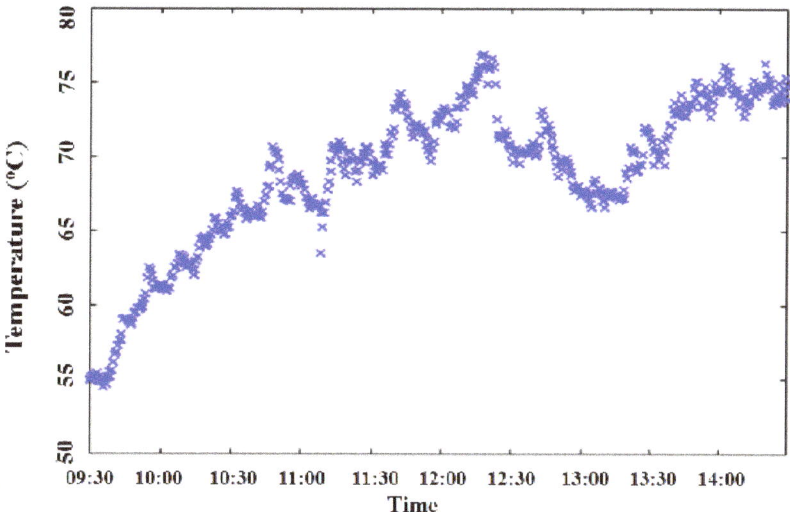

Figure 8. The surface temperature of the solar panel.

The hourly global solar radiation and ambient temperature on 12 April 2021, in the city of Dammam are shown in Figure 7. The ambient temperature was observed to remain above 30 °C during the period of 10 am to 3 pm. The measured surface temperature of the solar panel crossed the 75 °C mark during the same period (Figure 8).

The mean temperatures of the hot saline at the outlet of the main and connecting pipes are shown in Figure 9. The temperature of the water (Tw) at the back of the solar panel surrounding the main pipe was 70 °C. To account for the heat losses from the connecting pipe, the ambient temperature (Ta) in the numerical solution was kept at 35 °C. The effect of the heat losses from the connecting pipe was quite obvious in terms of the time taken to reach the steady temperature.

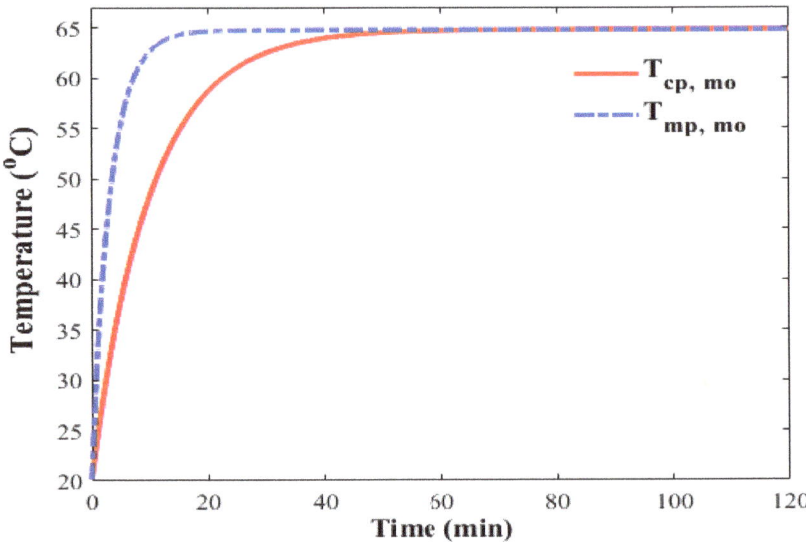

Figure 9. Mean outlet temperatures of the main pipe ($T_{mp,mo}$) and connecting pipe ($T_{cp,mo}$).

The effects of the feed rate of the hot saline (through the hot flow chamber) and cold fluid (through the cold flow chamber) on the permeate are shown in Figures 10 and 11.

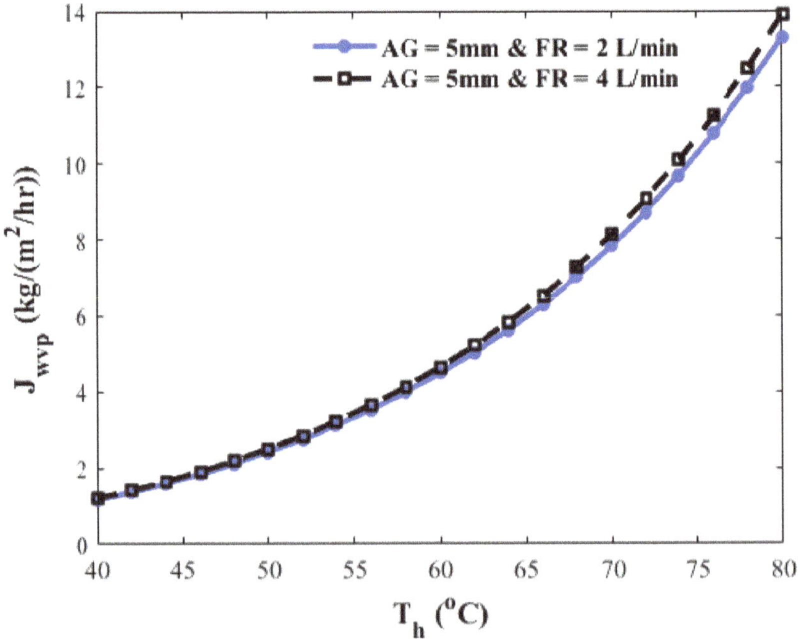

Figure 10. Effect of the hot feed (FR = 2 and 4 L/min) on the permeate (J_{wvp}) at different hot feed temperatures (T_h) (air gap thickness (AG) of 5 mm).

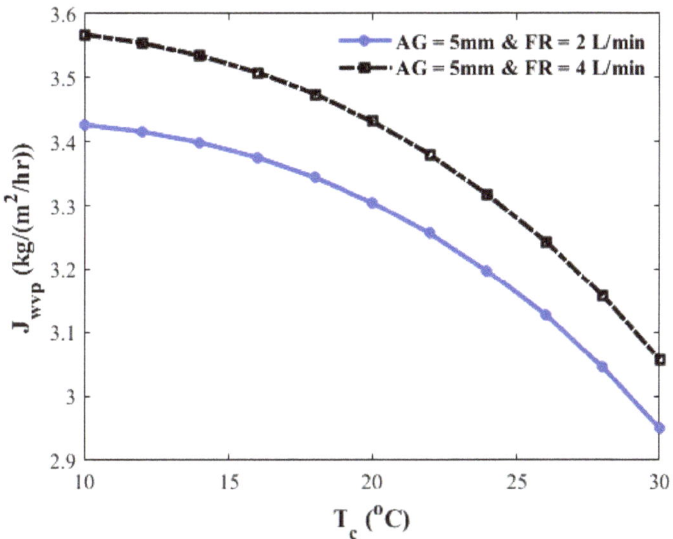

Figure 11. Effect of the hot feed (FR = 2 and 4 L/min) on the J$_{wvp}$ at different cold feed temperatures (T$_c$) (air gap thickness (AG) of 5 mm).

The effect of the hot feed temperature on the Jwvp at a constant cold feed temperature (20 °C) and fixed air gap thickness (5 mm) is presented in Figure 10. It is evident that when the hot feed temperature increased, the permeate increased. Increasing the feed of the hot solution from 2 L/min to 4 L/min produce a slim increase in the permeate corresponding to the hot feed temperature above 60 °C. Similarly, the effect of the cold feed temperature on the permeate at a constant hot feed temperature (55 °C) and fixed air gap thickness (5 mm) is presented in Figure 11. As expected, when the cold feed temperature increased, the permeate decreased. A visible rise in the permeate was obtained by boosting the hot feed rate from 2 L/min to 4 L/min.

The effect of the cold feed on the permeate flux is presented in Figures 12 and 13.

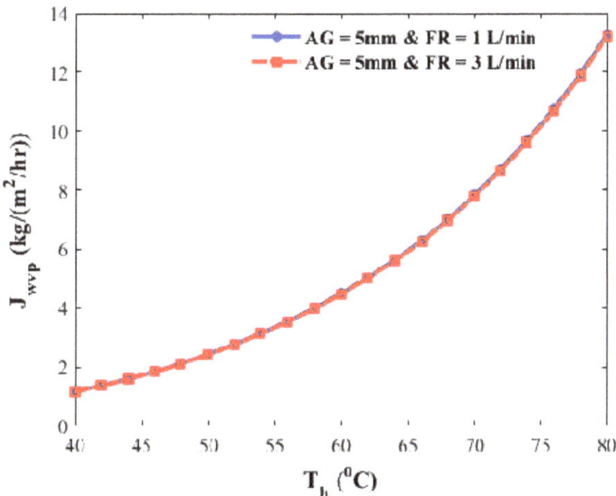

Figure 12. Effect of the cold feed (FR = 1 and 3 L/min) on the J$_{wvp}$ at different hot feed temperatures (T$_h$) (air gap thickness (AG) of 5 mm).

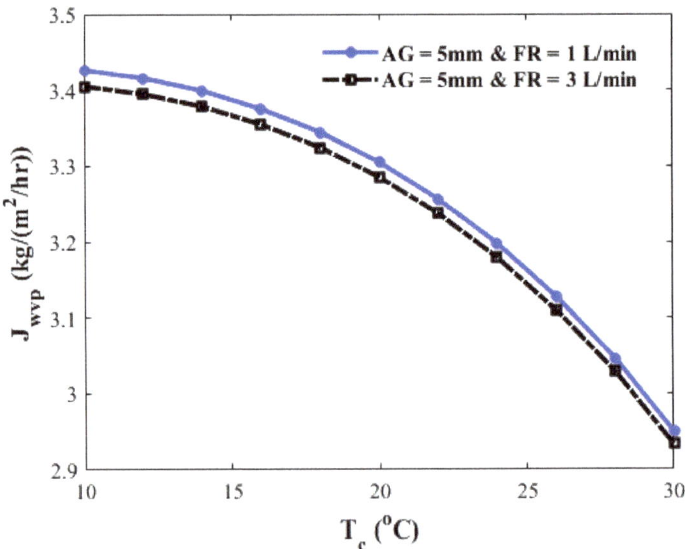

Figure 13. Effect of the cold feed (FR = 1 and 3 L/min) on the J_{wvp} at different cold feed temperatures (T_c) (air gap thickness (AG) of 5 mm).

While keeping the cold feed at a constant temperature (20 °C), the permeate corresponding to the cold feed of 1 L/min and 3 L/min at varying hot feed temperatures remained unchanged, as shown in Figure 12. In contrast, when keeping the hot feed temperature constant (55 °C), a drop in the flux was observed with the increase in the cold feed from 1 L/min to 3 L/min at different cold feed temperatures (Figure 13).

The effect of the air gap thickness (AG) on the permeate is presented in Figures 14 and 15.

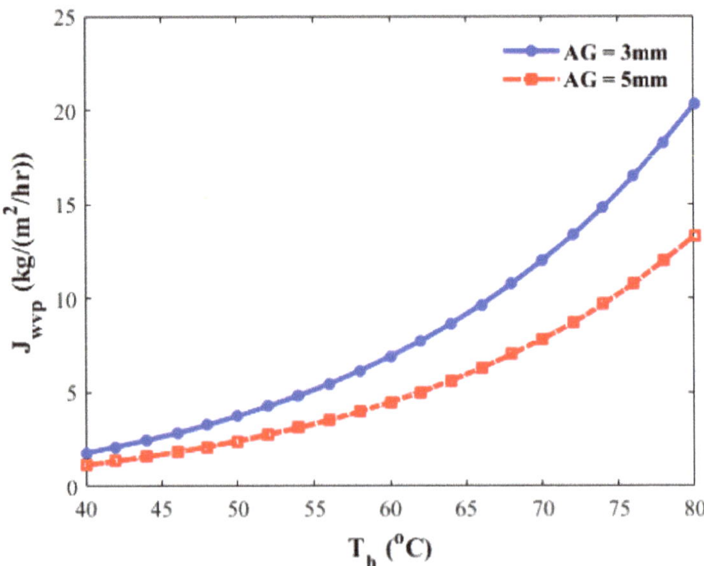

Figure 14. Effect of the air gap thickness (AG = 3 and 5 mm) on the J_{wvp} at different hot feed temperatures (T_h) (hot feed of 2 L/min and cold feed of 1 L/min).

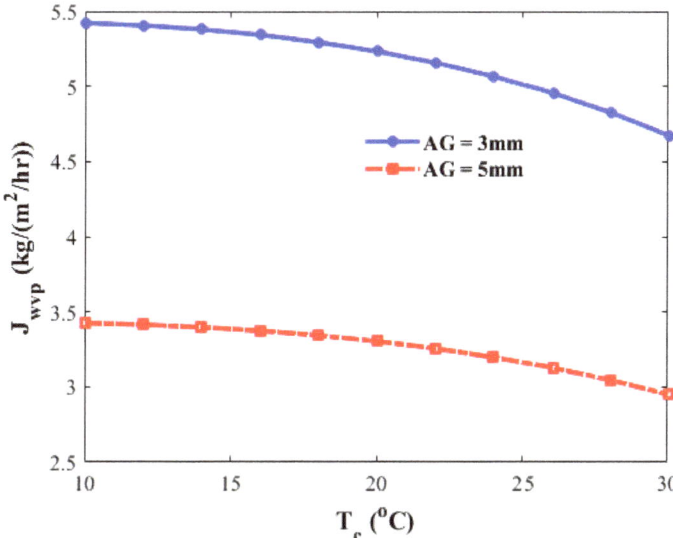

Figure 15. Effect of the air gap thickness (AG = 3 and 5 mm) on the J_{wvp} at different cold feed temperatures (T_c) (hot feed of 2 L/min and cold feed of 1 L/min).

The effect of the air gap thickness on the permeate is shown in Figure 14 at different hot feed temperatures. Increasing the air gap thickness had an adverse impact on the permeate, as it tended to decrease when the air gap thickness was raised from 3 mm to 5 mm. A similar effect, i.e., higher air gap thickness corresponded to lower permeate flux, was also observed in the case of varying cold feed temperature (Figure 15).

An analysis of the airgap membrane desalination parameters was also undertaken experimentally. The permeate flux was measured under different operating conditions. The effects of hot and cold temperatures on the permeate at a constant cold feed temperature (20 °C) are presented in Figures 16 and 17.

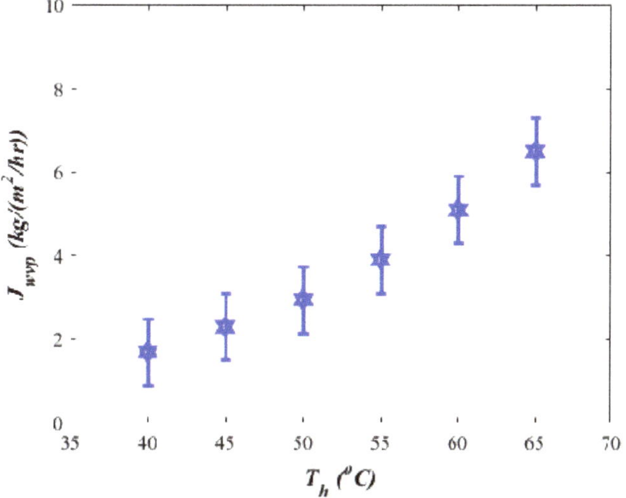

Figure 16. J_{wvp} at different hot feed temperatures (T_h).

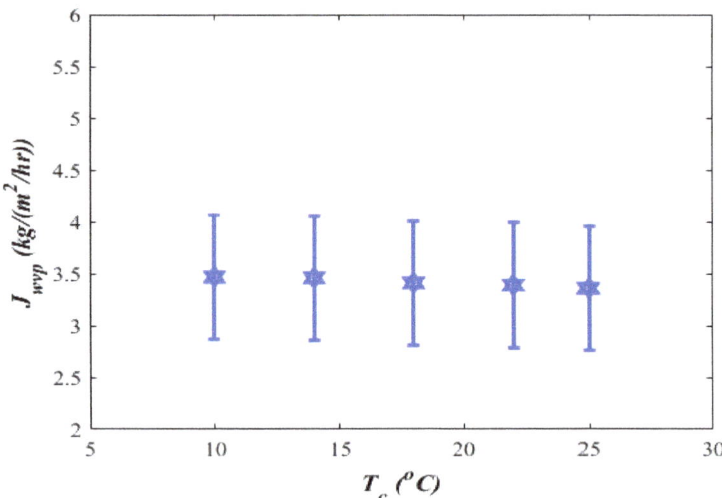

Figure 17. J_{wvp} at different cold feed temperatures (T_c).

The hot feed was kept at 4 L/min, while the cold feed was kept at 1 L/min. The experimental results were found to be in accord with the numerical results presented in the earlier section. Increasing the hot feed temperature resulted in an increase in the permeate (Figure 16) and increasing the cold feed temperature resulted in a decrease in the permeate (Figure 17). The permeate fluxes obtained from the experimental tests (EV) were compared with those obtained from the numerical simulations (NV) (Figure 18).

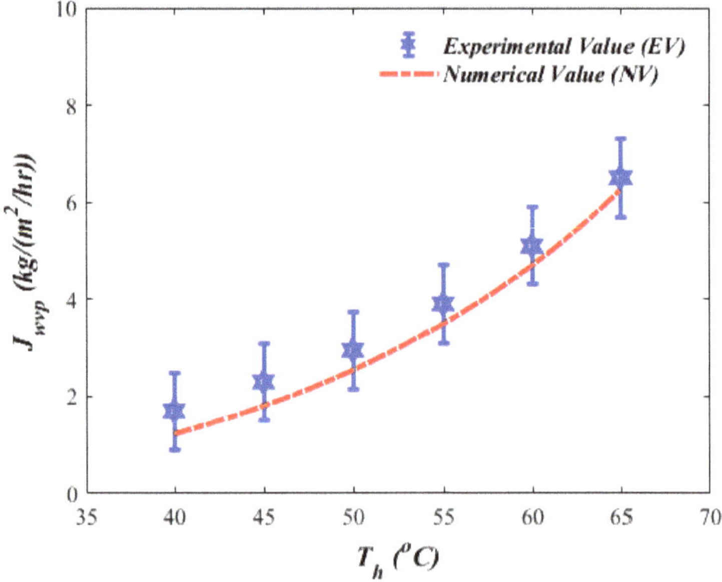

Figure 18. Permeate flux (J_{wvp}) comparison, experimental vs. numerical.

It is evident from the caparison that the experimental results (EV) and numerical values were quite comparable. The numerical results were well within the expected accuracy range (± 0.785 kg/m^2/h) of the experimental setup.

Figure 19 shows the energy consumption comparison as a result of photovoltaic thermal integration. The results showed that a 39% saving in energy consumption was afforded by the use of a photovoltaic thermal integrated membrane desalination system since the system provided electrical power to run the pump and the chiller, in addition to providing the thermal energy to heat the saline water, thus eliminating the use of a heater.

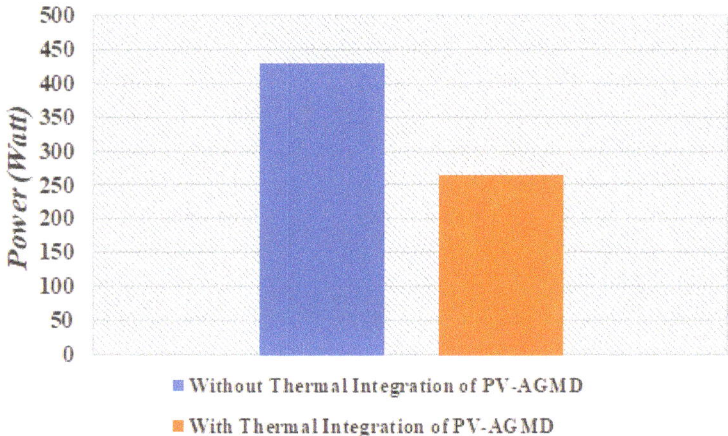

Figure 19. Energy consumption comparison.

5. Conclusions

A novel solar-assisted AGMD system was proposed that utilized solar energy as a heating source for the intake saline solution. A prototype of the proposed design was built and experimentally tested to validate the concept and the numerical model. Detailed thermodynamic, heat, and mass transfer analyses are presented. In the analysis, the temperature of water at the back of the solar panel that surrounded the main pipe was observed to reach above 70 °C. To account for the heat losses from the connecting pipe, the ambient temperature in the numerical solution was kept at 35 °C. The effect of the heat losses from the connecting pipe was noted in terms of the time taken to reach the steady temperature. It was observed that increasing the hot feed temperature on the permeate flux at a constant cold feed temperature (20 °C) and fixed air gap thickness (5 mm) increased the permeate flux. In addition, increasing the feed rate of hot solution from 2 L/min to 4 L/min resulted in a slight rise in the permeate flux corresponding to the hot feed temperature above 60 °C. In contrast, by increasing the cold feed temperature on the permeate flux at a constant hot feed temperature (55 °C) and fixed air gap thickness (5 mm), the permeate flux tended to decrease. Furthermore, it was observed that keeping a constant cold feed temperature (20 °C) at different cold feed flow rates did not have an effect on the permeate flux. On the other hand, when keeping the hot feed temperature constant (55 °C), a decrease in the permeate flux was observed with an increase in the cold feed rate from 1 L/min to 3 L/min at different cold feed temperatures. Increasing the air gap thickness resulted in a decrease in the permeate flux. The permeate fluxes obtained from the experimental tests were compared with those obtained from the numerical simulations. It was observed from the caparison that the experimental results and numerical values were quite comparable. The numerical results were well within the expected accuracy range (± 0.785 kg/m^2/h) of the experimental setup.

Author Contributions: Data curation, S.A.; Investigation, S.A., F.A.-A. and F.S.; Methodology, S.A., F.A.-A. and F.S.; Project administration, F.A.-A.; Resources, F.A.-A.; Software, S.A.; Supervision, F.A.-A.; Visualization, S.A.; Writing—original draft, S.A.; Writing—review and editing, F.A.-A. and F.S. All authors have read and agreed to the published version of the manuscript.

Funding: This research was funded by the Deputyship for Research & Innovation, Ministry of Education in Saudi Arabia grant project number IF-2020-024-Eng.

Institutional Review Board Statement: Not applicable.

Informed Consent Statement: Not applicable.

Data Availability Statement: The data that support the findings of this study are available on request from the corresponding author. The data are not publicly available due to privacy or ethical restrictions.

Acknowledgments: The authors extend their appreciation to the Deputyship for Research and Innovation, Ministry of Education in Saudi Arabia, for funding this research work through the project number IF-2020-024-Eng at Imam Abdulrahman bin Faisal University/College of Engineering.

Conflicts of Interest: The authors declare no conflict of interest.

List of Symbols

$A_{mp,s}$	Main pipe surface area
C_P	Specific heat capacity
$D_{cp,o}$	Outer diameter of connecting pipe
$D_{mp,o}$	Outer diameter of main pipe
J_{wvp}	Mass transferred
k_{mp}	Thermal conductivity of main pipe
L_{mp}	Length of main pipe
$\dot{m}$	Mass flow rate
P_{hm}	Vapor pressure on the feeding side
q_{cp}	Rate of heat transfer (connecting pipe)
R_{cp}	Thermal resistance of connecting pipe
$r_{cp,i}$	Inner radius of connecting pipe
$r_{mp,o}$	Outer radius of main pipe
T_a	Ambient air temperature
$T_{cp,mo}$	Mean outlet temperature (connecting pipe)
$T_{mp,mo}$	Mean outlet temperature (main pipe)
$\overline{U}_{cp}$	Overall heat transfer coefficient for connecting pipe
$\Delta T_{mp,lm}$	Log-mean temperature differences
$A_{cp,s}$	Connecting pipe surface area
$D_{cp,i}$	Inner diameter of connecting pipe
$D_{mp,i}$	Inner diameter of main pipe
h_T	Overall convection heat transfer coefficient
k_{cp}	Thermal conductivity of connecting pipe
L_{cp}	Length of connecting pipe
Mc	Mass transfer coefficient
P_{cm}	Vapor pressure on the condensing side
Q_{vp}	Enthalpy of vaporization
q_{mp}	Rate of heat transfer (main pipe)
R_{mp}	Thermal resistance of main pipe
$r_{cp,o}$	Outer radius of connecting pipe
$r_{mp,i}$	Inner radius of main pipe
$T_{cp,mi}$	Mean inlet temperature (connecting pipe)
$T_{mp,mi}$	Mean inlet temperature (main pipe)
T_w	Water temperature at the back of panel
$\overline{U}_{mp}$	Overall heat transfer coefficient for main pipe
ΔP	Vapor pressure difference

References

1. Pangarkar, B.L.; Sane, M.G.; Guddad, M. Reverse osmosis and membrane distillation for desalination of groundwater: A review. *Int. Sch. Res. Not.* **2011**, *2011*, 523124. [CrossRef]
2. Chen, Y.H.; Hung, H.G.; Ho, C.D.; Chang, H. Economic design of solar-driven membrane distillation systems for desalination. *Membranes* **2021**, *11*, 15. [CrossRef] [PubMed]
3. Shannon, M.A.; Bohn, P.W.; Elimelech, M.; Georgiadis, J.G.; Marinas, B.J.; Mayes, A.M. Science and technology for water purification in the coming decades. *Nature* **2008**, *452*, 301–310. [CrossRef] [PubMed]
4. Hameeteman, E. *Future Water (In)Security: Facts, Figures, and Predictions*; Global Water Institute: Columbus, OH, USA, 2013; pp. 1–16.
5. Elimelech, M.; Phillip, W.A. The future of seawater desalination: Energy, technology, and the environment. *Science* **2011**, *333*, 712–717. [CrossRef] [PubMed]
6. Duong, H.C.; Cooper, P.; Nelemans, B.; Cath, T.Y.; Nghiem, L.D. Evaluating energy consumption of air gap membrane distillation for seawater desalination at pilot scale level. *Sep. Purif. Technol.* **2016**, *166*, 55–62. [CrossRef]
7. Marni Sandid, A.; Bassyouni, M.; Nehari, D.; Elhenawy, Y. Experimental and simulation study of multichannel air gap membrane distillation process with two types of solar collectors. *Energy Convers. Manag.* **2021**, *243*, 114431. [CrossRef]
8. Shim, W.G.; He, K.; Gray, S.; Moon, I.S. Solar energy assisted direct contact membrane distillation (DCMD) process for seawater desalination. *Sep. Purif. Technol.* **2015**, *143*, 94–104. [CrossRef]
9. World Health Organization. *Safe Drinking-Water from Desalination*; World Health Organization: Geneva, Switzerland, 2011.
10. Al-Subaie, K.Z. Precise way to select a desalination technology. *Desalination* **2007**, *206*, 29–35. [CrossRef]
11. Al-Salmi, M.; Laqbaqbi, M.; Al-Obaidani, S.; Al-Maamari, R.S.; Khayet, M.; Al-Abri, M. Application of membrane distillation for the treatment of oil field produced water. *Desalination* **2020**, *494*, 114678. [CrossRef]
12. Qin, J.-J.; Lay, W.C.L.; Kekre, K.A. Recent developments and future challenges of forward osmosis for desalination: A review. *Desalin. Water Treat.* **2012**, *39*, 123–136. [CrossRef]
13. Petersen, R.J. Composite reverse osmosis and nanofiltration membranes. *J. Memb. Sci.* **1993**, *83*, 81–150. [CrossRef]
14. Mouiya, M.; Abourriche, A.; Bouazizi, A.; Benhammou, A.; El Hafiane, Y.; Abouliatim, Y.; Nibou, L.; Oumam, M.; Ouammou, M.; Smith, A. Flat ceramic microfiltration membrane based on natural clay and Moroccan phosphate for desalination and industrial wastewater treatment. *Desalination* **2018**, *427*, 42–50. [CrossRef]
15. Krzeminski, P.; Gil, J.A.; van Nieuwenhuijzen, A.F.; van der Graaf, J.H.J.M.; van Lier, J.B. Flat sheet or hollow fibre—comparison of full-scale membrane bio-reactor configurations. *Desalin. Water Treat.* **2012**, *42*, 100–106. [CrossRef]
16. Wang, P.; Chung, T.-S. Recent advances in membrane distillation processes: Membrane development, configuration design and application exploring. *J. Memb. Sci.* **2015**, *474*, 39–56. [CrossRef]
17. Chafidz, A.; Al-Zahrani, S.; Al-Otaibi, M.N.; Hoong, C.F.; Lai, T.F.; Prabu, M. Portable and integrated solar-driven desalination system using membrane distillation for arid remote areas in Saudi Arabia. *Desalination* **2014**, *345*, 36–49. [CrossRef]
18. Drioli, E.; Ali, A.; Macedonio, F. Membrane distillation: Recent developments and perspectives. *Desalination* **2015**, *356*, 56–84. [CrossRef]
19. Alkhudhiri, A.; Darwish, N.; Hilal, N. Membrane distillation: A comprehensive review. *Desalination* **2012**, *287*, 2–18. [CrossRef]
20. Rabie, M.; Elkady, M.F.; El-Shazly, A.H. Effect of channel height on the overall performance of direct contact membrane distillation. *Appl. Therm. Eng.* **2021**, *196*, 117262. [CrossRef]
21. Lawson, K.W.; Lloyd, D.R. Membrane distillation. *J. Memb. Sci.* **1997**, *124*, 1–25. [CrossRef]
22. Al-Obaidani, S.; Curcio, E.; Macedonio, F.; Di Profio, G.; Al-Hinai, H.; Drioli, E. Potential of membrane distillation in seawater desalination: Thermal efficiency, sensitivity study and cost estimation. *J. Memb. Sci.* **2008**, *323*, 85–98. [CrossRef]
23. Cheng, L.-H.; Lin, Y.-H.; Chen, J. Enhanced air gap membrane desalination by novel finned tubular membrane modules. *J. Memb. Sci.* **2011**, *378*, 398–406. [CrossRef]
24. Cath, T.Y.; Adams, V.D.; Childress, A.E. Experimental study of desalination using direct contact membrane distillation: A new approach to flux enhancement. *J. Memb. Sci.* **2004**, *228*, 5–16. [CrossRef]
25. Miao, Q.; Zhang, Y.; Cong, S.; Guo, F. Experimental investigation on floating solar-driven membrane distillation desalination modules. *Membranes* **2021**, *11*, 304. [CrossRef] [PubMed]
26. Attia, H.; Osman, M.S.; Johnson, D.J.; Wright, C.; Hilal, N. Modelling of air gap membrane distillation and its application in heavy metals removal. *Desalination* **2017**, *424*, 27–36. [CrossRef]
27. Gude, V.G.; Nirmalakhandan, N.; Deng, S. Renewable and sustainable approaches for desalination. *Renew. Sustain. Energy Rev.* **2010**, *14*, 2641–2654. [CrossRef]
28. Jönsson, A.-S.; Wimmerstedt, R.; Harrysson, A.-C. Membrane distillation-a theoretical study of evaporation through microporous membranes. *Desalination* **1985**, *56*, 237–249. [CrossRef]
29. Alklaibi, A.M.; Lior, N. Transport analysis of air-gap membrane distillation. *J. Memb. Sci.* **2005**, *255*, 239–253. [CrossRef]
30. Voutchkov, N. Energy use for membrane seawater desalination–current status and trends. *Desalination* **2018**, *431*, 2–14. [CrossRef]
31. Do Thi, H.T.; Pasztor, T.; Fozer, D.; Manenti, F.; Toth, A.J. Comparison of desalination technologies using renewable energy sources with life cycle, pestle, and multi-criteria decision analyses. *Water* **2021**, *13*, 3023. [CrossRef]
32. Al Washahi, M.; Gopinath, A.S. Techno Economical Feasibility Analysis of Solar Powered RO Desalination in Sultanate of Oman. In Proceedings of the 2017 9th IEEE-GCC Conference and Exhibition (GCCCE), Manama, Bahrain, 8–11 May 2017; pp. 1–9.

33. Cherif, H.; Belhadj, J. Environmental life cycle analysis of water desalination processes. In *Sustainable Desalination Handbook*; Elsevier: Amsterdam, The Netherlands, 2018; pp. 527–559.
34. Alquraish, M.M.; Mejbri, S.; Abuhasel, K.A.; Zhani, K. Experimental investigation of a pilot solar-assisted permeate gap membrane distillation. *Membranes* **2021**, *11*, 336. [CrossRef]
35. Duong, H.C.; Chivas, A.R.; Nelemans, B.; Duke, M.; Gray, S.; Cath, T.Y.; Nghiem, L.D. Treatment of RO brine from CSG produced water by spiral-wound air gap membrane distillation—A pilot study. *Desalination* **2015**, *366*, 121–129. [CrossRef]
36. Guillén-Burrieza, E.; Zaragoza, G.; Miralles-Cuevas, S.; Blanco, J. Experimental evaluation of two pilot-scale membrane distillation modules used for solar desalination. *J. Memb. Sci.* **2012**, *409–410*, 264–275. [CrossRef]
37. Guillén-Burrieza, E.; Blanco, J.; Zaragoza, G.; Alarcón, D.-C.; Palenzuela, P.; Ibarra, M.; Gernjak, W. Experimental analysis of an air gap membrane distillation solar desalination pilot system. *J. Memb. Sci.* **2011**, *379*, 386–396. [CrossRef]
38. Tian, R.; Gao, H.; Yang, X.H.; Yan, S.Y.; Li, S. A new enhancement technique on air gap membrane distillation. *Desalination* **2014**, *332*, 52–59. [CrossRef]
39. Elhenawy, Y.; Elminshawy, N.A.S.; Bassyouni, M.; Alanezi, A.A.; Drioli, E. Experimental and theoretical investigation of a new air gap membrane distillation module with a corrugated feed channel. *J. Memb. Sci.* **2020**, *594*, 117461. [CrossRef]
40. Singh, D.; Sirkar, K.K. Desalination by air gap membrane distillation using a two hollow-fiber-set membrane module. *J. Memb. Sci.* **2012**, *421*, 172–179. [CrossRef]
41. Geng, H.; Wu, H.; Li, P.; He, Q. Study on a new air-gap membrane distillation module for desalination. *Desalination* **2014**, *334*, 29–38. [CrossRef]
42. Bahar, R.; Hawlader, M.N.A.; Ariff, T.F. Channeled coolant plate: A new method to enhance freshwater production from an air gap membrane distillation (AGMD) desalination unit. *Desalination* **2015**, *359*, 71–81. [CrossRef]
43. Kumar, N.T.U.; Martin, A.R. Co-production performance evaluation of a novel solar combi system for simultaneous pure water and hot water supply in urban households of UAE. *Energies* **2017**, *10*, 481. [CrossRef]
44. Siddiqui, M.U.; Ali, S.; Khan, S.; Ali, S.; Horoub, M.M. Optimum Tilt Angles of Solar Collectors in Saudi Arabia. In Proceedings of the 3rd International Conference on Electrical, Communication, and Computer Engineering, Kuala Lumpur, Malaysia, 12–13 June 2021; pp. 12–13. [CrossRef]
45. Wu, S.-Y.; Guo, F.-H.; Xiao, L. A review on the methodology for calculating heat and exergy losses of a conventional solar PV/T system. *Int. J. Green Energy* **2015**, *12*, 379–397. [CrossRef]
46. Al-Amri, F.; Maatallah, T.S.; Al-Amri, O.F.; Ali, S.; Ali, S.; Ateeq, I.S.; Zachariah, R.; Kayed, T.S. Innovative technique for achieving uniform temperatures across solar panels using heat pipes and liquid immersion cooling in the harsh climate in the Kingdom of Saudi Arabia. *Alex. Eng. J.* **2022**, *61*, 1413–1424. [CrossRef]
47. Chang, H.; Liau, J.-S.; Ho, C.-D.; Wang, W.-H. Simulation of membrane distillation modules for desalination by developing user's model on Aspen Plus platform. *Desalination* **2009**, *249*, 380–387. [CrossRef]
48. Chang, H.; Wang, G.-B.; Chen, Y.-H.; Li, C.-C.; Chang, C.-L. Modeling and optimization of a solar driven membrane distillation desalination system. *Renew. Energy* **2010**, *35*, 2714–2722. [CrossRef]
49. Guijt, C.M.; Meindersma, G.W.; Reith, T.; De Haan, A.B. Air gap membrane distillation: 1. Modelling and mass transport properties for hollow fibre membranes. *Sep. Purif. Technol.* **2005**, *43*, 233–244. [CrossRef]

Article

Implementation and Validation for Multitasks of a Cost-Effective Scheme Based on ESS and Braking Resistors in PMSG Wind Turbine Systems

Thanh Hai Nguyen [1,*], **Asif Nawaz** [1], **Preetha Sreekumar** [1], **Ammar Natsheh** [1], **Vishwesh Akre** [2] **and Tan Luong Van** [3]

[1] Faculty of Engineering Technology and Science, Higher Colleges of Technology, Dubai P.O. Box 16062, United Arab Emirates

[2] Faculty of Computer Information Systems, Higher Colleges of Technology, Dubai P.O. Box 16062, United Arab Emirates

[3] Department of Electrical and Electronics Engineering, Ho Chi Minh City University of Food Industry, 140 Le Trong Tan, Ho Chi Minh City P.O. Box 760310, Vietnam

* Correspondence: nhai@hct.ac.ae; Tel.: +971-2-2064038

Abstract: This study deals with fault ride-through (FRT) capability and output power fluctuation suppression of wind turbine systems (WTS) having PMSG (permanent-magnet synchronous generator) for mitigating grid frequency variation and voltage flicker in the distribution system. The coordinated control of a cost-effective scheme based on energy storage supercapacitors (ESSs) and braking resistors (BR) is introduced to perform the multiple tasks of the WTS. In this hybrid scheme, the ESSs are initially used to absorb the fluctuated power component with the constraints of their ratings during the grid faults and wind speed variation conditions prior to the activation of the BRs when the ESSs cannot fully consume the mismatched power between the PMSG and grid during severe grid faults. With the additional BRs, the capacity of the costly ESSs is remarkably reduced, while the performance of the fault ride-through capability and power smoothening for the WTS are still kept satisfactory and in compliance with the requirements of advanced grid codes. Detailed experimental implementation and its results for a down-scaled prototype in a laboratory are shown to verify the effectiveness of the introduced scheme along with the simulation results with the high-power rating WTS.

Keywords: braking resistors; energy storage super-capacitors; fault ride-through; permanent-magnet synchronous generator; wind turbine systems

Citation: Nguyen, T.H.; Nawaz, A.; Sreekumar, P.; Natsheh, A.; Akre, V.; Van, T.L. Implementation and Validation for Multitasks of a Cost-Effective Scheme Based on ESS and Braking Resistors in PMSG Wind Turbine Systems. *Energies* **2022**, *15*, 8282. https://doi.org/10.3390/en15218282

Academic Editors: Sharul Sham Dol and Anang Hudaya Muhamad Amin

Received: 26 August 2022
Accepted: 29 October 2022
Published: 5 November 2022

Publisher's Note: MDPI stays neutral with regard to jurisdictional claims in published maps and institutional affiliations.

1. Introduction

Nowadays, renewable energy has attracted considerable attention since fossil fuels are being exhausted and environmental issues have become more serious. Wind energy is one of the most important renewable energy sources [1–5], where the significant penetration of wind power capacity may cause some problems in the power systems, such as grid instability, unbalance, and frequency variation [6,7].

The first issue is that a large amount of electrical power generated by the WTS may lead to adverse effects on the power quality of the network due to randomly varying wind speeds. In this condition, the output power fluctuation of the WTS incurs an imbalance of power between generation and consumption, which results in a deviation of the line frequency from its rated value. Therefore, suppressing the fluctuating components of WTS output power is essential. Another concern is the FRT capability of the WTS under grid fault conditions, where the WTS are required to remain in service and be able to provide reactive power to the network.

In the published literature, in order to mitigate the power fluctuation of the WTS, several power smoothening strategies have been suggested. By exploiting the high inertia effect of the WTS, the turbine power fluctuation can be smoothened to support the frequency regulation for the network [8,9], where no extra apparatus is required. However, its response is slow due to the high rotor inertia, and the smoothened power capacity is limited. Another method using a flywheel system was presented in [10,11], which utilizes simple control schemes. However, the flywheel system is bulky and costly. The pitch angles of the wind turbine blades with the rotor velocity can be regulated to smooth the output power of the WTS, but the WTS loses the maximum power-point tracking operation [12]. With the development and current availability of the energy storage systems, utilization of the energy storage systems to improve the power quality of the WTS has been introduced, where the battery energy storage systems or the hybrid system combining the batteries and supercapacitors are both suggested [13,14]. The main issues of using energy storage systems are high capital investment and maintenance costs.

In order to comply with the requirements of advanced grid codes under grid faults, a braking chopper has been considered a preferred and practical solution, which provides a lower cost and a simple control [15,16]. However, this scheme is used only for the FRT, which is not capable of improving the power quality. Other methods with modification of the control algorithms, such as the current feed-forward method, sliding mode control scheme, and robust adaptive PI controller, have been introduced to improve the response of the WTS during the grid faults [17,18]. Installing the energy storage systems at the connection point of the wind plants can achieve many objectives such as fluctuated power smoothing and providing the FRT capability etc. [19]. Other solutions with the ESSs distributed to single WTS at the back-to-back (B2B) converters through a bidirectional DC/DC converter (BDCC) were introduced [20], which can offer both an FRT capability and power fluctuation mitigation for the WTS. However, the required capacity of the ESS is still high to handle extremely severe grid faults, such as full interruptions of two or three phases of grid voltages, which rarely occur in practice compared to phase-to-phase or phase-to-ground short-circuit faults. Under such faults, the grid can partly absorb real power from the WTS. So, the mismatched power between the wind generator and grid under the common grid faults is less than the rated power of the system. For this scenario, employing the ESSs with a full rating is costly and unnecessary. The required rating of the ESS can be reduced by the de-loaded operation of the WTS, where the output power of the wind turbine generator is reduced [21]. This algorithm can be acceptable, but the WTS is not operating at the MPPT control. A hybrid scheme of the ESS and BR in the WTS was suggested [22,23], where the power capacity of the ESS was lowered. However, the improvement of power quality by installing ESS was not fully investigated, and a control mode switch for the grid-side converter and the BDCC between the normal and fault conditions may cause a current overshot in the system. These issues have not been presented in any of the literature so far.

This paper presents a coordinated control of the ESS and BR for improving the power quality as well as riding through the PMSG WTS, which guarantees a secure and smooth operation of the WTS during normal and severe grid fault conditions. The ESSs are utilized to absorb the fluctuation component of turbine power before injecting it into the grid under the wind speed variation, which smoothens the grid power. Under the grid sags, the ESSs consume a certain level of mismatched power between the PMSG and the network, and the rest is then dissipated by the BRs. A cascaded control of the outer power loop and inner current loop is applied for the BDCC, and a control strategy associated with the ESSs is used for the BRs, was compensation for a phase shift caused by the high-pass filter is employed to obtain an effective power reference of the ESS. With this control scheme, the required capacity of the ESSs is significantly reduced, while both the performances of the power fluctuation mitigation and the FRT under severe faults of the grid are still kept well.

The detailed experimental implementation for a down-scaled prototype in the laboratory is presented to validate the multiple tasks of PMSG WTS, where the squirrel-cage induction motor is used to simulate the characteristics of the wind turbine with the output power depending on the wind speed. The DSP TMS320VC33 chips are used as the main controller boards in the experimental test systems. The experimental results also demonstrate the simulation results for a 2-MW PMSG WTS. The additional contributions of this work are that a switching control mode of the grid-side converter and the BDCC between the normal and fault conditions are not required, and a phase-shift delay compensation for obtaining the power reference of the ESS is employed.

2. Control Scheme of PMSG Wind Turbine Systems

The ESS can be installed in the wind power plants by either a distributed connection to the single WTS or a centralized connection at the wind power plant terminal, where the distributed ESS can not only smoothen the output power of the WTS but also provide an FRT capability under the grid faults. A typical configuration of a direct-drive PMSG WTS with the hybrid scheme of the distributed ESS and BR is shown in Figure 1 [23]. The ESS in this work adopts electric double-layer capacitors (EDLC), which are linked to the DC link of the B2B converters through a BDCC. The BR is connected in parallel with the DC link through a switch. With this configuration, the hybrid scheme of the ESS and BR is well fit for the wind turbine systems with full-scale capacity converters such as PMSG-based wind turbine systems, while this scheme is not suitable to apply for the DFIG-based wind turbine systems because the rating of the converters in the rotor side is already as low as 30% of the system rating.

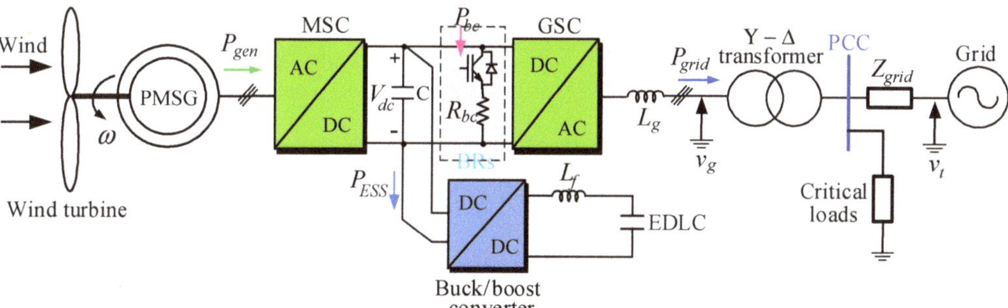

Figure 1. WTS with PMSG utilizes a hybrid scheme of ESSs and BRs in the distribution system [17].

2.1. Turbine Inertial Effect for Absorbing Power under Grid Faults

When grid faults occur, the real power injected into the grid will be restricted due to the limited current rating of the power converters. In order to reduce the mismatch of power between the PMSG and grid, which is to be absorbed by the ESS and BR, the wind turbine is accelerated to reduce the turbine output power since a part of the amount of energy due to inertial effect is required for the acceleration action.

A torque equation for a two-mass model of WTS is expressed as [20,24,25]

$$T_t - T_g = (J_t + J_g)\frac{d\omega}{dt} \tag{1}$$

where T_t, T_g, and J_t, J_g are the torques and inertias of the turbine and PMSG, respectively, and ω is the turbine speed.

For storing the inertial energy, the speed of the turbine is increased by controlling the PMSG. A time interval of the fault and a gradient of the speed is defined as ΔT and $\Delta k(\%)$, respectively. An inertial power, P_J, due to a speed change is given based on the constant inertia definition and (1) as [26]

$$P_J = 2 \cdot P_{rated}(H_T + H_G)\frac{\Delta k}{\Delta T} \tag{2}$$

where H_T and H_G are the turbine and generator inertia constants, respectively, and P_{rated} is the power rating.

2.2. Control of Grid-Side Converter

The grid-side converter (GSC) is used to control the DC-link voltage and the currents flowing into the network. Figure 2 depicts the control scheme of the GSC, where a cascaded control scheme with the main loop of DC-link voltage and an inner loop of grid currents is employed [27,28]. Initially, the positive- and negative-sequence components (PSC and NSC) of dq-axis grid voltages and currents, $E_{dqe}^{p,n}$ and $I_{dqe}^{p,n}$, respectively, are decomposed by digital all-pass filters and Park's transformation, which then a decouple synchronous reference frame PLL (phase-locked loop) for the three-phase system is applied to detect the grid phase angle [29,30]. The phase angle of the negative-sequence voltage component is set as the opposite of the positive-sequence one.

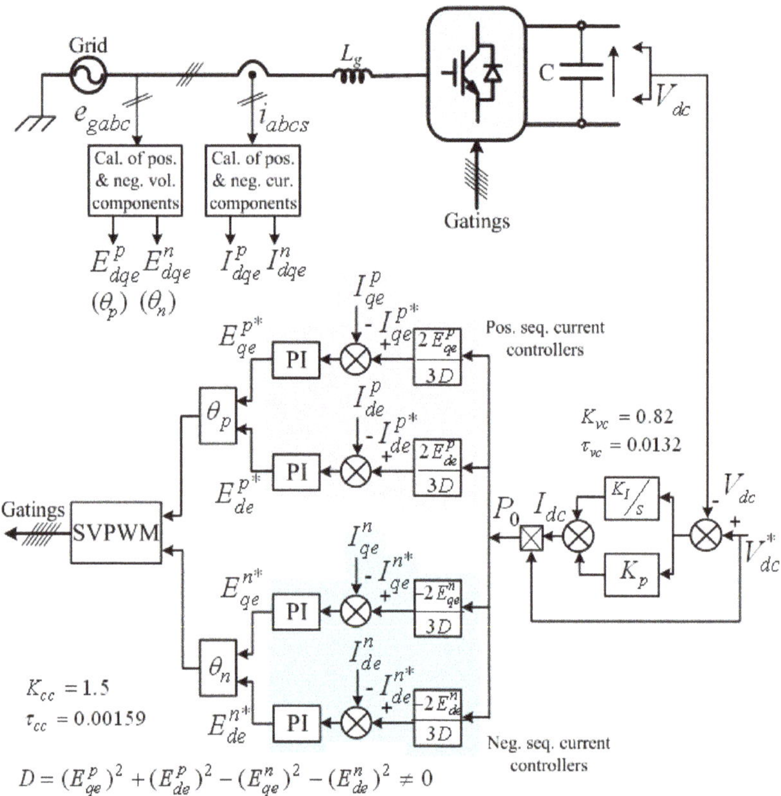

Figure 2. The control scheme of GSC [27,28].

The typical PI (proportional integral) regulators are employed for the DC-link voltage controllers, which determine the amount of real power delivered to the grid. The control algorithm is applied for both symmetrical and asymmetrical faults of grid voltages, where the PSC and NSC of the grid currents are regulated. To guarantee the active power of PMSG delivered to the grid fully, which keeps the DC-link voltage unchanged, along with the PSC of the grid currents, NSC is also required to be injected into the grid, where the dq-axes current reference of the PSC and NSC, I_{de}^{p*}, I_{qe}^{p*}, I_{de}^{n*}, I_{qe}^{n*}, respectively, are determined as below:

$$I_{qe}^{p*} = V_{dc}^* \times I_{dc}^* \frac{2E_{qe}^p}{3D} \text{ and } I_{de}^{p*} = V_{dc}^* \times I_{dc}^* \frac{2E_{de}^p}{3D}$$
$$I_{qe}^{n*} = -V_{dc}^* \times I_{dc}^* \frac{2E_{qe}^n}{3D} \text{ and } I_{de}^{n*} = -V_{dc}^* \times I_{dc}^* \frac{2E_{de}^n}{3D}$$

V_{dc}^* is the DC-link voltage reference, I_{dc}^* is the output of the DC-link voltage controller, E_{de}^p, E_{qe}^p, E_{de}^n, E_{qe}^n are the dq-axes voltage components of the PSC and NSC of the grid voltage, respectively and $D = \left(E_{de}^p\right)^2 + \left(E_{qe}^p\right)^2 - \left(E_{de}^n\right)^2 - \left(E_{qe}^n\right)^2$.

The PI regulators are also applied for the current controllers for all components, as shown in Figure 2. The controller gains are designed following the guidelines in [27,28], which are listed in Figure 2 as $K_{vc} = K_p^{vc} = 0.82$; $K_i^{vc} = 0.82$; $\tau_{vc} = 0.013$ for the DC-link voltage controller and $K_{cc} = K_p^{cc} = 1.5$; $K_i^{cc} = 930$; $\tau_{cc} = 0.0016$ for the grid current controllers. It is noted that by absorbing the PMSG power in the ESS and BR under the grid faults, the GSC remains at a margin to provide reactive power to the grid as required by the grid code [23,31,32], even though this issue is not investigated in this work.

2.3. Control of Machine-Side Converter (MSC)

Figure 3 shows a control scheme of PMSG under both normal and fault conditions. A vector control method is applied for the PMSG, where the *dq* axes of generator currents are regulated. An operation of maximum torque per ampere method for the PMSG is utilized, where the *d*-axis generator current is regulated to zero. So, the *q*-axis generator current is used to adjust the generator's real power, where the reference of the *q*-axis component is determined from the generator speed controller. The PI controller gains of the machine speed and currents are depicted in Figure 3 as $K_{sc} = K_p^{sc} = 10,125$; $K_i^{sc} = 81,000$; $\tau_{sc} = 0.125$ and $K_{cc} = K_p^{cc} = 0.518$; $K_i^{cc} = 388$; $\tau_{cc} = 0.00133$, respectively, which are designed based on the pole-placement technique [33]. Under the normal grid condition, the maximum power point tracking (MPPT) operation of the WTS is applied [34,35], which sets the generator speed reference at the optimal value for the speed controller, as shown in Figure 3. However, when the grid sag occurs, to reduce the output power of WTS, the MPPT operation is deactivated, and the turbine is accelerated. The acceleration rate of the turbine depends on the inertia constant of the system and the speed control parameters, where the selection procedure of the acceleration rate has been described in detail in [23].

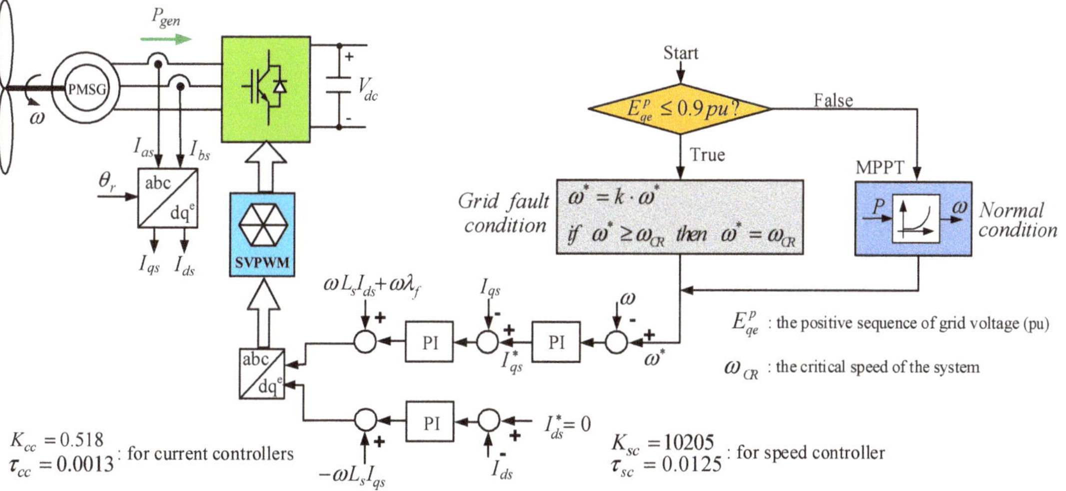

Figure 3. The control scheme of PMSG.

In this work, the acceleration rate, k, is chosen as 1.0005 [23]. The acceleration of the turbine causes a reduction in the output power of WTS according to two following reasons. Firstly, when the MPPT operation is not maintained due to increasing the turbine speed, the tip-speed ratio, λ, is not kept at the optimal value, λ_{opt}. So, the power conversion coefficient, C_p, is lower than the maximum one, C_{pmax}, which can be expressed as

$$\omega > \omega_{opt} \rightarrow \begin{cases} \lambda > \lambda_{opt} \\ C_p < C_{pmax} \\ P_t < P_{t,max} \end{cases} \tag{3}$$

where P_t is the turbine power and $P_{t,max}$ is the optimal turbine power at a certain speed.

Secondly, a portion of the power extracted from wind is stored in the turbine inertia as in (2) during the acceleration of the turbine, so the generator output power, P_{gen}, is reduced according to the acceleration of the turbine expressed as:

$$P_{gen} = P_t - P_J \tag{4}$$

3. Power Smoothing Operation and FRT Control

In a weak distribution system with a low X/R ratio, the fluctuation of the WTS output power may cause a voltage flicker at the point of common coupling, which has negative effects on critical loads [8]. By smoothing the output power of WTS associated with the reactive power injection, the voltage flicker, as well as the frequency variation, would be mitigated. In this work, the power smoothening operation and FRT control for the PMSG WTS in the distribution system are mainly investigated.

3.1. Calculation of Power References for the ESS

During the voltage dips, the real power injected into the network by the GSC is limited, which can be lowered than that of PMSG due to a reduction in grid voltage and a fixed current rating of GSC. The differential power is absorbed by the ESSs and BRs to avoid an overvoltage in the DC-link capacitors, which can be described as

$$P_{diff} = P_{gen} - P_{grid}\big|_{max\ posibility} \tag{5}$$

where P_{diff} is the surplus power of the PMSG and grid, and P_{grid} is the power injected into the power line through the GSC. In this operation, the GSC is controlled to inject the generator power into the network as high as possible with consideration of its current constraint.

It has been reported that the fluctuated power of the WTS in a frequency range of (0.05~1) Hz results in a deviation of the line frequency [8]. So, in order to improve the quality of the WTS power under normal conditions, the fluctuated component of the WTS output power should be suppressed, where the ESSs are utilized to absorb these components caused by random wind speed conditions. The high-frequency power component of PMSG, P_{fluc}, is extracted by the high-pass filter (HPF), which is expressed as

$$P_{fluc} = \frac{s^2}{s^2 + 2\zeta\omega_c s + \omega_c^2} P_{gen} \tag{6}$$

where in this work, the lowest frequency or cut-off frequency of the HPF is chosen as 0.05 Hz and $\zeta = 0.707$ and $\omega_c = 2 \times \pi \times 0.05 = 0.314$ rad/s, which makes sure that the higher frequency components are also extracted for compensating by the ESS.

To improve the real power injected into the grid and the performance of the ESS, the phase shift of the second order HPF should be compensated. A lead-lag compensator is utilized for a phase-shift compensation at the cut-off frequency, where its transfer function, $G_{comp}(s)$, is expressed as

$$G_{comp}(s) = K_{comp}\frac{1 + a_1 s}{1 + a_2 s} \tag{7}$$

where the pole of the lead-lag compensator is selected according to the cut-off frequency of the HPF as $a_2 = \frac{1}{\omega_c^2} = \frac{1}{0.314} \approx 10$, and the zero is chosen by placing it 10 times further from the pole as $a_1 = 1$. K_{comp} is selected as 10 to obtain a unity gain for high-frequency operation range.

Finally, the power reference of the ESSs is determined as

$$P_{fluc} = 10\frac{1+s}{1+10s} \cdot \frac{s^2}{s^2 + 2\zeta\omega_c s + \omega_c^2} P_{gen} \tag{8}$$

Utilizing the above power reference of the ESSs, the main objective of controlling the ESSs is to suppress the fluctuation component of the WTS output power. The optimal charge and discharge of the ESSs for minimizing the operational cost and power losses and prolonging the lifetime of the ESSs are beyond the scope of this work.

3.2. Control of the ESS

The real power is stored or released from the ESSs by controlling the BDCC, where its control scheme diagram is depicted in Figure 4. Under both normal and voltage sag conditions, the real power is a control target of the ESSs, where the power references are determined in (8) and (5) for the two cases, respectively. A cascaded control scheme is adopted for the BDCC, which consists of an outer PI loop for the power and an inner PI loop for the inductor current. From Figure 4, the output of the power controller, which is the inductor current reference, I_{ESS}^*, is given as

$$I_{ESS}^* = K_p^{pc}(P_{ESS}^* - P_{ESS}) + \frac{K_i^{pc}}{s}(P_{ESS}^* - P_{ESS}) \tag{9}$$

$$\text{and } P_{ESS} = V_{ESS} I_{ESS} \tag{10}$$

where P_{ESS} is the ESS power, V_{ESS} and I_{ESS} are the ESS voltage and current, respectively, K_p^{pc} and K_i^{pc} are the power controller gains.

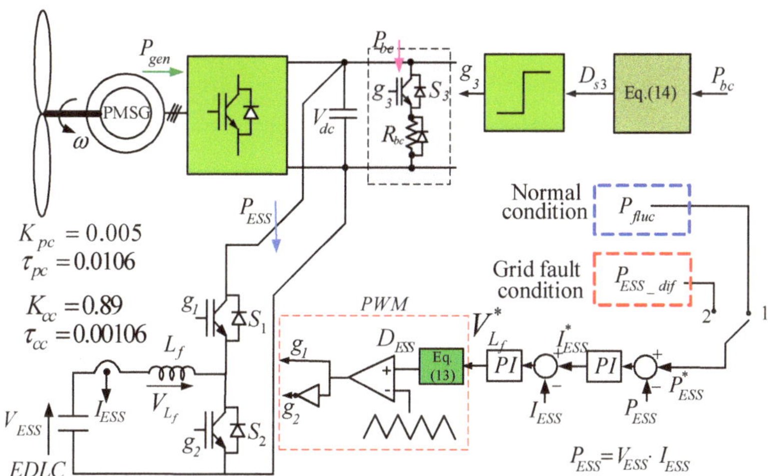

Figure 4. Power and current controllers of the ESSs and the braking chopper operation.

Then, the transfer function for the power controller is expressed as

$$\frac{P_{ESS}}{P_{ESS}^*} = \frac{K_p^{pc}s + K_i^{pc}}{(\frac{1}{V_{ESS}} + K_p^{pc})s + K_i^{pc}} \approx 1 \quad \text{for all } s \tag{11}$$

In this work, the power controller gains are selected by a trial-and-error method as $K_p^{pc} = 0.005$, $K_i^{pc} = 0.472$, and $\tau_{pc} = 0.0106$. Note that the bandwidth of the power controller is low, which is about 100 Hz.

It is seen from Figure 4 the output of the current controller, $V_{L_f}^*$, is expressed as [36]

$$V_{L_f}^* = K_p^{cc}(I_{ESS}^* - I_{ESS}) + \frac{K_i^{cc}}{s}(I_{ESS}^* - I_{ESS}) \tag{12}$$

where K_p^{cc} and K_i^{cc} are the ESS current controller gains, which are selected following the pole-placement technique as $K_p^{cc} = 0.89$, $K_i^{cc} = 840$, and $\tau_{cc} = 0.00106$ [36].

Based on the output of the current controller, which is the inductor voltage reference, the duty cycle, D_{ESS}, for the DC/DC converter is determined by

$$D_{ESS} = \frac{V_{ESS} + V_{L_f}^*}{V_{dc}} \tag{13}$$

where V_{dc} is the DC-link voltage.

The duty cycle is compared to a carrier signal to generate the gating signals for switches S_1 and S_2 of the BDCC. In this work, the carrier frequency is selected as 2 kHz.

3.3. Control of the BRs

The braking chopper is only operated when the ESSs cannot absorb fully a surplus power between the PMSG and the grid, which is shown in Figure 4. The switch S_3 is allocated to adjust the power consumed by the BR, where its duty cycle is determined from the required power and the braking resistance, R_{bc}, expressed as

$$D_{S3} = \frac{R_{bc}}{V_{dc}^2} P_{bc} \tag{14}$$

where P_{bc} is the power command for the BRs, which is calculated as

$$P_{bc} = \begin{cases} P_{diff} - P_{ESS\_rated} \geq 0 & : \quad \text{fault condition} \\ P_{fluc} - P_{ESS\_rated} \geq 0 & : \quad \text{normal condition} \end{cases} \cdot \tag{15}$$

It is worth noting that the pole-placement technique is adopted to design the controller gains, where the PI gains of all controllers have been obtained for the studied systems. For the field deployment system, this method can be used to obtain the gain with fine-tuning to achieve good performance.

4. Ratings of the ESSs and the BRs

The ESS is utilized to consume the fluctuated power components of the WTS in the normal grid condition, where the capacity of the ESS is selected appropriately in terms of power and energy ratings to mitigate the output power fluctuation for a certain frequency range. It is noted that this work does not target to forecast or model the accurate wind speed patterns, so a simple time-period model of randomly varying wind speed, which causes the fluctuated power of WTS; is used as expressed below [37]:

$$v(t) = A_0 + \sum \Delta A_i \sin(2\pi f_i t) \tag{16}$$

where A_0 is the mean wind speed, ΔA_i is the magnitudes of turbulences of wind speed, and f_i is the frequencies ($f_i = 0.05 \sim 10$ Hz).

For a short-term fluctuation of wind power, the output power variation is low, where the power fluctuation is mostly within 30% of its average [8]. So, the power rating of the ESSs is selected as

$$P_{ESS\_rated} = 0.3 P_{rated}. \tag{17}$$

Then, the energy rating of the ESS is selected depending on turbulent components of wind speed and frequency range, where the ESSs can absorb the fluctuated power component. The energy stored in the ESSs is expressed as

$$E_{ESS\_rated} = \int_0^{1/2f_i} P_{ESS\_rated} \sin(2\pi f_i t). \tag{18}$$

The EDLC capacitance, C, is selected from its ratings of the voltage and energy as

$$C = \frac{2 \cdot E_{ESS\_rated}}{\Delta V_{cap} \cdot V_{cap}^{rated}} \tag{19}$$

where V_{cap}^{rated} and ΔV_{cap} are the ratings of EDLC voltage and its allowable variation, respectively.

A decreased amount of the grid power under grid voltage sag conditions, P_{LVRT}, for the worst case, can be determined from the grid codes as

$$P_{LVRT} = (1 - V_{\min,pu})P_{rated} \tag{20}$$

where $V_{\min,pu}$ is a lower limit level of the grid voltage determined from the grid codes.

Finally, the power rating and resistance of the BRs are determined as

$$P_{bc\_rated} = P_{LVRT} - P_{ESS\_rated} - P_J \text{ and } R_{bc} = \frac{V_{dc}^2}{P_{bc\_rated}}. \tag{21}$$

Based on the rating design above, an approximate cost comparison between using only ESS and the hybrid scheme of ESS and braking resistors is carried out applying to 2 MW wind turbine system, which is shown in Table 1. The power rating of the ESS is selected as 2 MW if only ESS is used, while the power ratings of the ESS and braking resistor are chosen as 0.6 MW and 1.4 MW, respectively, for the hybrid scheme. It is shown in Table 1 that the total cost of the hybrid system is about 36.24% compared to that of the system using only the ESS.

Table 1. Cost comparison of the ESS only and hybrid scheme.

System Specification	Only ESS	Hybrid Scheme	
		ESS	Braking Resistor
Power rating	2 MW	0.6 MW	1.4 MW
Unit rating	4.8 kW [38]	4.8 kW [38]	125 kW [39]
Unit price	$526 [38]	$526 [38]	$1145 [39]
Number of units	417	125	12
Total cost	$219,342	$79,490	

It is stated that the main concerns of the research are on the FRT capability and output power fluctuation suppression. However, it is worth noting that the PMSG-based WTSs with the hybrid scheme of the ESSs and the BRs, as shown in Figure 1, are able to operate in a standalone mode instead of grid connection, where the control targets of the GSC and the buck/boost DC-DC converter need to be changed, and the MSC is still controlled to track the maximum power from wind. For the standalone mode, the BDCC is allocated to maintain the DC-link voltage, while the GSC is used to regulate the AC voltages of the loads or the PCC.

5. Simulation Results

PSIM simulation tests were carried out for a PMSG WTS with a rating of 2 MW to verify the effectiveness of the presented scheme. The specifications of the WTS and PMSG are listed in Table 2, and the parameters of the ESSs and the BRs are listed in Table 3, which are designed as per the guidelines described in Section 4. The terminal voltage of BTB converters is 0.69-kV/60-Hz.

Table 2. Specifications of turbine and PMSG.

Parameters	Values
Power rating	2 MW
Turbine rotor radius	42 m
Turbine inertia constant	4.2 s
Generator voltage	690 V
Resistance/inductance	8.56 mΩ/3.59 mH
Number of pole pairs	60
Generator inertia	0.75 s

Table 3. Specifications of ESS and BR.

Parameters	Values
Ratings	6.37 MJ/0.6 MW
Capacitance of EDLC	200 F
Inductor (L_f)	0.5 mH
Operating voltage (V_{ESS})	0.4 kV
$P_{bc\text{-}rated}$	1.133 MW
R_{bc}	1.49 Ω

5.1. Test for a Fault Ride-Through Capability

The control performance of the GSC under the unbalanced voltage sag condition is illustrated in Figure 5. Figure 5a shows the three-phase grid voltages, which drop to 60%, 60%, and 25%, respectively, for three-phase voltages, for 250 ms. Figure 5b shows the DC-link voltage, which is maintained close to its rating of 1.2 kV with an overshoot value of less than 1.5%. It is illustrated in Figure 5c–f that the control performances of dual-current controllers for the P_NSC of the *dq*-axis grid currents, respectively, are satisfactory. Under unbalanced sag, the NSC of grid currents exists, as shown in Figure 5c,d. It is also demonstrated in Figure 5 that before and after the sag, the GSC works well.

Figure 6 shows the response of the WTS and the control response of PMSG, ESS, and BR under the fault. The speed of WTS is shown in Figure 6a, where the turbine speed starts increasing at the instant of the fault detected, which causes a reduction in the turbine output power. The PMSG and grid powers are shown in Figure 6b, where the grid power is lower than the generator power. The mismatch power is absorbed by the ESSs and BRs. The *dq*-axis PMSG currents are shown in Figure 6c,d, respectively, which show that the current controllers work satisfactorily. The operation of the ESSs and BRs is also illustrated in Figure 6e,f, where the performances of the ESS controls for its power and current, respectively, are good, in which the actual values follow their references well. Figure 6g shows the voltage of EDLC, which is increased due to a charge of the ESS during the fault. The current of the BRs is shown in Figure 6h.

5.2. Power Fluctuation Mitigation Tests

Figure 7 demonstrates the performances of the whole WTS in the case of varying wind speeds but under normal grid conditions. The varying wind speed patterns applied to the turbine blades are shown in Figure 7a, which contains 8 terms as expressed in (16). The terms consist of DC, 0.05 Hz, 0.1 Hz, 0.2 Hz, 0.5 Hz, 1 Hz, 5 Hz, and 10 Hz components. Figure 7b shows the PMSG speed. It is apparent in Figure 7c–e that the controllers of grid currents and DC-link voltage, respectively, are satisfactory. Figure 7f,g shows the *dq*-axis components of the generator currents, respectively, which are also controlled well. Figure 7h demonstrates that even though the power of PMSG fluctuates, the grid power does not contain high-frequency components. By operating the ESSs and the BRs to absorb the variation power component, the power injected into the network is much mitigated, where the less than 0.1-Hz frequency components appear only. The response of the ESSs and BRs for improving the power quality of WTS is also demonstrated, in which the high-frequency fluctuation components of the PMSG power are absorbed by the ESSs, as shown in Figure 7h. Figure 7i shows the ESS currents, whose waveforms are almost the same as the power waveform. The control performances of the ESS power and current are also verified. Figure 7j shows the EDLC voltage, which is either increased or reduced depending on the ESS power direction. The BR current is shown in Figure 7k, in which the BR operates shortly.

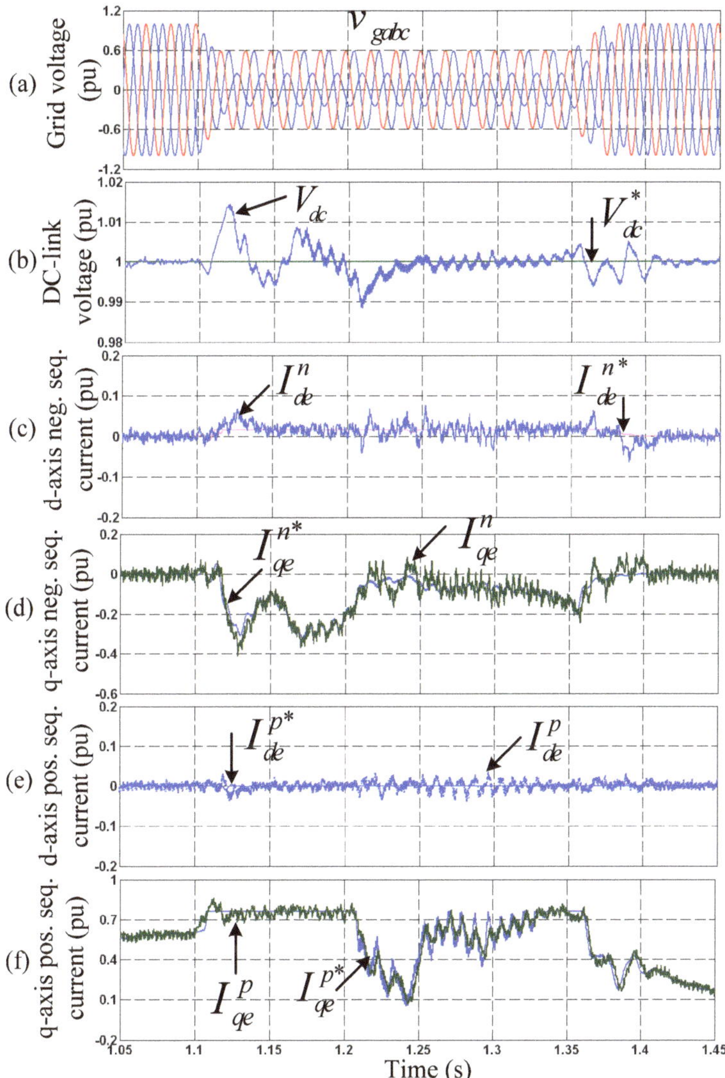

Figure 5. GSC performance at unbalanced voltage dip. (**a**) Grid phase voltages. (**b**) The DC-link voltage of B2B converters. (**c**) NSC d-axis currents. (**d**) NSC q-axis currents. (**e**) PSC d-axis currents. (**f**) PSC q-axis currents.

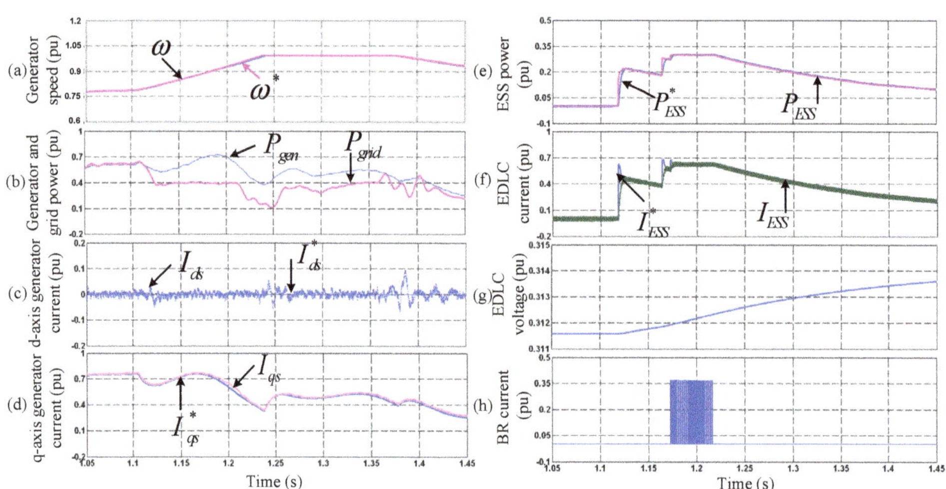

Figure 6. Control performances of the PMSG, ESS, and BRs at the fault. (**a**) WTS speed. (**b**) Grid and PMSG powers. (**c**) d-axis components of PMSG current. (**d**) q-axis components of PMSG current. (**e**) ESS powers. (**f**) EDLC currents. (**g**) EDLC voltage. (**h**) Current of BRs.

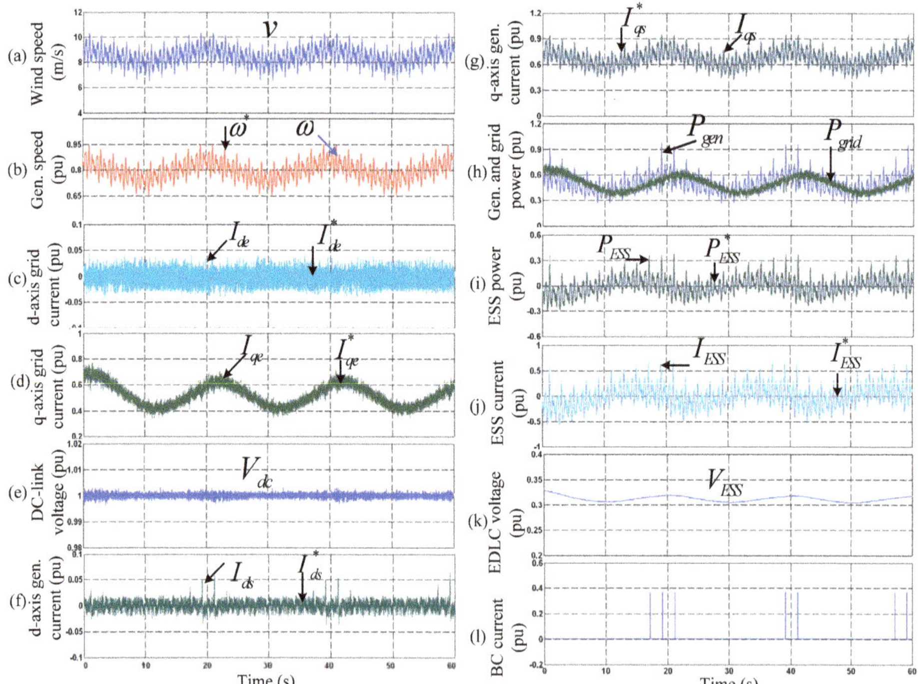

Figure 7. Power fluctuation mitigation performance. (**a**) Wind velocity pattern. (**b**) PMSG speed. (**c**) d-axis grid currents. (**d**) q-axis grid currents. (**e**) DC-link voltages. (**f**) d-axis currents of PMSG. (**g**) q-axis currents of PMSG. (**h**) Generator and grid powers. (**i**) ESS power. (**j**) EDLC current. (**k**) EDLC voltage. (**l**) BR current.

6. Experimental Implementation and Results

6.1. Experimental Implementation of the Studied System

The experimental bench for a down-scale prototype was developed in the laboratory to validate the introduced scheme. Figure 8 shows an experimental setup where an M-G set was built. A 3-kW SCIM (squirrel-cage induction motor) was emulated as a wind turbine system, which adopts a motor torque control according to the turbine characteristic depending on given wind speed patterns. A separate B2B converter was used to drive the SCIM. The investigated PMSG was coupled with the SCIM, where another B2B converter associated with a bidirectional DC/DC converter and a braking resistor controlled by an IGBT was used to control the PMSG. The specifications of the experimental system are listed in Table 4. Supercapacitor, which is a product of LS Mtron, was adopted for the ESSs. A grid simulator was used to generate voltage conditions.

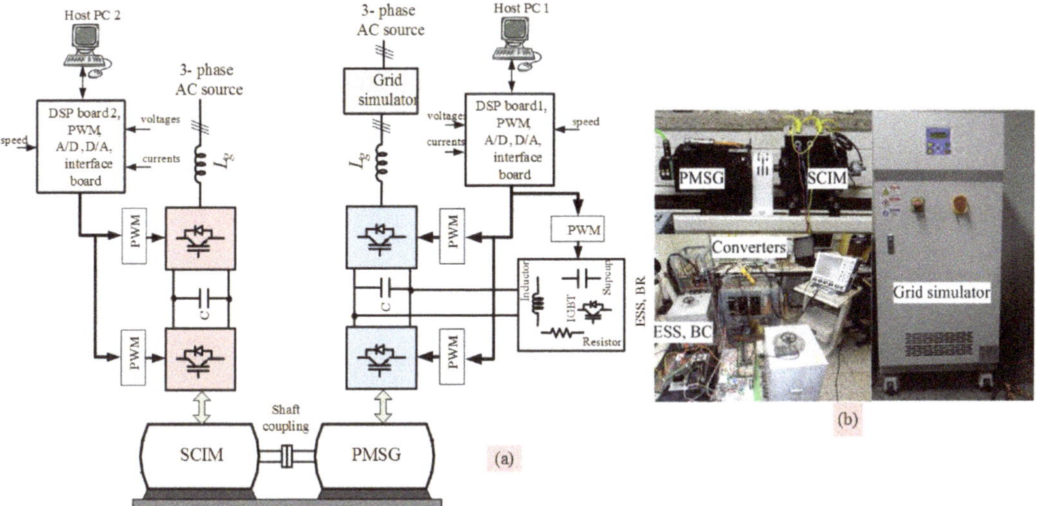

Figure 8. Experimental setup. (**a**) Configuration. (**b**) Apparatus.

Table 4. Specifications of PMSG, ESSs, and BRs.

Parameters	Values
Power rating	2.68 kW
Number of pole pairs	3
Speed rating	1200 rpm
Inertia of WTS	0.071 kg.m^2
Stator resistance/inductance	0.49 Ω/5.35 mH
Boost inductance	3.17 mH
Capacitance of ESSs	2.92 F
Operating voltage	120 V
BR resistance	23 Ω

Semikron IGBT dual modules (SKM75GB128D) with the ratings of 1200 V and 75 A are realized for all the switches of the GSC, MSC, ESSs, and BRs, where the gate drivers of Semikron SKHI 22 are used to control the switches. Transducers using the Hall effects are employed for the current and voltage measurements, which are LA-25 NP and LV-25.

NP, respectively. Incremental encoders with 1024 ppr (pulse per revolution) are used for obtaining the machine speed and rotor position. DSP TMS320VC33 chips are used to implement the digital controllers for the converters [40], where the count/comparator unit of the pulse-width modulator is implemented in an erasable programmable logic device (EP1K50-EPLD). The gate pulse generation scheme is shown in Figure 9, where the digital controllers are implemented on the DSP. The output of controllers is the voltage references, which are the inputs of the space vector pulse-width modulation (SVPWM). The switching times calculated from the SVPWM technique are transferred into the numbers of pulses, which then the gating signals are generated by the count/comparator unit of EPLD.

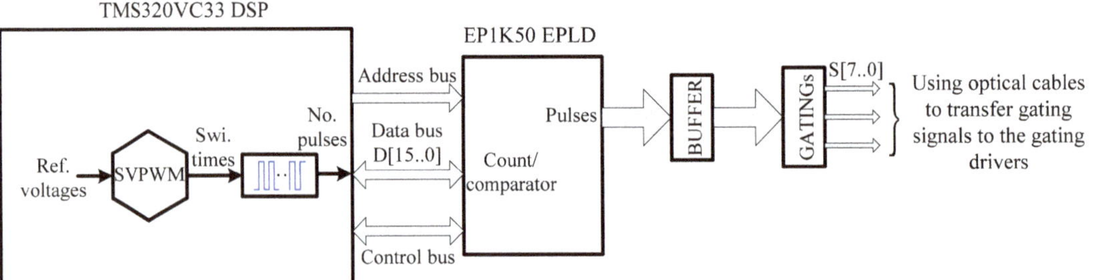

Figure 9. Gate pulse generation scheme based on the DSP and EPLD.

In this work, a bilinear transform is applied to discretize the controllers, filters, and measurements for digital implementation. The current controller sampling time is 100 µs, and the sampling time of the speed controller is 2 ms. The switching frequency of converters is 5 kHz. The gains of the controllers for the GSC, MSC, and BDCC in the experimental system are listed in Table 5, which are designed following the guidelines mentioned in Sections 2 and 3.

Table 5. Controller gains for the experimental system.

Converters and Controllers		**Gains**
GSC	DC-link voltage controller	$K_p^{vc} = 0.605$ and $K_i^{vc} = 44.1$
	Current controllers	$K_p^{cc} = 2.7$ and $K_i^{cc} = 430$
MSC	Speed controller	$K_p^{sc} = 1.052$ and $K_i^{sc} = 8.42$
	Current controllers	$K_p^{cc} = 2.41$ and $K_i^{cc} = 220.5$
DC/DC converter	Power controller	$K_p^{pc} = 0.005$ and $K_i^{pc} = 0.472$
	Current controller	$K_p^{cc} = 2.67$ and $K_i^{cc} = 1141.2$

6.2. FRT Tests for an Unbalanced Voltage Sag

Firstly, the FRT capability of the WTS is investigated under the unbalanced voltage sag, where the three grid-phase voltages are reduced to 80%, 60%, and 40%, respectively, applying for the test as shown in Figure 10a–c for the phase voltages and magnitudes of its P_NSC, respectively. The control performance of the GSC is mainly demonstrated in Figure 10. From Figure 10d,e, it is known that the performance of the current controllers is good for the dq-axis grid currents of PSC, respectively, in which the actual values follow reference one well. It is apparent from Figure 10f,g the q- and d-negative-sequence grid current components exist under the unbalanced sag condition, and the performance control is good. The transients for these components appear at the beginning and the end of fault but within the allowable region. Figure 10h shows the DC-link voltage, which is kept constant at 340 V as in the normal grid condition.

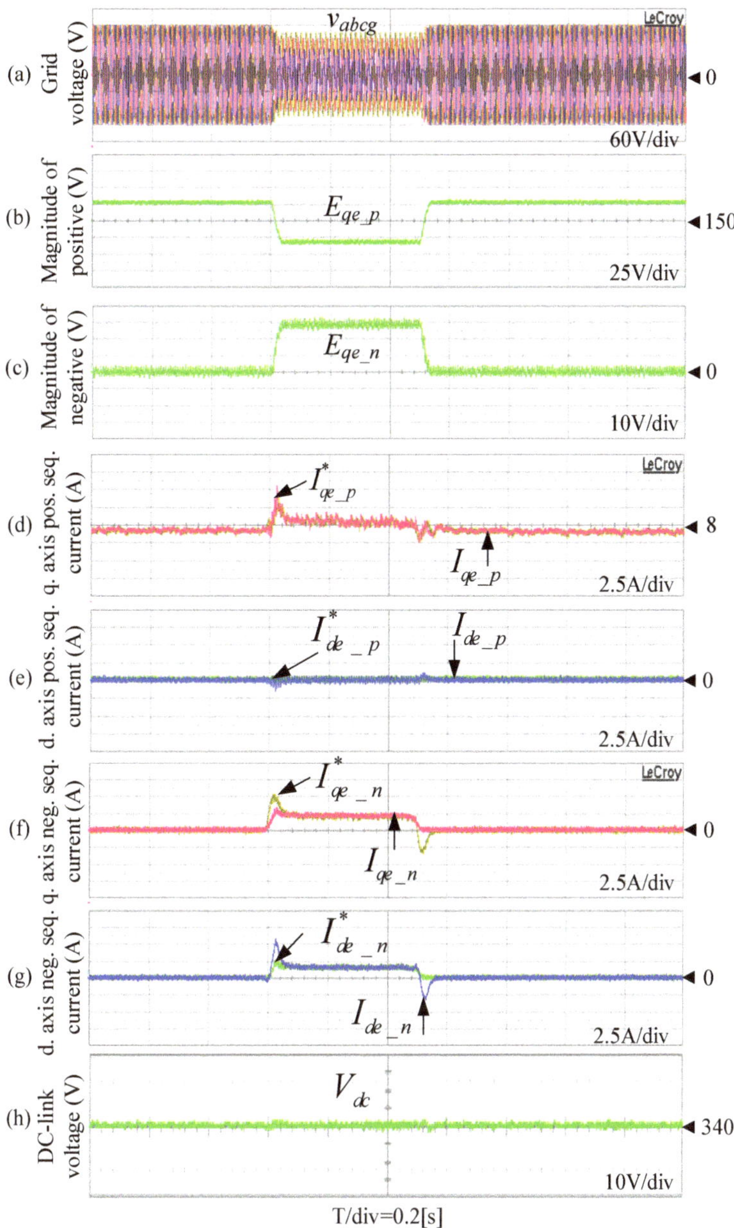

Figure 10. Performance of the grid-side converter at unbalanced voltage dip. (**a**) Grid phase voltages. (**b**) The magnitude of PSC. (**c**) The magnitude of NSC. (**d**) q-axis currents of PSC. (**e**) d-axis currents of PSC. (**f**) q-axis currents of NSC. (**g**) d-axis currents of NSC. (**h**) The DC-link voltage of B2B converters.

A response of the turbine simulator is shown in Figure 11, which also demonstrates the turbine inertia effect. When a sag occurs, the power conversion coefficient and turbine power are decreased, as shown in Figure 11a,b, respectively, since the turbine speed is increased, as seen in Figure 11c. The power match of the whole WTS is investigated, in which the acceleration of the turbine and the operation of ESSs and BRs are activated. Due to a decrease in turbine power, where kinetic energy is stored in the system inertia, the generator power is reduced, as shown in Figure 11d. Figure 11e shows the power flowing into the grid, which is much reduced due to a deep drop in the grid voltage. The amount of mismatch power between the PMSG and network is consumed by the ESSs and BRs, where the ESSs absorb the amount of 400 W equal to its rating as shown in Figure 11f, while the rest is dissipated on the BR as shown in Figure 11g. According to Figure 11h, the EDLC voltage is increased due to charging, and the effect of ESR (equivalent series resistance) of the super-capacitor is also shown in this figure. At the beginning of voltage sag, the EDLC voltage is increased fast due to an increase in EDLC current from zero. Inversely, the EDLC voltage is decreased at the end of sag due to a decrease in EDLC current down to zero. The control performance of the ESS is also demonstrated. Figure 11i shows that the power control performance is good, in which the ESS power follows its reference closely. Under normal conditions, the ESS is deactivated where its power is zero. Figure 11j shows the boost inductor currents, which demonstrates that the controller works satisfactorily.

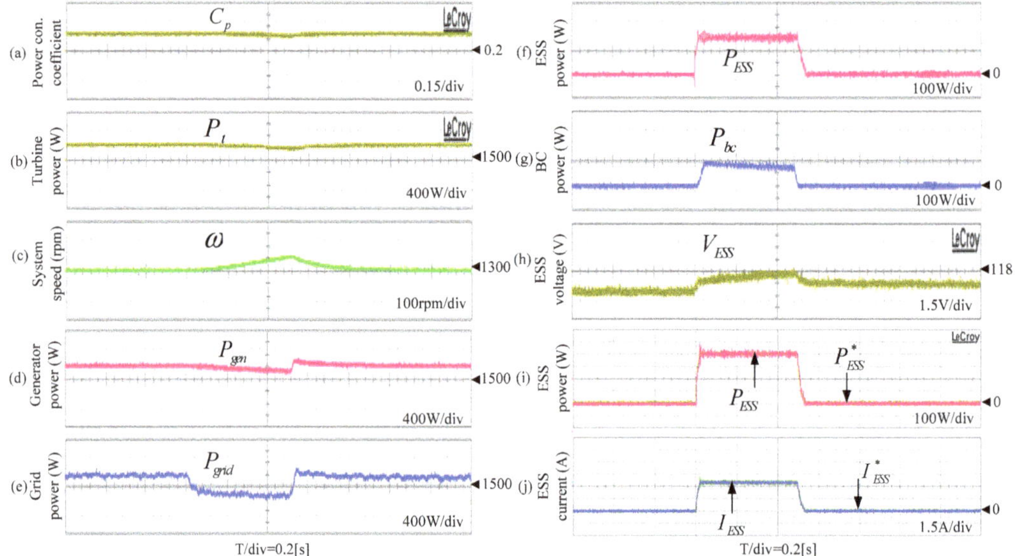

Figure 11. Response of WTS under unbalanced sag. (**a**) Power coefficient. (**b**) Turbine power. (**c**) Turbine speed. (**d**) PMSG power. (**e**) Power flows into the grid. (**f**) Power charging to EDLC. (**g**) Power consuming by BRs. (**h**) EDLC voltage. (**i**) Power controller. (**j**) Current controller.

6.3. Power Fluctuation Mitigation Tests

Improvement of the output power quality of the WTS in the condition of varying wind speed is investigated in these tests. Figure 12a shows the wind speed, where its peak value is about 17% higher than the mean value of 9.5 m/s. The generator speed varies similarly to the wind speed pattern, as shown in Figure 12b. According to the MPPT control, the generator power also fluctuates as shown in Figure 12c, where the peak power is about 45% higher than the average power. It is shown in Figure 12d that the grid power can be smoothened with less than 10% variation due to the operation of the ESS and the BR to extract the fluctuated component of the generator power. When the ripple component of PMSG power exceeds the ESS rating, the BR is activated and dissipates the extra power, as illustrated in Figure 12e. As seen from Figure 12f,g, the controllers of the ESS power and current work satisfactorily, respectively. Figure 12h shows the super-capacitor voltage, which varies depending on their charging or discharging powers. Figure 12i shows the DC-link voltage, which is regulated well at 340 V.

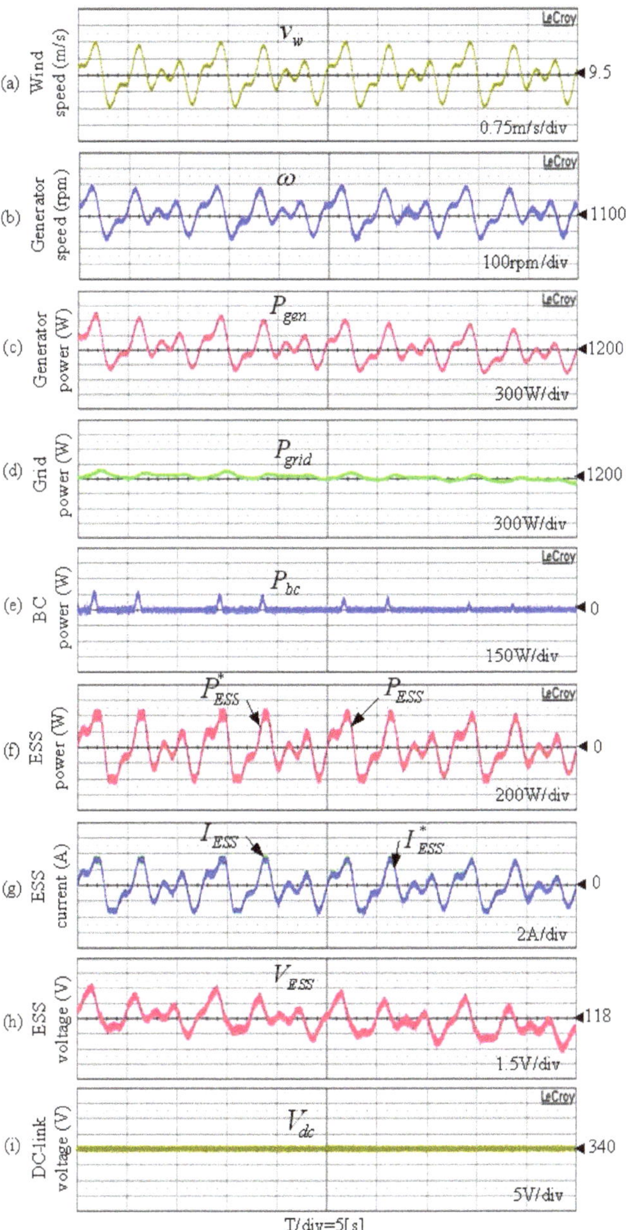

Figure 12. Power fluctuation mitigation for varying wind velocity. (**a**) Wind velocity. (**b**) Generator speed. (**c**) PMSG power. (**d**) Power flows into the grid. (**e**) Power of BRs. (**f**) Power of supercapacitor. (**g**) Supercapacitor current. (**h**) Voltage of supercapacitor. (**i**) DC-link voltage.

7. Conclusions

This paper demonstrates the multiple task performance of a coordinated control scheme utilizing the hybrid of the ESS and BR for the PMSG-based WTS in distribution systems, where a fault ride-through capability under both symmetrical and non-symmetrical grid sags and output power fluctuation mitigation were achieved. Adding the BRs helps to reduce the capacity of the costly ESSs, while the performance of FRT capability is still kept well even under the most serious condition of three-phase grid voltage corruption, and the ESSs are able to perform the power smoothening. In the hybrid scheme, a trade-off between the rating of the ESSs and grid power fluctuation mitigation capacity can be made depending on the system operators, and the rating of the BRs can be further reduced when proper control of the PMSG speed is performed. The detailed experimental implementation for the laboratory-scaled PMSG-based WTS utilizing the SCIM drive as the wind turbine has been presented, where the experimental results for the FRT capability and output power fluctuation suppression have been shown to validate the effectiveness of the scheme and demonstrate the simulation results for the 2-MW PMSG-based WTS.

This work has sufficiently covered its salient objectives of applying a cost-effective scheme of the ESS and braking resistor in the PMSG-based WTS for the LVRT and power smoothing capability, but the state-of-charge and lifetime effect of the ESS has not been discussed. In further research, an optimally coordinated control scheme of the ESS with braking resistors and PMSG wind turbine will be studied, which considers the ESS lifetime and optimal power management for the whole system.

Author Contributions: Conceptualization, T.H.N. and A.N. (Asif Nawaz); methodology, T.H.N. and P.S.; software, T.H.N., A.N. (Asif Nawaz), P.S., A.N. (Ammar Natsheh), V.A. and T.L.V.; validation, T.H.N., P.S. and A.N. (Ammar Natsheh); formal analysis, T.H.N., A.N. (Ammar Natsheh) and V.A.; investigation, T.H.N., A.N. (Asif Nawaz), P.S., A.N. (Ammar Natsheh), V.A. and T.L.V.; resources, A.N. (Asif Nawaz) and P.S.; data curation, V.A. and T.L.V.; writing—original draft preparation, T.H.N.; writing—review and editing, A.N. (Asif Nawaz), P.S., A.N. (Ammar Natsheh), V.A. and T.L.V.; visualization, A.N. (Asif Nawaz), P.S. and A.N. (Ammar Natsheh); supervision, T.H.N.; project administration, T.H.N.; funding acquisition, T.H.N. All authors have read and agreed to the published version of the manuscript.

Funding: This research was funded by Higher Colleges of Technology under Interdisciplinary Grant_212294-Fund code 113166 dated 24 February 2022.

Informed Consent Statement: Not applicable.

Conflicts of Interest: The authors declare no conflict of interest.

References

1. Huynh, P.; Tungare, S.; Banerjee, A. Maximum power point tracking for wind turbine using integrated generator-rectifier systems. *IEEE Trans. Power Electron.* **2021**, *36*, 504–512. [CrossRef]
2. Nguyen, T.H.; Van, T.L.; Nawaz, A.; Natsheh, A. Feedback Linearization-Based Control Strategy for Interlinking Inverters of Hybrid AC/DC Microgrids with Seamless Operation Mode Transition. *Energies* **2021**, *14*, 5613. [CrossRef]
3. Nguyen, T.H.; Quach, N.T. A hybrid HVDC converter based on M2C and diode rectifiers without DC capacitors for offshore wind farm integration. *Int. J. Electr. Power Energy Syst.* **2021**, *133*, 107260. [CrossRef]
4. Ramasamy, T.; Abdul Basheer, A.; Tak, M.-H.; Joo, Y.-H.; Lee, S.-R. An Effective DC-Link Voltage Control Strategy for Grid-Connected PMVG-Based Wind Energy Conversion System. *Energies* **2022**, *15*, 2931. [CrossRef]
5. Prince, M.K.K.; Arif, M.T.; Gargoom, A.; Oo, A.M.T.; Haque, E. Modeling, parameter measurement, and control of PMSG-based grid-connected wind energy conversion system. *J. Mod. Power Syst. Clean Energy* **2021**, *9*, 1054–1065. [CrossRef]
6. do Nascimento, T.F.; Nunes, E.A.D.F.; Villarreal, E.R.L.; Pinheiro, R.F.; Salazar, A.O. Performance analysis of an electromagnetic frequency regulator under parametric variations for wind system applications. *Energies* **2022**, *15*, 2873. [CrossRef]
7. Yuan, H.; Xin, H.; Huang, L.; Wang, Z.; Wu, D. Unified power quality conditioner with advanced dual control for performance improvement of DFIG-based wind farm. *IEEE Trans. Energy Convers.* **2019**, *34*, 838–848. [CrossRef]
8. Luo, C.; Banakar, H.; Shen, B.; Ooi, B.T. Strategy to smooth wind power fluctuation of wind turbine generator. *IEEE Trans. Energy Convers.* **2007**, *22*, 341–349. [CrossRef]
9. Wang, D. Utilisation of kinetic energy from wind turbine for grid connections: A review paper. *IET Renew. Power Gener.* **2018**, *12*, 615–624. [CrossRef]

10. Carrasco, J.M.; Franquelo, L.G.; Bialasiewicz, J.T.; Galvan, E.; Guisado, R.C.P.; Prats, M.A.M.; Leon, J.I.; Alfonso, M. Power-electronic system for the grid integration of renewable energy sources: A survey. *IEEE Trans. Ind. Electron.* **2006**, *53*, 1002–1016. [CrossRef]

11. Cardenas, R.; Asher, G.; Pena, R.; Clare, J. Power smoothing control using sensorless flywheel drive in wind–diesel generation system. In Proceedings of the IEEE 28th Annual IEEE IECON, Seville, Spain, 5–8 November 2002; pp. 3303–3308.

12. Magnus, D.; Scharlau, C.C.; Pfitscher, L.L.; Costa, G.C.; Silva, G.M. A novel approach for robust control design of hidden synthetic inertia for variable speed wind turbines. *Electr. Power Syst. Res.* **2021**, *196*, 107267. [CrossRef]

13. Barra, P.; de Carvalho, W.C.; Menezes, T.S.; Fernandes, R.A.S.; Coury, D.V. A review on wind power smoothing using high-power energy storage systems. *Renew. Sustain. Energy Rev.* **2021**, *137*, 110455. [CrossRef]

14. Eydi, M.; Alishahi, M.; Zarif, M. A novel output power determination and power distribution of hybrid energy storage system for wind turbine power smoothing. *IET Electric Power Appl.* **2022**, 1–17. [CrossRef]

15. Gencer, A. Analysis and Control of Fault Ride Through Capability Improvement PMSG Based on WECS Using Active Crowbar System During Different Fault Conditions. *Elektron. Elektrotechnika* **2018**, *24*, 64–69. [CrossRef]

16. Nasiri, M.; Arzani, A. Roust control scheme for the braking chopper of PMSG-based wind turbines—A comparative assessment. *Int. J. Electr. Power Energy Syst.* **2022**, *134*, 107322. [CrossRef]

17. Jeong, D.H.; Kim, C.H.; Gui, Y.H.; Chung, C.C. Sliding mode control for LVRT of a PMSG wind turbine using stored energy in rotor inertia. In Proceedings of the IEEE Power and Energy Society General Meeting, Boston, MA, USA, 17–21 July 2016; pp. 1–5.

18. Alhejji, A.; Bouzid, Y. Robust Adaptive PI Controller of Low Voltage Ride-Through For PMSG-Based Wind Turbine. In Proceedings of the 2019 6th International Conference on Control, Decision and Information Technologies (CoDIT), Paris, France, 23–26 April 2019; pp. 1233–1237.

19. Kim, C.; Kim, W. Coordinated Fuzzy-Based Low-Voltage Ride-Through Control for PMSG Wind Turbines and Energy Storage Systems. *IEEE Access* **2020**, *8*, 105874–105885. [CrossRef]

20. Shen, Y.-W.; Ke, D.-P.; Qiao, W.; Sun, Y.-Z.; Kirschen, D.S.; Wei, C. Transient reconfiguration and coordinated control for power converters to enhance the LVRT of a DFIG wind turbine with an energy storage device. *IEEE Trans. Energy Convers.* **2015**, *30*, 1679–1690. [CrossRef]

21. Kim, C.; Kim, W. Low-Voltage Ride-Through Coordinated Control for PMSG Wind Turbines Using De-Loaded Operation. *IEEE Access* **2021**, *9*, 66599–66606. [CrossRef]

22. Nguyen, T.H.; Lee, D.-C.; Song, S.-H.; Kim, E.-H. Improvement of power quality for PMSG wind turbine systems. In Proceedings of the IEEE-ECCE, Atlanta, GA, USA, 12–16 September 2010; pp. 2763–2770.

23. Nguyen, T.H.; Lee, D.-C. Advanced fault ride-through technique for PMSG wind turbine systems using line-side converter as STATCOM. *IEEE Trans. Ind. Electron.* **2013**, *60*, 2842–2850. [CrossRef]

24. Lubosny, Z. *Wind Turbine Operation in Electric Power System*; Springer-Verlag: Berlin/Heidelberg, Germany, 2003; Chapter 5.

25. Shaker, M.S.; Kraidi, A.A. Robust fault-tolerant control of wind turbine systems against actuator and sensor faults. *Arab. J. Sci. Eng.* **2017**, *42*, 3055. [CrossRef]

26. Akhmatov, V. *Induction Generators for Wind Power*; Multi-Science Publishing Company: Essex, UK, 2005; Chapter 3.

27. Jang, J.-I.; Lee, D.-C. High performance control of three-phase PWM converters under non-ideal source voltage. In Proceedings of the IEEE ICIT, Mumbai, India, 15–17 December 2006; pp. 2791–2796.

28. Song, H.-S.; Nam, K. Dual current control scheme for PWM converter under unbalanced input voltage conditions. *IEEE Trans. Ind. Appl.* **1999**, *46*, 953–959.

29. Rodriguez, P.; Pou, J.; Bergas, J.; Candela, J.I.; Burgos, R.P.; Boroyevich, D. Decoupled double synchronous reference frame PLL for power converters control. *IEEE Trans. Power Electron.* **2007**, *22*, 584–592. [CrossRef]

30. Rodriguez, P.; Luna, A.; Aguilar, R.S.M.; Otadui, I.E.; Teodorescu, R.; Blaabjerg, F. A stationary reference frame grid synchronization system for three-phase grid-connected power converters under adverse grid conditions. *IEEE Trans. Power Electron.* **2012**, *27*, 99–112. [CrossRef]

31. Ibrahim, A.O.; Nguyen, T.H.; Lee, D.-C.; Kim, S.-C. A fault ride-through technique of DFIG wind turbine systems using dynamic voltage restorers. *IEEE Trans. Energy Convers.* **2011**, *26*, 871–882. [CrossRef]

32. Fernández, G.; Martínez, A.; Galán, N.; Ballestín-Fuertes, J.; Muñoz-Cruzado-Alba, J.; López, P.; Stukelj, S.; Daridou, E.; Rezzonico, A.; Ioannidis, D. Optimal D-STATCOM Placement Tool for Low Voltage Grids. *Energies* **2021**, *14*, 4212. [CrossRef]

33. Sul, S.-K. *Control of Electrical Machine Drive Systems*; Wiley: Hoboken, NJ, USA, 2011.

34. Data, R.; Ranganathan, V.T. A method of tracking the peak power points for a variable speed wind energy conversion system. *IEEE Trans. Energy Convers.* **2003**, *18*, 163–168. [CrossRef]

35. Sharma, S.; Chandra, A.; Saad, M.; Lefebvre, S.; Asber, D.; Lenoir, L. Voltage flicker mitigation employing smart loads with high penetration of renewable energy in distribution systems. *IEEE Trans. Sust. Energy* **2017**, *8*, 414–424. [CrossRef]

36. Camara, M.B.; Gualous, H.; Gustin, F.; Berthon, A. Design and new control of DC/DC converters to share energy between supercapacitors and batteries in hybrid vehicles. *IEEE Trans. Veh. Tech.* **2008**, *57*, 2721–2735. [CrossRef]

37. Wang, J.; Xiong, X.; Li, H.; Lu, X. Time-periodic model of wind speed and its application in risk evaluation of wind-power-integrated power systems. *IET Renew. Power Gener.* **2019**, *13*, 46–54. [CrossRef]

38. Available online: https://www.amazon.ae/Maxwell-DuraBlue-Capacitor-3000farad-ultracapacitor/dp/B081LBMNH2/ref=sr_1_4?crid=S8K48G97H0M1&keywords=super+capacitor&qid=1665136325&qu=eyJxc2MiOiI0LjQyIiwicXNhIjoiMy43NSIsInFzcCI6IjAuMDAifQ%3D%3D&sprefix=supercapacito%2Caps%2C234&sr=8-4 (accessed on 8 October 2022).
39. Available online: https://ae.wiautomation.com/siemens/drives-motors-circuits-protection/sinamics/6SL30001BE313AA0?utm_source=shopping_free&utm_medium=organic&utm_content=AE50617&gclid=Cj0KCQjwnP-ZBhDiARIsAH3FSReMDOWHKVbC6VEdM0KUlGzyPw3MqflDOLICsMR2KF62a50nwJ3P2EwaAngdEALw_wcB (accessed on 8 October 2022).
40. Texas Instruments. *TMS320C3x User's Guide*; Texas Instruments: Dallas, TX, USA, 1997.

Article

Optimal Design of Hybrid Renewable Energy Systems Considering Weather Forecasting Using Recurrent Neural Networks

Alfonso Angel Medina-Santana and Leopoldo Eduardo Cárdenas-Barrón *

Tecnológico de Monterrey, School of Engineering and Sciences, Ave. Eugenio Garza Sada 2501, Monterrey 64849, Mexico
* Correspondence: lecarden@tec.mx; Tel.: +52-8183582000

Abstract: Lack of electricity in rural communities implies inequality of access to information and opportunities among the world's population. Hybrid renewable energy systems (HRESs) represent a promising solution to address this situation given their portability and their potential contribution to avoiding carbon emissions. However, the sizing methodologies for these systems deal with some issues, such as the uncertainty of renewable resources. In this work, we propose a sizing methodology that includes long short-term memory (LSTM) cells to predict weather conditions in the long term, multivariate clustering to generate different weather scenarios, and a nonlinear mathematical formulation to find the optimal sizing of an HRES. Numerical experiments are performed using open-source data from a rural community in the Pacific Coast of Mexico as well as open-source programming frameworks to allow their reproducibility. We achieved an improvement of 0.1% in loss of load probability in comparison to the seasonal naive method, which is widely used in the literature for this purpose. Furthermore, the RNN training stage takes 118.42, 2103.35, and 726.71 s for GHI, wind, and temperature, respectively, which are acceptable given the planning nature of the problem. These results indicate that the proposed methodology is useful as a decision-making tool for this planning problem.

Keywords: RNN; LSTM; long-term forecasting; GHI; non-linear optimization; solar energy; renewable energy; deep learning; optimal sizing; time-series forecasting

Citation: Medina-Santana, A.A.; Cárdenas-Barrón, L.E. Optimal Design of Hybrid Renewable Energy Systems Considering Weather Forecasting Using Recurrent Neural Networks. *Energies* **2022**, *15*, 9045. https://doi.org/10.3390/en15239045

Academic Editors: Sharul Sham Dol and Anang Hudaya Muhamad Amin

Received: 21 October 2022
Accepted: 18 November 2022
Published: 29 November 2022

Publisher's Note: MDPI stays neutral with regard to jurisdictional claims in published maps and institutional affiliations.

1. Introduction

Global warming is a global phenomenon with hard consequences, such as increasing the frequency of damaging disasters, melting of glaciers and polar ice caps, and modification of the water cycle [1]. Among the factors which increase global warming is energy generation based on fossil sources due to the emission of carbon dioxide (CO_2) and methane (CH_4) in the atmosphere [2]. This represents a huge challenge given that energy demand is projected to grow between 4% and 9% between the years 2019 and 2030 [3].

The use of hybrid renewable energy systems (HRESs) to supply electricity is a convenient solution due to the decrease in cost of renewables and to mitigate carbon emissions [4]. The adoption of these systems and other alternatives is fomented by multiple policies and global agreements [5]. In the case of Mexico, short-term and medium-term goals have been fixed for electricity generation based on clean energies (30% for 2021 and 35% for 2024) [6].

In particular, using HRESs for supplying rural areas with electricity requires special attention [7]. It is worth noting that lack of electricity comes in the vast majority from rural areas [8] and it is in rural areas where natural resources are available, thus HRESs constitute an affordable solution [9]. However, HRES sizing is a complex problem: on the one hand, it is essential to cover a minimum of the electric demand of rural inhabitants so that the entire system should not be undersized; on the other hand, it is not desirable to oversize the system since it increases the capital costs, making a rural project infeasible [10].

Many studies have addressed HRES design as an optimization problem [11]. There exist in the literature classical and modern techniques [12]. Classical methodologies entail using linear programming (LP) [13] problems, mixed-integer linear programming (MILP) problems [14], analytical [15], and numerical [16] methods. Metaheuristics involve the use of a single [17] or a hybrid [18] algorithm. Finally, HOMER and ViPOR are widely used software for HRES design [19,20]. For instance, [21] compares a baseline PV/battery system hybridized with combustion-based prime movers. The effects of startup thresholds and the type of prime mover on the cost of energy (COE), waste heat, and lifecycle carbon emissions are studied. Formulating HRES design as an optimization problem requires selecting among the available models of component behavior available for each technology [22,23].

The models of components that make up HRES can also be found in the literature of the energy management system problem, which is related to the sizing problem we are addressing in this work but whose differences should be noticed. The home energy management problem is addressed in [24] by integrating PV and ESS and using genetic algorithms (GAs) to determine the optimal scheduling of appliances considering user preferences and home-to-grid energy exchange. In Ref. [25], an optimization method to schedule three types of household appliances was proposed, taking into account dynamic behavior of customers electricity prices, and weather conditions. Some modern technologies, such as electric vehicles (EVs), used as storage units are incorporated in [26], which presents a linear programming formulation. Literature reviews are presented in [27,28] to compare energy management in campus microgrids. In Ref.[29], real-time pricing, time of use, and critical peak have been studied. In general, modifying the demand is aimed at using two approaches: demand side management and demand response [30]. The former is commonly used by utilities to improve a system's reliability, whereas the latter is focused on encouraging end users to reduce electricity bills. However, it should be noticed that, in general, load scheduling aims to change the demand given that the configuration of a microgrid has already been set, constituting an operational problem, whereas the sizing problem deals with the allocation of space, budget, and other resources when buying or designing the HRES, constituting a planning problem.

One of the particular issues of HRES design is taking into account renewable resources uncertainty, which leads to different approaches. For instance, chance-constrained programming is employed in [31]. In Ref. [32] day-ahead energy management is addressed, considering uncertainties in energy generation and demand and using robust optimization. In contrast, [33] includes in the robust energy management a quadratic pricing model instead of the widely used linear cost function for energy purchase from the grid. Some other works have employed stochastic programming to deal with renewables uncertainty, as detailed in [34]. Modeling uncertainty with SP approaches has proven useful; however, when using SP in general, assumptions with respect to the distribution where the uncertain parameters come from are required. These assumptions are avoided in this work by using RNN. For instance, the current work does not need to assume a probability density function (PDF) for weather parameters.

Some works have addressed this using past data as forecast corresponding to the Seasonal Naive approach. For instance, in Refs. [35] and [36] the Water-Energy nexus is considered in the design of an off-grid system for low-income communities. However, there exist other more complex methodologies to estimate weather conditions for planning purposes.

Most weather forecasting works have addressed only this task for the short term [37]. Among these works, deep learning methodologies have proved to be useful [38]. However, long-term weather forecasting in a suitable manner is required for the purpose of microgrids design [39].

Only a few works have used Artificial Neural Networks (ANN) to predict weather conditions along with an optimization model. For instance, in Ref. [40], weather parameters along with power demand are predicted hourly for a whole year employing ANN. In that work, a heuristic based on tabu search is proposed for the optimization procedure of the

Life-Cycle cost of the Independent Hybrid Power System (IHPS). A novel framework is proposed by [41] which includes a new heuristic as a combination of chaotic search, harmony search, and simulated annealing to find the optimal sizing of a stand-alone hybrid renewable energy system. Again, weather and load forecasting is incorporated using ANN to improve the accuracy of the size-optimized design. In Ref. [42], the Biogeography-Based Optimization (BBO) algorithm is proposed with ANN as an optimization algorithm and solar-wind forecasting model, respectively. Furthermore, the authors of that work analyze the advantages of using forecasts instead of past data to improve the optimization of the Small Autonomous Hybrid Power System. Harmony-search and a combination of harmony-search and chaotic-search is proposed in [43]. This work analyses an HRES which aims to fulfill the water demand of households including weather forecasting using iterative neural networks.

To the best of our knowledge, none of the works in the literature have employed RNN for this purpose, which has been proven to overcome traditional forecasting techniques in global competitions ([44,45]). For that reason, in this work, we propose a full methodology to find the optimal sizing of an HRES using RNN to predict long-term weather conditions. Furthermore, we employ multivariate time series clustering to build scenarios that are fed to the model. Additionally, we propose a nonlinear model formulation compatible with the two previous steps that takes into account renewable and nonrenewable sources. Finally, this model also incorporates the triple bottom of sustainability which includes the economic, environmental, and social dimensions.

The rest of the paper is structured as follows. Section 2 presents the background for the three techniques: RNN for long-term time series forecasting, multivariate clustering, and nonlinear optimization. Section 3 summarizes the data for the case study. Section 4 explains the full methodology for optimal HRES sizing proposed in this work. Section 5 shows and discusses the results for every step explained in Section 4. Finally, Section 6 provides the conclusions and future work.

2. Background Framework

The mathematical models for the Recurrent Neural Network (RNN) models, clustering, benchmark metrics and optimization models, which are the tools that made up the proposed methodology, are proposed in this section.

2.1. Recurrent Neural Networks

Sequence-to-sequence RNN architectures are made up of an encoder and a decoder, as depicted in Figure 1. This deep learning architecture has been widely studied for time series forecasting tasks, as described in [45].

Long Short Term Memory (LSTM) cells are an improved version of the vanilla recurrent neural network unit since they deal better with vanishing gradients and are proven more powerful in conveying historical information throughout the sequences using three gates: input, output, and forget gate. Furthermore, they transmit two states: the hidden state, and the cell state. The LSTM formulation is stated as follows:

$$i_t = \sigma(W_i \cdot h_{t-1} + V_i \cdot x_t + P_i \cdot C_{t-1} + b_i) \tag{1a}$$

$$o_t = \sigma(W_0 \cdot h_{t-1} + V_0 \cdot x_t + P_0 \cdot C_t + b_0) \tag{1b}$$

$$f_t = \sigma(W_f \cdot h_{t-1} + V_f \cdot x_t + P_f \cdot C_{t-1} + b_f) \tag{1c}$$

$$\tilde{C}_t = tanh(W_c \cdot h_{t-1} + V_c \cdot x_t + b_c) \tag{1d}$$

$$C_t = i_t \odot \tilde{C}_t + f_t \odot C_{t-1} \tag{1e}$$

$$h_t = o_t \odot tanh(C_t) \tag{1f}$$

$$z_t = h_t \tag{1g}$$

where the subindex t indicates the timestep. i, o, and f correspond to the input, output, and forget gate. W denotes the weight matrices of the three gates and the cell state whereas

V denotes the weight matrices of the current inputs. C corresponds to the cell state. x, b, and σ are the input values, the biases, and the activation function, respectively. Finally, $\odot$ corresponds to the element-wise multiplication.

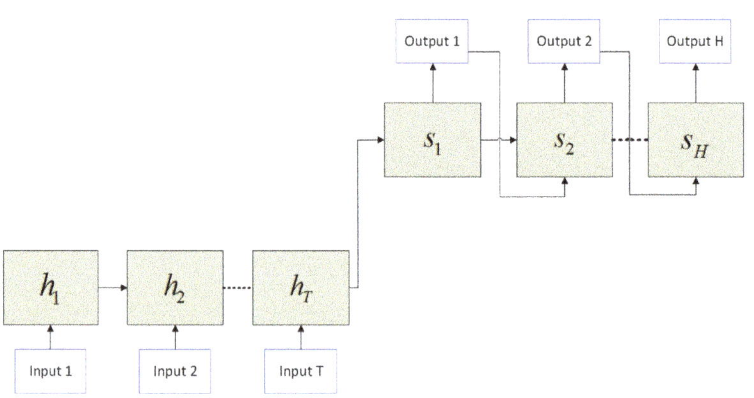

Figure 1. Sequence-to-sequence RNN architecture.

The accumulated error (E) for the back-propagation process is computed using only the forescasts generated by the decoder as expressed in (2):

$$E = \sum_{t=1}^{H} A_t - F_t \qquad (2)$$

where A is the observed value and F is the forecast.

2.2. Benchmark Methods

The evaluation of our forecasts is based on the following well-known metrics: Mean Absolute Scaled Error (MASE), Mean Absolute Error (MAE), Root Mean Squared Error (RMSE), Weight Average Percentage Error (WAPE), Absolute Percentage Bias (APB).

$$MASE = \frac{\frac{\sum_{t=1}^{H} |A_t - F_t|}{H}}{\frac{\sum_{t=m+1}^{H} |A_t - A_{t-m}|}{n-m}} \qquad (3)$$

$$MAE = \sum_{t=1}^{H} \frac{|A_t - F_t|}{H} \qquad (4)$$

$$RMSE = \sqrt{\frac{\sum_{t=1}^{H} (A_t - F_t)^2}{H}} \qquad (5)$$

$$WAPE = \frac{\sum_{t=1}^{H} |A_t - F_t|}{\sum_{t=1}^{H} |A_t|} \qquad (6)$$

$$APB = \frac{\sum_{t=1}^{H} (A_t - F_t)}{\sum_{t=1}^{H} A_t} \times 100 \qquad (7)$$

where A is the observed value, F is the forecasted value, n is the train set length, H is the forecasting horizon, and m is the seasonal period.

2.3. Hierarchical Agglomerative Clustering

This algorithm does not require us to specify the number of clusters in advance. At the beginning, each datum is treated as a singleton cluster and then the algorithm agglomerates pairs of them. The algorithm ends when there is only one single cluster.

This method requires selecting the metric employed as a measure of affinity between clusters and the way in which pairs of clusters are compared. For this work, the Euclidean distance is used as a measure of affinity using the Ward method, which consists of the use of the Ward variance minimization algorithm. Given that the clusters x and y have been combined to generate cluster z, the distance between z and another cluster v is calculated with the following formula [46]:

$$d(z,v) = \sqrt{\frac{|v|+|x|}{T}d(v,x)^2 + \frac{|v|+|y|}{T}d(v,y)^2 - \frac{|v|}{T}d(x,y)^2} \tag{8}$$

where z is the newly joined cluster consisting of clusters x and y, z is an unused cluster in the forest, $T = |v| + |x| + |y|$, and $|*|$ represents the cardinality operator of a set. Moreover, for internal validation, in the absence of truth ground, the Calinski-Harabasz index is used to select the number of clusters.

The Calinski–Harabasz index, also known as the variance ratio criterion, is the ratio of the sum of between-clusters dispersion and of intercluster dispersion for all clusters. The formula to compute this index is the following:

$$CH = \left[\frac{\sum_{k=1}^{K} n_k \|c_k - c\|^2}{K-1}\right] / \left[\frac{\sum_{k=1}^{K} \sum_{i=1}^{n_k} \|d_i - c_k\|^2}{N-K}\right] \tag{9}$$

where n_k and c_k are the number of points and centroid of the kth cluster, respectively, c is the global centroid, N is the total number of data points, and d_i is the ith element associated with a cluster k.

2.4. Optimization

A nonlinear optimization problem has the following form [47]:

$$\min_{\vec{x}} \quad f(\vec{x})$$
$$\text{s.t.} \quad \vec{g}(\vec{x}) = 0 \tag{10}$$
$$\vec{h}(\vec{x}) \le 0$$

where $\vec{x}$ is the decision variable vector, f is an scalar function, and g and h are vector-valued functions.

3. NLP Model Formulation

This sections describes in detail the optimization to find the optimal sizing and operational level of the HRES depicted in Figure 2. The economic, social, and environmental aspects are considered, which involve the three dimensions of sustainability [48]. The notation is described in Section 7. The required sets to formulate the multi-period, multi-scenario optimization model are the following:

$$\tau = \{1,...,T\} \tag{11a}$$

$$\varsigma = \{1,...,S\} \tag{11b}$$

$$\Psi = \{PV, WT, DG\} \tag{11c}$$

The set τ entails the periods of time t in which the each day is partitioned. Given that the cardinality of τ is T, the duration for the corresponding elements of this set is calculated as $\Delta t = \frac{24}{T}$ (in hours). Furthermore, the set ς includes the different scenarios in which the 365 days were clustered using the method explained in Section 2. Then, the number of days represented by each scenario s is different and is expressed as Θ_s. Moreover, each element of set Ψ corresponds to one energy generation technology.

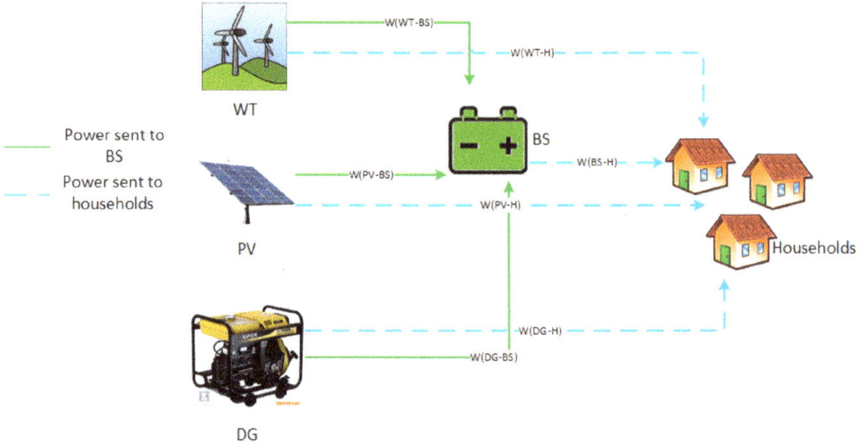

Figure 2. Superstructure to supply electricity to the rural inhabitants.

3.1. Objective Function

The objective function of this model is to minimize the total annualized cost whereas the social and environmental dimensions of sustainability are included as constraints.

3.1.1. Economic Dimension

The minimization objective is the total annualized cost (TAC). It is made up by the sum of the incurred costs to install, maintain, and operate the energy facilities. Therefore, it includes the capital cost ($CapCost$), cost of fuel ($FuelCost$), and operation and maintenance cost ($Cost^{O\&M}$) as shown in Equation (12a).

$$TAC = k_f \cdot (CapCost + FuelCost) + Cost^{O\&M} \tag{12a}$$

Capital cost

The capital cost for power generation units ($CapCost^\psi$) is calculated using the installed capacity, the variable cost associated to the size (v^ψ), and the fixed cost associated to installation (ϖ^ψ) as stated in Equations (12b)–(12d):

$$CapCost^{PV} = \varpi^{PV} + v^{PV} \cdot A^{PV} \tag{12b}$$

$$CapCost^{WT} = \varpi^{WT} + v^{WT} \cdot A^{WT} \tag{12c}$$

$$CapCost^{DG} = \varpi^{DG} + v^{DG} \cdot W^{DG-MAX} \tag{12d}$$

For the battery system (BS) and inverter (INV), the expressions to calculate the capital costs are presented in Equations (12e) and (12f):

$$CapCost^{BS} = \varpi^{BS} + v^{BS} \cdot E^{BS-MAX} \tag{12e}$$

$$CapCost^{INV} = \varpi^{INV} + v^{INV} \cdot W^{INV-MAX} \tag{12f}$$

The total capital cost ($CapCost$) is expressed in Equation (12g).

$$CapCost = CapCost^{WT} + CapCost^{PV} + CapCost^{DG} + CapCost^{BS} + CapCost^{INV} \quad (12g)$$

Operation and maintenance cost

Equation (13a) formulates the annual production of energy by each technology whereas Equation (13b) indicates the annual amount of energy stored by the BS:

$$E^{\psi} = \sum_{s=1}^{S} \left(\sum_{t=1}^{T} W_{t,s}^{\psi} \cdot \Theta_s \right) \cdot \Delta t, \forall \psi \in \Psi \quad (13a)$$

$$E^{BS} = \sum_{s=1}^{S} \left(\sum_{t=1}^{T} W_{t,s}^{BS-H} \cdot \Theta_s \right) \cdot \Delta t \quad (13b)$$

The annual energy production and the maximum capacity of equipment are the drivers for operation and maintenance costs (Equations (13c)–(13e)):

$$O\&MCost^{\psi} = \chi^{O\&M-\psi} \cdot E^{\psi}, \forall \psi \in \Psi \quad (13c)$$

$$O\&MCost^{BS} = \chi^{O\&M-BS} \cdot E^{BS} \quad (13d)$$

$$O\&MCost^{INV} = \chi^{O\&M-INV} \cdot W^{INV-MAX} \quad (13e)$$

The total operation and maintenance cost ($Cost^{O\&M}$) is expressed as follows:

$$Cost^{O\&M} = O\&MCost^{PV} + O\&MCost^{WT} + O\&MCost^{BS} + O\&MCost^{DG}$$
$$+ O\&MCost^{INV} \quad (13f)$$

Cost of fuel

The cost of fuel associated with the DG unit is computed as the total cost for the fuel employed by the diesel generator over the whole year. This is expressed by Equation (14):

$$FuelCost = \sum_{s=1}^{S} \left(\sum_{t=1}^{T} F_{t,s} \cdot \Theta_s \right) \cdot \lambda \quad (14)$$

where Θ_s is the number of days in each scenario s.

3.1.2. Environmental Dimension

The system's reliability is aimed to be improved by adding a diesel generator. However, pollution generated by primary-sources-based generation units is highly undesirable. For that reason, we employ the renewable factor (RF) to set a lower bound (LB_{RF}) for the proportion of energy produced by renewable energy with respect to the energy produced by the diesel generator [18]. The renewable factor metric is formulated in Equation (15):

$$RF = 1 - \frac{\sum_{s=1}^{S} \left(\sum_{t=1}^{T} W_{t,s}^{DG} \cdot \Theta_s \right)}{\sum_{s=1}^{S} \left(\sum_{t=1}^{T} \left(W_{t,s}^{WT} + W_{t,s}^{PV} \right) \cdot \Theta_s \right)} \geq LB_{RF} \quad (15)$$

3.1.3. Social Dimension

For rural inhabitants, the HRES provides energy supply so that they can meet their information, education, and communication primary needs. In that sense, reliability constitutes a desirable feature for the optimized system. Loss of load occurs whenever the system load exceeds the available generating capacity [49]. In this work, the social dimension is expressed as minimizing the ratio corresponding to the Loss of Load Probability (LOLP) because of insufficient installed capacity which is expressed in Equation (16) [50]. In other words, we are setting an upper bound (UB_{LOLP}) for the proportion of unsatisfied demand throughout the whole year:

$$LOLP = \frac{\sum_{s=1}^{S}\left(\sum_{t=1}^{T}\left(W_{t,s}^{H} - \eta^{INV} \cdot (W_{t,s}^{WT-H} + W_{t,s}^{PV-H} + W_{t,s}^{DG-H} + W_{t,s}^{BS-H})\right) \cdot \Theta_s\right)}{\sum_{s=1}^{S}\left(\sum_{t=1}^{T} W_{t,s}^{H} \cdot \Theta_s\right)}$$

$$\leq UB_{LOLP}$$

(16)

3.2. Operational Constraints

The behavior of the system components under steady-state conditions is formulated in this subsection.

3.2.1. Power Generation

Power generation technologies include a photovoltaic system (PV), wind turbines (WT), and a diesel generator (DG). For each technology, the equations to calculate the power generation and efficiency are expressed below.

Photovoltaic system (PV)

The PV output power produced is calculated using the allocated area for solar panels (A^{PV}), the solar irradiance (α), and the efficiency of the photovoltaic system (η^{PV}).

$$W_{t,s}^{PV} = A^{PV} \cdot \eta_{t,s}^{PV} \cdot \alpha_{t,s}, \forall t \in \tau, \forall s \in \varsigma \tag{17a}$$

where the corresponding efficiency depends on the difference between ambient temperature (T^{amb}) and the reference temperature (T_{Ref}).

$$\eta_{t,s}^{PV} = \eta_{Ref}^{PV} \cdot \left(1 - \beta \cdot \left(T_{t,s}^{amb} - T_{Ref}\right)\right), \forall t \in \tau, \forall s \in \varsigma \tag{17b}$$

Wind turbine (WT)

The eolic output power is addressed in Equation (17c):

$$W_{t,s}^{WT} = \frac{1}{2} \cdot \frac{1}{1000} \cdot \rho^{air} \cdot A^{WT} \cdot \eta^{WT} \cdot v_{t,s}^{3}, \forall t \in \tau, \forall s \in \varsigma \tag{17c}$$

The efficiency of the aerogenerators η^{WT} is calculated using Equation (17d):

$$\eta^{WT} = \eta^{rotor} \cdot \eta^{generator} \cdot \eta^{transmission} \tag{17d}$$

Diesel generator (DG)

The power generation of the diesel generator is expressed in Equation (17e):

$$\eta_{t,s}^{DG} = \frac{W_{t,s}^{DG} \cdot \Delta t}{F_{t,s}}, \forall t \in \tau, \forall s \in \varsigma \tag{17e}$$

The efficiency of the diesel generator is affected by the partial load during each period t. The mathematical relationships corresponding to this technology are formulated in Equations (17f)–(17i):

$$PL_{t,s}^{DG} = \frac{W_{t,s}^{DG}}{W^{DG-MAX}}, \forall t \in \tau, \forall s \in \varsigma \tag{17f}$$

$$PL^{DG-MIN} \geq PL_{t,s}^{DG} \leq PL^{DG-MAX}, \forall t \in \tau, \forall s \in \varsigma \tag{17g}$$

$$W_{t,s}^{DG} \leq W^{DG-MAX}, \forall t \in \tau, \forall s \in \varsigma \tag{17h}$$

$$\eta_{t,s}^{DG} = \varphi^{DG}(PL_{t,s}) \cdot \eta_0^{DG}, \forall t \in \tau, \forall s \in \varsigma \tag{17i}$$

3.2.2. Energy Storage

BS acts as a backup when renewable resources are scarce and allow to meet the gap between energy generation and demand.

Battery system (BS)

We are considering Lithium batteries for the energy storage system. The entries of the BS come from PV, WT, and DG units (see Equation (18)):

$$E_{t,s}^{BS} - E_{t-1,s}^{BS} = \Delta t \cdot (\eta_{t,s}^{BS} \cdot \left(W_{t,s}^{PV-BS} + W_{t,s}^{WT-BS} + W_{t,s}^{DG-BS}\right) - W_{t,s}^{BS-H}), \forall t \in \tau, \forall s \in \varsigma \tag{18}$$

The BS efficiency affects the energy inputs and depends on the State of Charge (SoC) using the Equations (19)–(21):

$$SoC_{t,s} = \frac{E_{t,s}^{BS}}{E^{BS-MAX}}, \forall t \in \tau, \forall s \in \varsigma \tag{19}$$

$$E_{t,s}^{BS} \leq E^{BS-MAX}, \forall t \in \tau, \forall s \in \varsigma \tag{20}$$

$$\eta_{t,s}^{BS} = \phi^{BS}(SoC_{t,s}), \forall t \in \tau, \forall s \in \varsigma \tag{21}$$

where ϕ^{BS} is a non-linear function that relates the state of charge (SoC) and the BS efficiency η^{BS}. A lower bound for SoC is considered to prevent early deterioration:

$$SoC_{LB} \leq SoC_{t,s} \leq 1, \forall t \in \tau, \forall s \in \varsigma \tag{22}$$

3.2.3. Power Supply to Households

Household demand of electricity is supplied using the PV, WT, and DG units. The excess pf power supply is stored in the BS:

$$W_{t,s}^{PV} = W_{t,s}^{PV-H} + W_{t,s}^{PV-BS}, \forall t \in \tau, \forall s \in \varsigma \tag{23a}$$

$$W_{t,s}^{WT} = W_{t,s}^{WT-H} + W_{t,s}^{WT-BS}, \forall t \in \tau, \forall s \in \varsigma \tag{23b}$$

$$W_{t,s}^{DG} = W_{t,s}^{DG-H} + W_{t,s}^{DG-BS}, \forall t \in \tau, \forall s \in \varsigma \tag{23c}$$

Inverter operation (INV)

The inverter capacity is constrained by the maximum amount of energy sent to the households as expressed in Equation (23d):

$$W^{INV-MAX} \geq W_{t,s}^{H}, \forall t \in \tau, \forall s \in \varsigma \tag{23d}$$

Land use

The elements of the superstructure occupy the available land (see Figure 2):

$$A_T \geq SF \cdot (A^{PV} + A^{WT} + A^{BS} + A^{INV} + A^{DG}) \tag{24}$$

where A_T is the total area and SF is the proportion of the total area that can effectively occupied by the system. The use of land of each element is expressed in Equations (25)–(27):

$$\mu^{BS} \cdot A^{BS} = E^{BS-MAX} \tag{25}$$

$$\mu^{INV} \cdot A^{INV} = W^{INV-MAX} \tag{26}$$

$$\mu^{DG} \cdot A^{DG} = W^{DG-MAX} \tag{27}$$

Initialization of storage units

The formulation for BS presented in Equation (18) requires an strategy to deal with the balance of energy for $t = 1$. In this work, we propose to define the following decision variable: $E_{0,s}^{BS}$, such that $E_{0,s}^{BS} = E_{t-1,s}^{BS}$ when $t = 1$. Moreover, the formulation incorporates the linkage between the end of one day and the beginning of the next one, such that $E_{0,s}^{BS} \leq E_{t,s}^{BS}$, when $t = T$.

Non-linear relationships for efficiencies [36]

The correlation ϕ^{BS} between BS efficiency and SoC is given by Equation (28):

$$\eta_{t,s}^{BS} = -2.77 \cdot SoC_{t,s}^2 + 2.361 \cdot SoC_{t,s} + 0.498, \forall t \in \tau, \forall s \in \varsigma \tag{28}$$

The expression φ^{DG} to relate the partial load with the efficiency of the DG unit is expressed in Equation (29):

$$\varphi_{t,s}^{DG}(PL) = -0.4324 \cdot PL_{t,s}^2 + 0.8847 \cdot PL_{t,s} + 0.5385, \forall t \in \tau, \forall s \in \varsigma \tag{29}$$

4. Case Study

As a case study, we use data from a rural community located in the Pacific coast of Mexico. Technical data was obtained through direct measures, geographic data, and surveys [51].

Weather historical information was obtained from the National Renewable Energy Laboratory (NREL) through an API to obtain data collected from specific coordinates in this region. The National Solar Radiation Database (NSRDB) is one the most accepted datasets for solar energy and meteorological data collected with high temporal (30 min) and spatial resolution (4 km) [52,53]. Information for the specific station we take the measurements from is shown in Table 1.

Table 2 contains data about efficiencies, land use, and costs. Moreover, a lower bound of 15% for the state of charge of the BS (SoC LB) is defined to avoid early wearing out. Finally, the security factor SF is defined as 0.75. Therefore, only 75% of the total land can be occupied to allow some space among units of the superstructure. Tables 2 and 3 present the parameter values that should be introduced into the model explained in Section 3. The reference value for the temperature of the PV technology is 20 °C. Finally, we use a lower bound for renewable factor (LB_{RF}) and an upper bound for LOLP (UB_{LOLP}) of 0.5.

Table 1. Station information.

ID	571501
Latitude	19.65
Longitude	−101.66

Table 2. Parameters of the rural area [54].

Description	Value
Price of fuel (USD/kWh)	0.023
Total area (m^2)	5000
Annualization factor	0.117
Air density (kg/m^3)	1.225

Table 3. Technical parameters for the technologies [55,56].

	WT	PV	DG	BS	INV
Nominal efficiency η_0, η	0.112	0.21	0.35		0.8
Fixed Cost ω (USD)	100	800	1250	30	30
Variable cost ν (USD/kW, USD/kWh)	950	200	745	25	530
O&M Cost , $\chi^{O\&M}$ (USD/kWh, USD/kW)	0.03	0.003	0.01	0.0012	0.3
Land usage, μ (kW/m^2, kWh/m^2)	-	-	0.025	25.36	31.07

5. Methodology

In this section, we outline the overall methodology proposed in this work to find the optimal sizing of an HRES.

The first step is data collection. In this work, we acquire weather and technical information. The former was obtained from NREL API, and more details are provided in Section 4. Three weather variables were obtained: solar irradiance, wind speed, and ambient temperature. Regarding solar irradiance, we consider the global horizontal irradiance (GHI) given that it is widely used to compute PV power generation [57].

The second step was to describe the dataset. In this work, our dataset is made up of fourteen years of historical half-hourly data (from 2007 to 2020). We employ a timeseries decomposition to observe the patterns related to trend, seasonality and residual components. The analysis of daily distribution of weather parameters is also carried out which, in particular, support getting rid of zero or neglectable values of GHI.

The third step is to predict the three weather variables mentioned earlier with two methodologies: seasonal naive and RNN. For RNN, we take the mean of ten runs using different random seeds to avoid the bias that seed selection might have in our outputs. Furthermore, each weather variable is predicted separately by a single RNN model. Moreover, in the case of the GHI long-term forecast, the Solis clear-sky model is incorporated as an additional continuous feature. To grasp more details, the interested reader can consult [58]. Then, we end up with two datasets with forecast values for the weather variables for the year 2020.

The fourth step is to create two scenarios using multivariate time series clustering. The analysis of the number of clusters is based on the Calinski–Harabasz index, which is explained in Section 2.3.

The fifth step is to find an optimal sizing for each set of scenarios. For this purpose, the nonlinear mathematical formulation is described in Section 3. An extension of this model suitable for the WEFN analysis can be studied in [59].

Finally, the evaluation of both designs is performed by carrying out a simulation in which we compute the loss of load probability (LOLP) for both designs as a metric for comparison.

All these steps are illustrated in Figure 3.

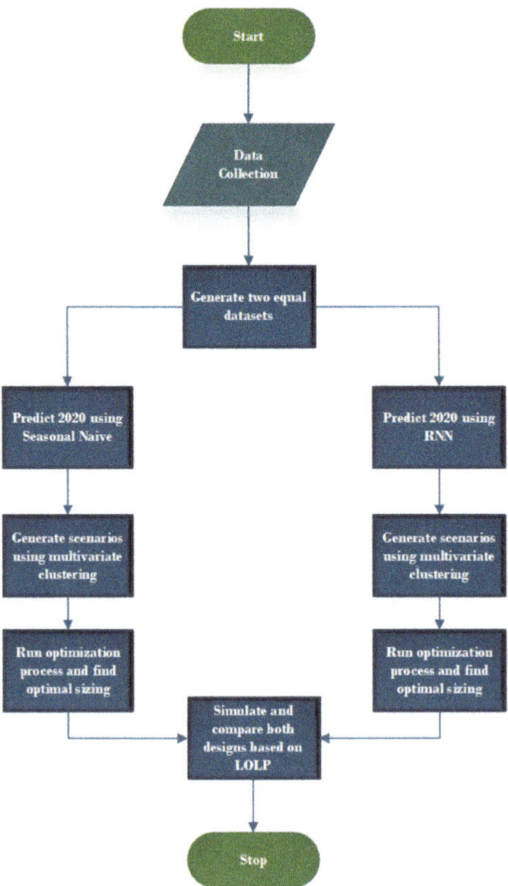

Figure 3. Flowchart of the methodology.

6. Analysis of Results

In this section, we show the results for each step explained in Section 5.

6.1. Data Exploration

Figure 4 shows boxplots for GHI data, and zero values are found before 8:00 am and after 17:30 pm. We then only consider values among these hours since the RNN model will not extract significant information besides these limits.

A time series decomposition using the moving average method was performed to identify patterns in the historical GHI, wind speed, and temperature data. December 2019 data is used to illustrate this procedure since this is the last month of our training set. To this aim, the additive model is chosen since the variability of the seasonal fluctuations do not depend on the level of the time series [60].

The additive model is stated as follows:

$$y_t = S_t + T_t + R_t \tag{30}$$

where y_t is the observed value, and S_t, T_t, and R_t are the seasonal, trend, and residual components, respectively. Observed values for GHI are depicted in Figure 5a.

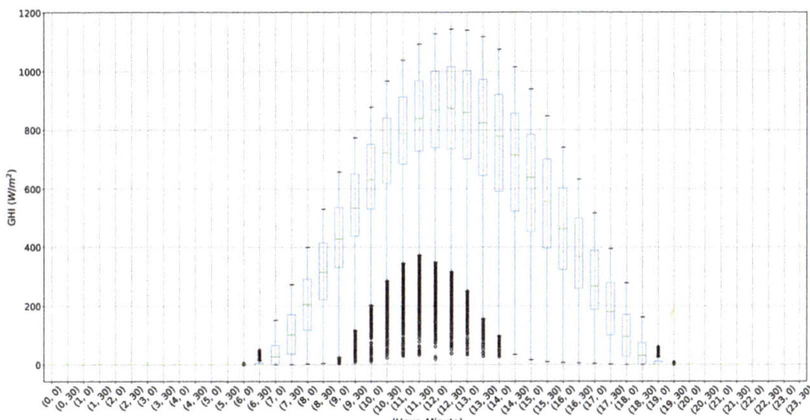

Figure 4. Boxplot of GHI for every intra-day time step from 2007 to 2019.

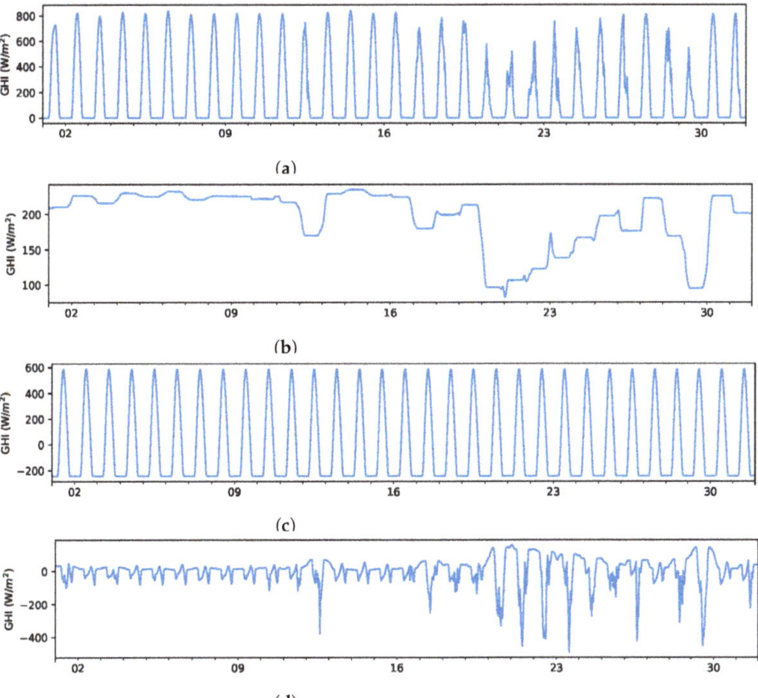

Figure 5. GHI time series decomposition. (**a**) Observed values. (**b**) Trend component. (**c**) Seasonal component. (**d**) Residual component.

Although no clear pattern for the trend is observed (see Figure 5b), a much more stronger seasonal component is observed in Figure 5c. The residual component is depicted in Figure 5d.

In contrast, the boxplot for wind speed in Figure 6 indicates that there are no such hourly intervals with only zero values. Furthermore, the trend component observed in Figure 7b is weak or nonexistent. Nevertheless, the seasonal component shown in Figure 7c has stronger results.

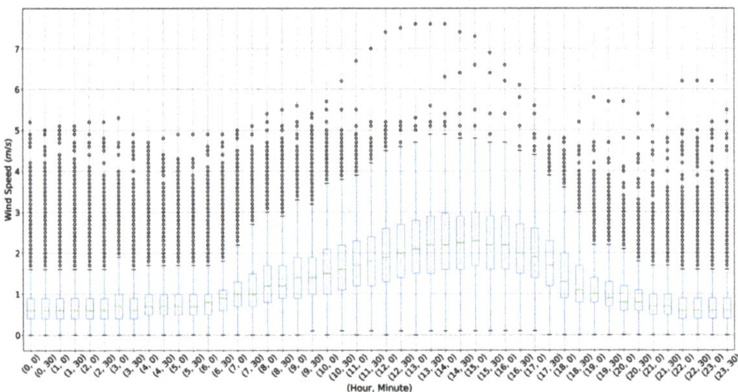

Figure 6. Boxplot of wind speed for every intraday timestep from 2007 to 2019.

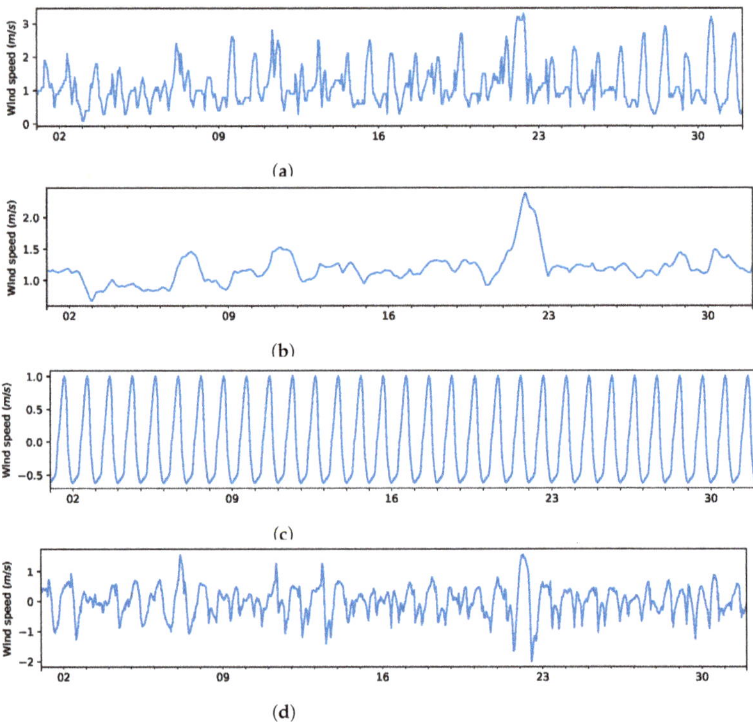

Figure 7. Wind time series decomposition. (**a**) Observed values. (**b**) Trend component. (**c**) Seasonal component. (**d**) Residual component.

The boxplot of temperature time series in Figure 8 indicates that it follows a daily seasonal pattern similar to the GHI variable. Regarding the time series decomposition, there is a downward part of the trend from the beginning until day 23, which indicates the importance of this component inside each month (see Figure 9b). However, the scale is much lower than the seasonal component, as shown in Figure 9c.

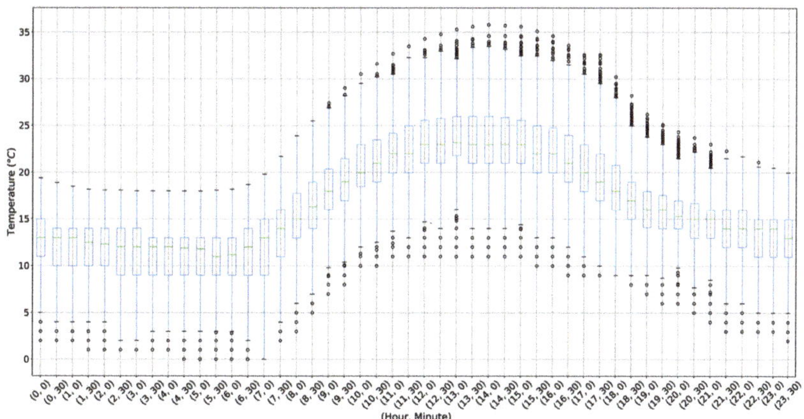

Figure 8. Boxplot of temperature for every intraday timestep from 2007 to 2019.

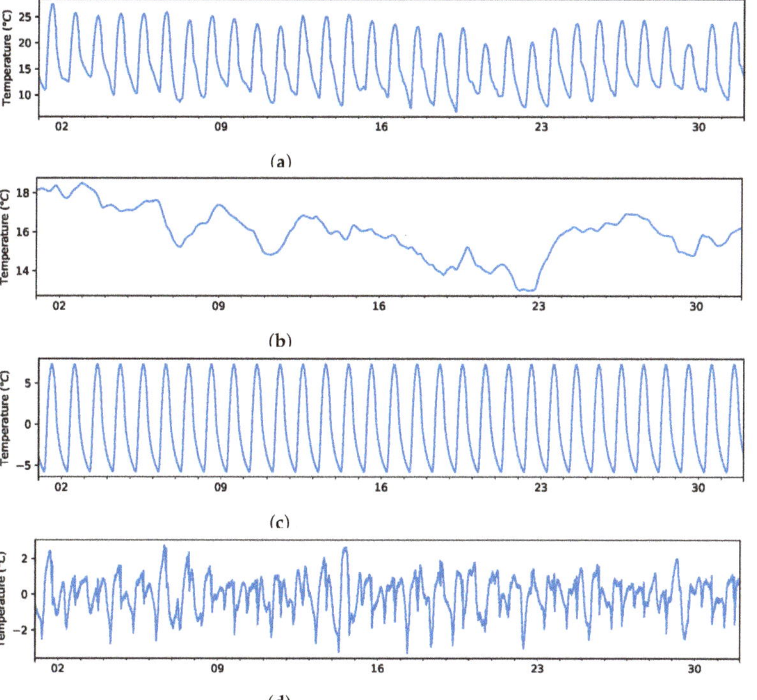

Figure 9. Temperature time series decomposition. (**a**) Observed values. (**b**) Trend component. (**c**) Seasonal component. (**d**) Residual component.

6.2. Data Preparation

In this work, to predict each variable, seven days of data (with 20 and 48 timesteps for GHI and wind/temperature, respectively) from seven consecutive years are fed to the model as inputs.

In the case of GHI, the first sample in the training set is constituted as follows: it includes 20 timesteps of 1 January for seven years (2007–2013). Two continuous features are included for each timestep: the observed GHI and the clear-sky value using the Solis model. According to data exploration carried out in Section 6.1, GHI presents daily seasonality, and given that the declination angle varies during a year between $-23.45°$ and $23.45°$, there is also a strong yearly seasonal component in this time series. For that reason, we incorporate dummy variables related to the month and hour of each timestep. In this way, we aim to increase the model capacity to grasp the daily and yearly seasonality from data.

For wind speed and temperature, only historical data and one dummy variable (for months) were included. This is due to the fact that for these two time series, each day entails 48 timesteps and therefore each sample has more data. Thus, including dummy variables for the hour leads to out-of-memory problems.

RNN requires 3D shape data where the first axis corresponds to the dimension of the sample. In our case, the input for the proposed RNN model is a matrix with 140 rows and 22 columns (for GHI), and 12 columns (for wind speed and temperature). It is worth noting that we dropped the first column of the dummy variables for months (all the variables) and hours (only GHI) to avoid collinearity issues. This is illustrated in Figure 10. Moreover, the three training samples are depicted in Figure 11. The output of each model is a 2D-rank tensor where the output linked to each sample is a 20-dimension vector and 48-dimension vector for the case of GHI and wind/temperature, respectively.

The model proposed for time series forecasting adopts the multiple input multiple output (MIMO) approach detailed in [61]. For each variable, the proposed method yields one forecasting model to predict one day of data. Then, for our long-term forecasting task of predicting a whole year, the obtained model has to be fed 365 times to predict one year of each weather parameter.

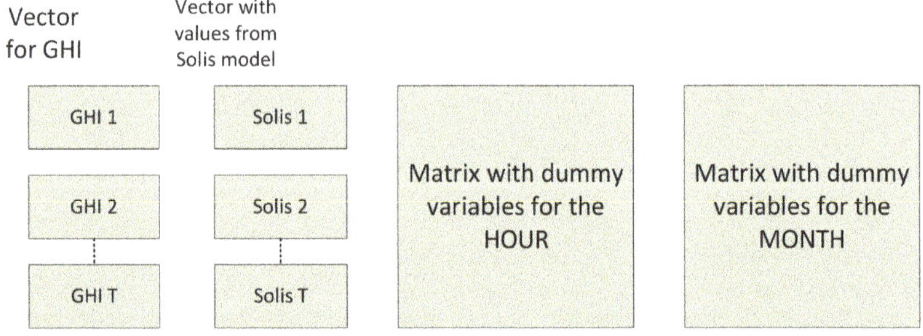

Figure 10. Composition of an input matrix in GHI training set for the RNN model.

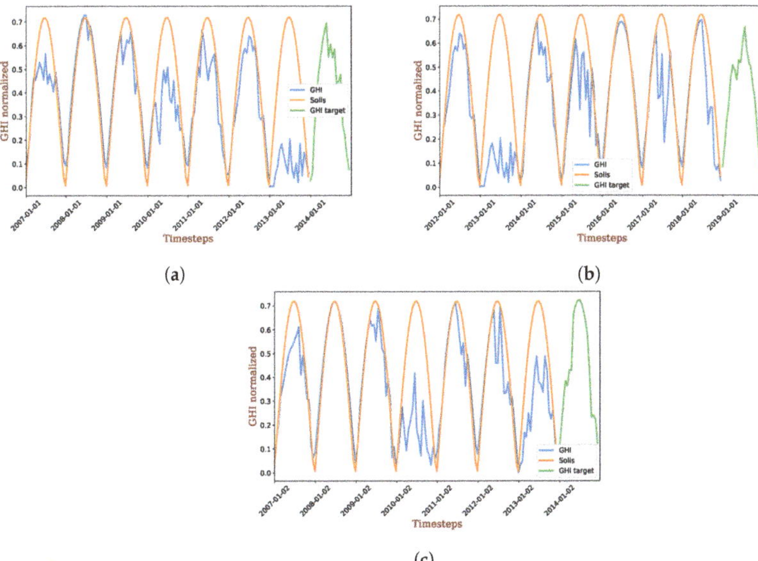

Figure 11. Three different samples to feed RNN model after data preparation [58]. (**a**) First sample in the training set. (**b**) Sample 101 in the training set. (**c**) Sample 102 in the training set.

6.3. Evaluation

The computational framework used in this work for modeling and evaluation purposes is Keras which employs a seed to randomly generate the initial values of the weight parameters. For this reason, our forecasts to be evaluated are the average of 20 forecasts to isolate the effect of the seed from the performance evaluation. Figures 12–14 illustrate the performance of RNN in this forecasting task for the parameters GHI, wind speed, and temperature, respectively. To illustrate the performance of our models, we plot weeks 16, 34, and 51 to contrast the difference between forecast and observed values of year 2020. Figure 12 shows that although more noise is found in week 33 (see Figure 12b), a similar level of the seasonality of the global irradiance is depicted by the RNN throughout the whole year. In contrast, for wind speed, different amplitudes in the cycles are predicted by the RNN (see Figure 13b,c). These results indicate the model's capacity to learn not only daily but yearly patterns. Finally, Figure 14 illustrates that the model learns quite accurately the patterns in temperature data.

To define and train our model, we use Keras from Tensorflow 2.7 (developed by researchers from Google) which is an open-source deep learning API for python. Training our RNN model requires taking care of computational time. To this aim, we build the sequence-to-sequence architecture according to CUDA requirements to keep our model parallelizable [62]. Regarding the hardware, the model was run using a GPU NVIDIA GeForce GTX 960M to accelerate the training process. Before testing, a validation stage is carried out using a grid search procedure. Thererfore, the training set for this stage entails only from 2007 up to 2018, so that the 2019 data is used as validation set. Table 4 depicts the values employed for the grid search procedure where the bold text specifies the chosen values found based on the average MSE of ten runs per combination.

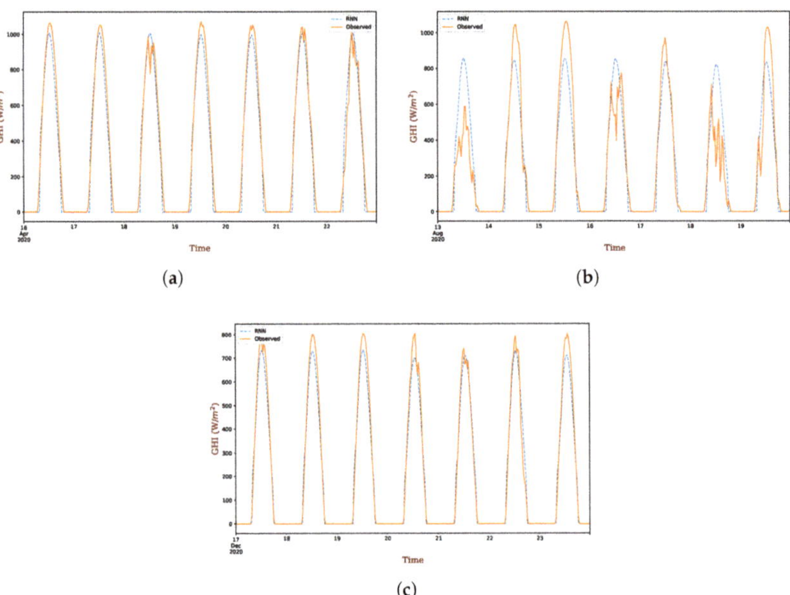

Figure 12. GHI prediction for three weeks of 2020 with RNN. (**a**) Week 16. (**b**) Week 33. (**c**) Week 51.

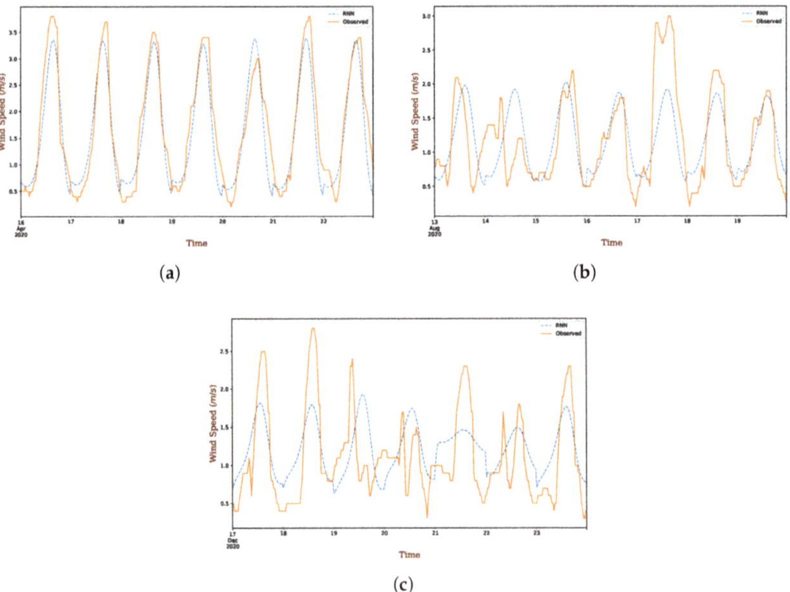

Figure 13. Wind speed prediction for three weeks of 2020 with RNN. (**a**) Week 16. (**b**) Week 33. (**c**) Week 51.

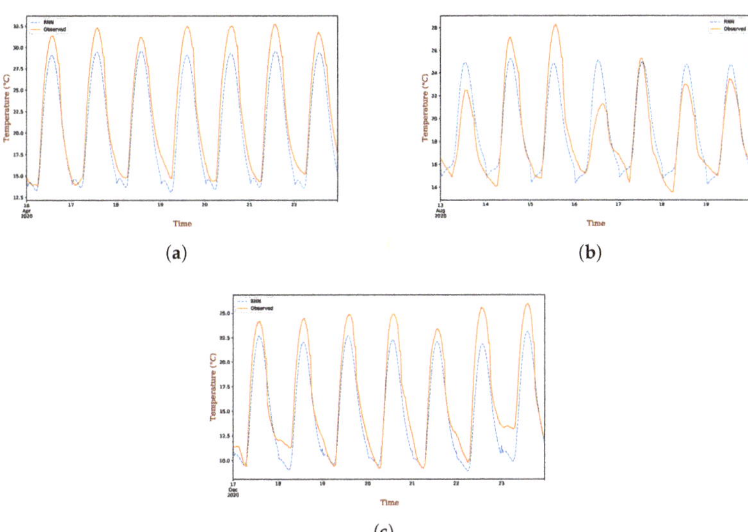

Figure 14. Temperature prediction for three weeks of 2020 with RNN. (**a**) Week 16. (**b**) Week 33. (**c**) Week 51.

Table 4. Hyperparameter values for grid-search.

	Epochs	Batch Size	Units Encoder	Units Decoder	Learning Rate
GHI	40, **80**, 120	256, 512, **1024**	**16**, 32, 64	**16**, 32, 64	**0.0001**, 0.001, 0.01
Wind velocity	40, **80**, 120	256, 512, **1024**	16, **32**, 64	16, **32**, 64	**0.0001**, 0.001, 0.01
Temperature	40, **80**, 120	256, 512, **1024**	16, **32**, 64	16, **32**, 64	**0.0001**, 0.001, 0.01

The numbers in bold correspond to the chosen values for the hyperparameters.

Moreover, the medians of the forecasting metrics are shown in Table 5. Table 6 shows the average computational time for each model in seconds. It is worth noting that training time for GHI is lower than the other two variables because for GHI we remove zero values in the preprocessing step, ending up with only 20 timesteps per day (whereas 48 are used for temperature and wind speed).

Table 5. Evaluation of weather forecasting.

Metric	GHI	Wind Speed	Temperature
RMSE	170.033	0.668	2.057
WAPE	0.224	0.354	0.084
MASE	0.815	0.767	0.820
MAE	131.144	0.482	1.616
APB	2.795	5.566	3.041

Table 6. Training and testing time in seconds.

Computational Time	GHI	Wind Speed	Temperature
Training	118.42	2103.35	726.71
Testing	2.03	1.92	3.09

6.4. Hierarchical Clustering

Special care should be taken regarding the dimensionality of the problem. A balance of energy equation such as the one presented in Equation (18) at the current time resolution

(48 timesteps per day) would mean adding 17,520 (48 times 365) constraints only for that equation, which would be unnecessarily hard to manage by the solver in an affordable time. For that reason, we propose an NLP formulation where a scenario is a typical day weighted according to the number of days similar to it. To find these scenarios, in this work we employ multivariate clustering, specifically hierarchical agglomerative clustering, to agglomerate similar days into time series clusters for four parameters of the NLP model: GHI, wind, temperature, and power consumption. To this aim, we use the Euclidean distance as a similarity function, and the Ward method to agglomerate and update the distance matrix at each iteration as is explained in Section 2. Our search space for the number of clusters is between 2 and 48, and we examine all the possibilities based on the CH index for each forecasting methodology, as is illustrated in Figure 15. We select 30 scenarios since it allows us to increase the CH index from 48 and at the same time maintain a number of scenarios near our search space upper bound. Figures 16 and 17 show 5 out of the 30 scenarios obtained by this method using the seasonal naive forecast and RNN, respectively. It is worth noting that RNN allows us to retrieve a smoother forecast and therefore smoother scenarios.

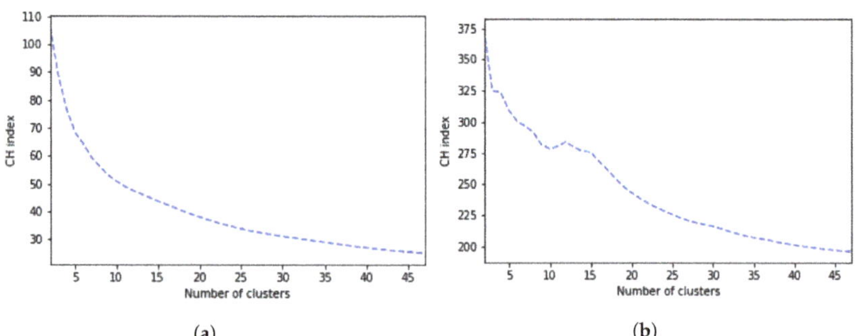

(**a**)   (**b**)

Figure 15. Search for the number of clusters. (**a**) SN forecast. (**b**) RNN forecast.

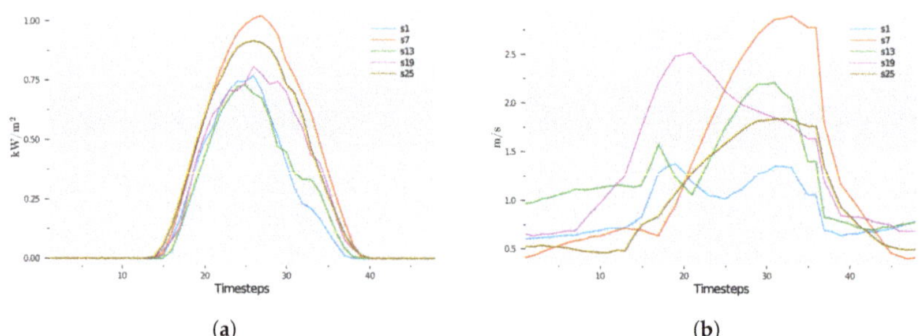

(**a**)   (**b**)

Figure 16. *Cont.*

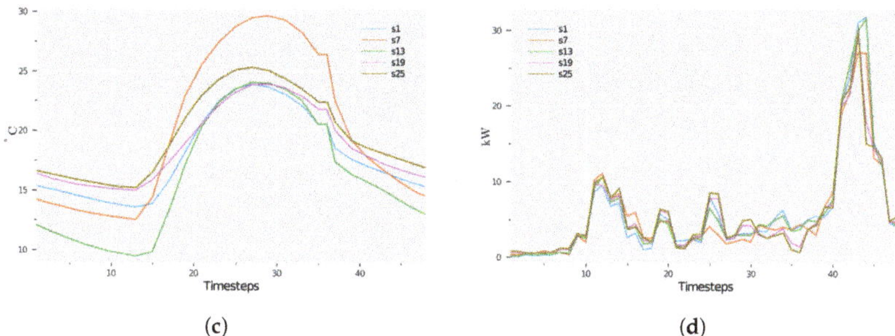

(c)
(d)

Figure 16. Scenarios created from seasonal naive forecast. (**a**) GHI. (**b**) Wind speed. (**c**) Temperature. (**d**) Power demand.

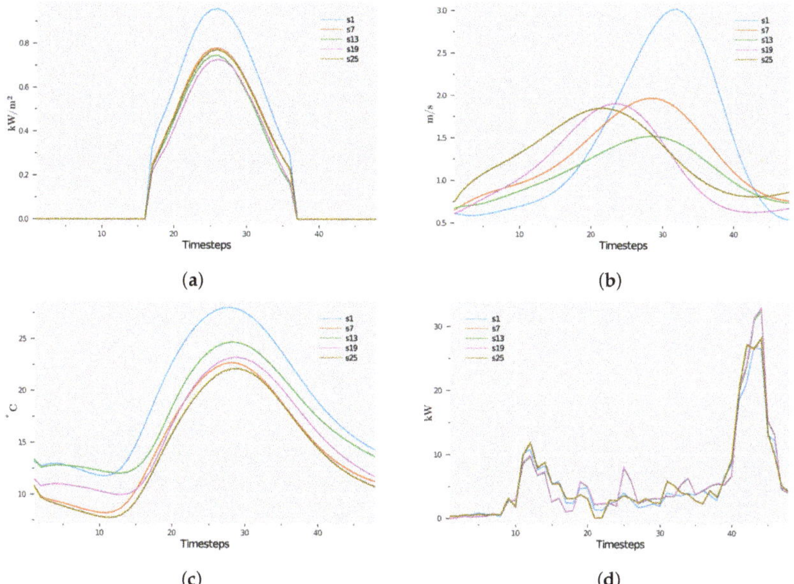

(a)
(b)

(c)
(d)

Figure 17. Scenarios created from RNN forecast. (**a**) GHI. (**b**) Wind speed. (**c**) Temperature. (**d**) Power demand.

6.5. Optimization Results

Numerical experiments for this optimization task were performed using a CPU with the following characteristics: Intel(R) Core(TM) i5-6300HQ @ 2.30 GHz, 8 GB of RAM. The optimization model was programmed in Julia using the JuMP library [63]. Our main goal from the numerical experiments is to yield the optimal decision variables value corresponding to the areas allocated for each technology as well as the nominal capacity required. These values are obtained using two sets of parameters (scenarios): the ones obtained using the RNN forecast and the ones obtained using the seasonal naive forecast. Table 7 presents the areas as well as the optimal costs for both cases. It is worth noting that RNN scenarios demand a high level of area for PV panels and that the first approach requires a higher dimension for the inverter. The objective function, total annual cost, is USD 4526.536 and USD 4424.052 for RNN and SN, respectively. Figures 18 and 19 present six scenarios of PV output and BS energy level for the RNN and SN approach, respectively. We can observe that the peak of PV power output has higher variability in the latter case, whereas the energy level in BS does not vary, to keep it at the point of maximum efficiency.

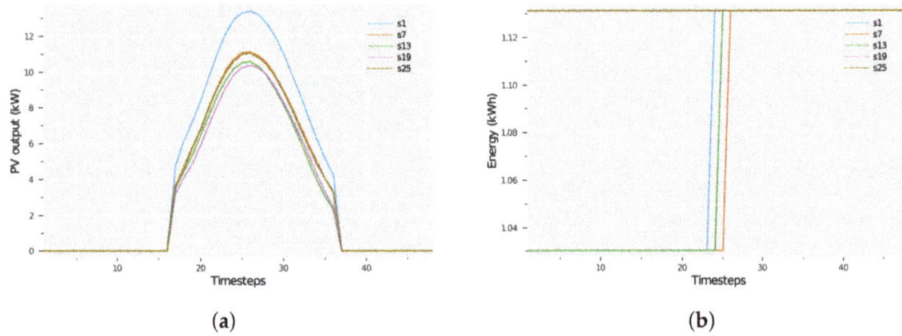

(a)   (b)

Figure 18. Optimization results for PV and BS using RNN forecast. (a) PV output. (b) Energy stored in BS.

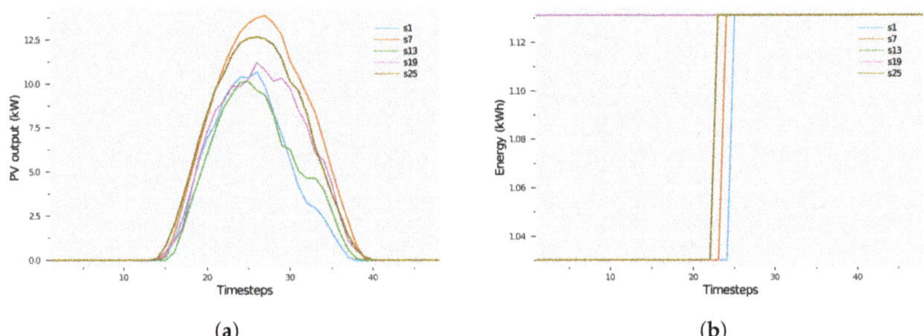

(a)   (b)

Figure 19. Optimization results for PV and BS using seasonal naive forecast. (a) PV output. (b) Energy stored in BS.

Table 7. Results of the optimization run for both approaches.

Variable	PV		WT		DG		BS		INV	
	RNN	SN	RNN	SN	RNN	SN	RNN	SN	RNN	SN
Area, A (m^2)	68.724	67.299	0.090	0.100	67.267	67.267	0.100	0.100	1.079	1.043
Energy, E (kWh)	29,462.760	29,462.698	0.231	0.293	14,731.495	14,731.495	0.100	0.100	99,911.352	99,916.457
Capital cost, $CapCost$ (USD)	14,544.881	14,259.820	194.999	195.000	2502.849	2502.849	93.400	93.400	18,274.549	17,686.509
O&M cost $OMcost$ (USD)	88.388	88.388	0.006	0.008	147.314	147.314	0.000	0.000	10.061	9.728
Total Cost, TAC(USD)	RNN:	4526.536	SN:	4424.052						

6.6. Simulation

In order to establish a fair comparison and have a good sense of the performance of the proposed forecasting methodology hierarchical clustering, and the optimization model, we propose a logic to control the HRES system depicted in Figure 20. The flow diagram is based on the HRES operation description described in some works, such as [64]. In the beginning, the controller reads the current information regarding the BS unit. Then, it retrieves the information about PV and WT generation. If there is any excess of load with respect to the demand, it charges the battery, or even dumps the excess of power in case it is necessary. However, in case this renewable power generation is not sufficient at any given point, then the BS unit is used to meet the current power demand, and if this is just not possible, the DG unit is used for that purpose. For this instance, we obtain a simulation result in which the LOLP for RNN and SN are 0.06008 and 0.6014, respectively, which means that a reduction of the LOLP metric in 0.1% was achieved.

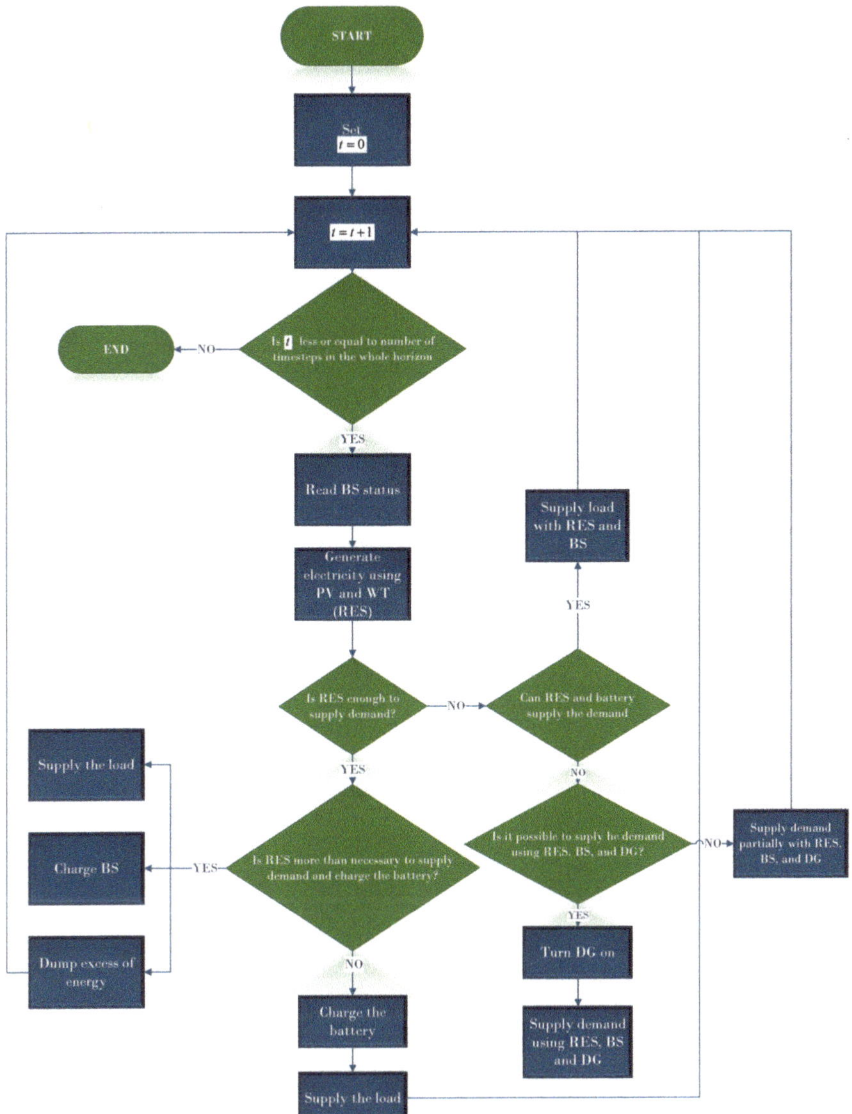

Figure 20. Flow diagram for the HRES management strategy.

7. Conclusions and Future Work

The optimal design of HRES requires finding a balance between undersizing and oversizing, taking into account sustainability concerns. Very few jobs have incorporated the weather forecast into the HRES design as an optimization problem. In this work, we propose a full methodology to find the optimal configuration of an HRES using RNN to predict three variables: global solar irradiance, wind speed, and ambient temperature.

The RNN model in this work employs the RMSprop optimizer for training the parameters. Furthermore, a grid search procedure was carried out to find appropriate values for the hyper-parameters in the validation stage. The three models for the weather variables: GHI, wind, and temperature took 118.42, 2103.35, and 726.71 seconds for training purposes, respectively. Moreover, the MASE for them were 0.815, 0.767, and 0.820 so they proved better than the Seasonal Naive method.

Before feeding the forecast into the model, multivariate hierarchical agglomerative time series clustering was applied to create 30 scenarios that represent the whole year of data. This allowed us to reduce the problem dimension to avoid out-of-memory issues. This step allowed us to solve the entire problem in 32.02 min.

A nonlinear mathematical formulation for the optimization problem was proposed, taking into account three power generation units: solar photovoltaics, wind turbines, and diesel generator. Moreover, a battery system was also available as a backup to ensure HRES autonomy. Renewable factor (RF) and loss of load probability (two important metrics for HRES design) were included as constraints. Our approach was then compared with the use of past data, which entails the seasonal naive method to predict the whole year of 2020. As a case study, we ran numerical experiments using data collected from a rural community on the Pacific Coast of Mexico. Our results show that the total annual cost to supply electricity to the rural inhabitants is USD 4526.536 and USD 4424.052 for RNN and SN, respectively. Additionally, the resulting HRES design is mainly made up of PV and DG units in both cases.

A simulation process was then carried out to evaluate both approaches, and for the instance studied, prediction with the RNN model was proved to be 0.1% superior based on the LOLP metric using real data collected for the year 2020.

For future work, it is interesting to apply meta-heuristics such as BBO and PSO algorithms to solve the optimization problem so that the application of this methodology is entirely cost-free. Furthermore, it could be beneficial to study the use of train tensor neural networks as a top-edge methodology to predict weather parameters. Additionally, comparing different methods for hierarchical clustering could lead to an increase in the benefit of using complex forecasting methods. Finally, given that there are several HRES sizing approaches in the literature, a comparison of the latest and more important ones in terms of numerical solutions using the same superstructure and case study for all of them would be beneficial to identify research directions.

Author Contributions: Conceptualization, A.A.M.-S. and L.E.C.-B.; data curation, A.A.M.-S.; formal analysis, A.A.M.-S. and L.E.C.-B.; investigation, A.A.M.-S. and L.E.C.-B.; methodology, A.A.M.-S. and L.E.C.-B.; software, A.A.M.-S.; supervision, L.E.C.-B.; validation, A.A.M.-S. and L.E.C.-B.; visualization, A.A.M.-S.; writing—original draft, A.A.M.-S.; writing—review and editing, L.E.C.-B. All authors have read and agreed to the published version of the manuscript.

Funding: This research received no external funding.

Institutional Review Board Statement: Not applicable.

Informed Consent Statement: Not applicable.

Data Availability Statement: Data employed in this article can be found in https://nsrdb.nrel.gov/ (accessed on 1 February 2022).

Acknowledgments: To Angel Medina De Paz for your unconditional support during the realization of this work.

Conflicts of Interest: The authors declare no conflict of interest.

Abbreviations

Sets

Symbol	Description	Units
t—τ	Time	30 min period
s—ς	Scenario	Period of days
ψ—Ψ	Power generation technologies	

Acronyms

Symbol	Description	Units
amb	Ambient conditions	
BS	Battery system	
H	Housing—households	
INV	Inverter	
LB	Lower bound	
MAX	Maximum	
MIN	Minimum	
NLP	Nonlinear programming	
O&M, OM	Operation and maintenance	
PV	Photovoltaic system	
SO	Social objective	
TAC	Total annual cost	
UB	Upper bound	
W	Electricity	
WT	Wind turbine	

Variables

Symbol	Description	Units
A	Area allocated for a unit	m^2
$Cost$	Cost	$
$CapCost$	Capital cost	$
E	Energy	kWh
f	Objective function	
$O\&MCost$	Operation and maintenance cost	$
SoC	State of charge	
SOp	Social objective	
T	Temperature	°C
W	Electric flux	kW
TAC	Total annual cost	$/year
η	Efficiency	
ϕ	Operative function	

Parameter

Symbol	Description	Units
k_f	Annualization factor	$
SF	Security factor	
α	Solar irradiance	kW/m^2
β	Temperature coefficient of the PV system	
χ	Unit cost	$/kWh, $/kW
μ	Factor for land usage	kW/m^2, kWh/m^2
Δt	Duration of a time period t	h
ν	Variable cost	$/kW, $/m^2, $/kWh
ω	Fixed cost	$
ρ	Density	kg/m^3
Θ	Duration of each scenario s	days
υ	Wind speed	m/s

Abbreviation

NL	Nonlinear	
LSTM	Long short-term memory	
FFNN	Feedforward neural network	
RNN	Recurrent neural network	
HRES	Hybrid renewable energy system	

References

1. Ameur, H.B.; Han, X.; Liu, Z.; Peillex, J. When did the global warming start? A new baseline for carbon budgeting. *Econ. Model.* **2022**, *116*, 106005. [CrossRef]
2. Lu, Z.Q.; Wu, C.G.; Wu, N.Y.; Lu, H.L.; Wang, T.; Xiao, R.; Liu, H.; Wu, X.H. Change trend of natural gas hydrates in permafrost on the Qinghai-Tibet Plateau (1960–2050) under the background of global warming and their impacts on carbon emissions. *China Geol.* **2022**, *5*, 475–509.
3. IEA. World Energy Outlook 2020. Available online: https://www.iea.org/reports/world-energy-outlook-2020 (accessed on 1 October 2021).
4. Kahwash, F.; Barakat, B.; Taha, A.; Abbasi, Q.H.; Imran, M.A. Optimising Electrical Power Supply Sustainability Using a Grid-Connected Hybrid Renewable Energy System—An NHS Hospital Case Study. *Energies* **2021**, *14*, 7084. [CrossRef]
5. Lim, B.; Hong, K.; Yoon, J.; Chang, J.I.; Cheong, I. Pitfalls of the eu's carbon border adjustment mechanism. *Energies* **2021**, *14*, 7303. [CrossRef]
6. SENER. *Reporte de Avance de Energías Limpias Primer Semestre 2018*; SENER: Getxo, Spain, 2018; p. 21.
7. Erdinc, O. *Optimization in Renewable Energy Systems: Recent Perspectives*; Butterworth-Heinemann: Oxford, UK, 2017.
8. Bank, W. Sustainable Energy for all Database. 2017. Available online: https://data.worldbank.org/indicator/EG.ELC.ACCS.RU.ZS (accessed on 27 April 2020).
9. Ferrer-Martí, L.; Domenech, B.; García-Villoria, A.; Pastor, R. A MILP model to design hybrid wind-photovoltaic isolated rural electrification projects in developing countries. *Eur. J. Oper. Res.* **2013**, *226*, 293–300. [CrossRef]
10. Eriksson, E.; Gray, E.M. Optimization of renewable hybrid energy systems–A multi-objective approach. *Renew. Energy* **2019**, *133*, 971–999. [CrossRef]
11. Emad, D.; El-Hameed, M.; Yousef, M.; El-Fergany, A. Computational methods for optimal planning of hybrid renewable microgrids: A comprehensive review and challenges. *Arch. Comput. Methods Eng.* **2019**, *27*, 1297–1319. [CrossRef]
12. Al-Falahi, M.D.; Jayasinghe, S.; Enshaei, H. A review on recent size optimization methodologies for standalone solar and wind hybrid renewable energy system. *Energy Convers. Manag.* **2017**, *143*, 252–274. [CrossRef]
13. Kusakana, K.; Vermaak, H.; Numbi, B. Optimal sizing of a hybrid renewable energy plant using linear programming. In Proceedings of the IEEE Power and Energy Society Conference and Exposition in Africa: Intelligent Grid Integration of Renewable Energy Resources (PowerAfrica), Bari, Italy, 7–10 September 2021; IEEE: Piscataway, NJ, USA, 2012; pp. 1–5.
14. Domenech, B.; Ranaboldo, M.; Ferrer-Martí, L.; Pastor, R.; Flynn, D. Local and regional microgrid models to optimise the design of isolated electrification projects. *Renew. Energy* **2018**, *119*, 795–808. [CrossRef]
15. Khatod, D.K.; Pant, V.; Sharma, J. Analytical approach for well-being assessment of small autonomous power systems with solar and wind energy sources. *IEEE Trans. Energy Convers.* **2009**, *25*, 535–545. [CrossRef]
16. Luna-Rubio, R.; Trejo-Perea, M.; Vargas-Vázquez, D.; Ríos-Moreno, G. Optimal sizing of renewable hybrids energy systems: A review of methodologies. *Sol. Energy* **2012**, *86*, 1077–1088. [CrossRef]
17. Nadjemi, O.; Nacer, T.; Hamidat, A.; Salhi, H. Optimal hybrid PV/wind energy system sizing: Application of cuckoo search algorithm for Algerian dairy farms. *Renew. Sustain. Energy Rev.* **2017**, *70*, 1352–1365. [CrossRef]
18. Suman, G.K.; Guerrero, J.M.; Roy, O.P. Optimisation of solar/wind/bio-generator/diesel/battery based microgrids for rural areas: A PSO-GWO approach. *Sustain. Cities Soc.* **2021**, *67*, 102723. [CrossRef]
19. Bahramara, S.; Moghaddam, M.P.; Haghifam, M. Optimal planning of hybrid renewable energy systems using HOMER: A review. *Renew. Sustain. Energy Rev.* **2016**, *62*, 609–620. [CrossRef]
20. Lambert, T.W.; Hittle, D. Optimization of autonomous village electrification systems by simulated annealing. *Sol. Energy* **2000**, *68*, 121–132. [CrossRef]
21. Das, B.K.; Al-Abdeli, Y.M.; Kothapalli, G. Optimisation of stand-alone hybrid energy systems supplemented by combustion-based prime movers. *Appl. Energy* **2017**, *196*, 18–33. [CrossRef]
22. Lan, H.; Wen, S.; Hong, Y.Y.; David, C.Y.; Zhang, L. Optimal sizing of hybrid PV/diesel/battery in ship power system. *Appl. Energy* **2015**, *158*, 26–34. [CrossRef]
23. Abd el Motaleb, A.M.; Bekdache, S.K.; Barrios, L.A. Optimal sizing for a hybrid power system with wind/energy storage based in stochastic environment. *Renew. Sustain. Energy Rev.* **2016**, *59*, 1149–1158. [CrossRef]
24. Iqbal, M.M.; Sajjad, I.A.; Khan, M.F.N.; Liaqat, R.; Shah, M.A.; Muqeet, H.A. Energy management in smart homes with pv generation, energy storage and home to grid energy exchange. In Proceedings of the 2019 International Conference on Electrical, Communication, and Computer Engineering (ICECCE), Swat, Pakistan, 24–25 July 2019; IEEE: Piscataway, NJ, USA, 2019; pp. 1–7.
25. Rasheed, M.B.; Javaid, N.; Ahmad, A.; Jamil, M.; Khan, Z.A.; Qasim, U.; Alrajeh, N. Energy optimization in smart homes using customer preference and dynamic pricing. *Energies* **2016**, *9*, 593. [CrossRef]
26. Nasir, T.; Raza, S.; Abrar, M.; Muqeet, H.A.; Jamil, H.; Qayyum, F.; Cheikhrouhou, O.; Alassery, F.; Hamam, H. Optimal scheduling of campus microgrid considering the electric vehicle integration in smart grid. *Sensors* **2021**, *21*, 7133. [CrossRef]
27. Muqeet, H.A.; Munir, H.M.; Javed, H.; Shahzad, M.; Jamil, M.; Guerrero, J.M. An energy management system of campus microgrids: State-of-the-art and future challenges. *Energies* **2021**, *14*, 6525. [CrossRef]
28. Muqeet, H.A.; Javed, H.; Akhter, M.N.; Shahzad, M.; Munir, H.M.; Nadeem, M.U.; Bukhari, S.S.H.; Huba, M. Sustainable Solutions for Advanced Energy Management System of Campus Microgrids: Model Opportunities and Future Challenges. *Sensors* **2022**, *22*, 2345. [CrossRef]

29. Fei, L.; Shahzad, M.; Abbas, F.; Muqeet, H.A.; Hussain, M.M.; Bin, L. Optimal Energy Management System of IoT-Enabled Large Building Considering Electric Vehicle Scheduling, Distributed Resources, and Demand Response Schemes. *Sensors* **2022**, *22*, 7448. [CrossRef] [PubMed]
30. Balouch, S.; Abrar, M.; Abdul Muqeet, H.; Shahzad, M.; Jamil, H.; Hamdi, M.; Malik, A.; Hamam, H. Optimal Scheduling of Demand Side Load Management of Smart Grid Considering Energy Efficiency. Front. *Energy Res.* **2022**, *10*, 861571.
31. Kamjoo, A.; Maheri, A.; Dizqah, A.M.; Putrus, G.A. Multi-objective design under uncertainties of hybrid renewable energy system using NSGA-II and chance constrained programming. *Int. J. Electr. Power Energy Syst.* **2016**, *74*, 187–194. [CrossRef]
32. Hosseini, S.M.; Carli, R.; Dotoli, M. Robust optimal energy management of a residential microgrid under uncertainties on demand and renewable power generation. *IEEE Trans. Autom. Sci. Eng.* **2020**, *18*, 618–637. [CrossRef]
33. Hosseini, S.M.; Carli, R.; Dotoli, M. Robust energy scheduling of interconnected smart homes with shared energy storage under quadratic pricing. In Proceedings of the 2019 IEEE 15th International Conference on Automation Science and Engineering (CASE), Vancouver, BC, Canada, 22–26 August 2019; IEEE: Piscataway, NJ, USA, 2019; pp. 966–971.
34. Zakaria, A.; Ismail, F.B.; Lipu, M.H.; Hannan, M.A. Uncertainty models for stochastic optimization in renewable energy applications. *Renew. Energy* **2020**, *145*, 1543–1571. [CrossRef]
35. Fuentes-Corté, L.F.; Ortega-Quintanilla, M.; Flores-Tlacuahuac, A. Water–Energy Off-Grid Systems Design Using a Dominant Stakeholder Approach. *ACS Sustain. Chem. Eng.* **2019**, *7*, 8554–8578. [CrossRef]
36. Fuentes-Cortés, L.F.; Ponce-Ortega, J.M. Optimal design of energy and water supply systems for low-income communities involving multiple-objectives. *Energy Convers. Manag.* **2017**, *151*, 43–52. [CrossRef]
37. Singla, P.; Duhan, M.; Saroha, S. A comprehensive review and analysis of solar forecasting techniques. *Front. Energy* **2021**, *16*, 187–223. [CrossRef]
38. Aslam, M.; Seung, K.H.; Lee, S.J.; Lee, J.M.; Hong, S.; Lee, E.H. Long-term Solar Radiation Forecasting using a Deep Learning Approach-GRUs. In Proceedings of the 2019 IEEE 8th International Conference on Advanced Power System Automation and Protection (APAP), Xi'an, China, 21–24 October 2019; IEEE: Piscataway, NJ, USA, 2019; pp. 917–920.
39. Kumar, N.M.; Subathra, M. Three years ahead solar irradiance forecasting to quantify degradation influenced energy potentials from thin film (a-Si) photovoltaic system. *Results Phys.* **2019**, *12*, 701–703. [CrossRef]
40. Zhang, W.; Maleki, A.; Rosen, M.A. A heuristic-based approach for optimizing a small independent solar and wind hybrid power scheme incorporating load forecasting. *J. Clean. Prod.* **2019**, *241*, 117920. [CrossRef]
41. Zhang, W.; Maleki, A.; Rosen, M.A.; Liu, J. Sizing a stand-alone solar-wind-hydrogen energy system using weather forecasting and a hybrid search optimization algorithm. *Energy Convers. Manag.* **2019**, *180*, 609–621. [CrossRef]
42. Gupta, R.; Kumar, R.; Bansal, A.K. BBO-based small autonomous hybrid power system optimization incorporating wind speed and solar radiation forecasting. *Renew. Sustain. Energy Rev.* **2015**, *41*, 1366–1375. [CrossRef]
43. Maleki, A.; Khajeh, M.G.; Rosen, M.A. Weather forecasting for optimization of a hybrid solar-wind–powered reverse osmosis water desalination system using a novel optimizer approach. *Energy* **2016**, *114*, 1120–1134. [CrossRef]
44. Smyl, S. A hybrid method of exponential smoothing and recurrent neural networks for time series forecasting. *Int. J. Forecast.* **2020**, *36*, 75–85. [CrossRef]
45. Hewamalage, H.; Bergmeir, C.; Bandara, K. Recurrent neural networks for time series forecasting: Current status and future directions. *Int. J. Forecast.* **2021**, *37*, 388–427. [CrossRef]
46. Müllner, D. Modern hierarchical, agglomerative clustering algorithms. *arXiv* **2011**, arXiv:1109.2378.
47. Otto, K.N. *Product Design: Techniques in Reverse Engineering and new Product Development*; Prentice Hall: Hoboken, NJ, USA, 2003.
48. Askarzadeh, A. Optimisation of solar and wind energy systems: A survey. *Int. J. Ambient. Energy* **2017**, *38*, 653–662. [CrossRef]
49. Čepin, M. *Assessment of Power System Reliability: Methods and Applications*; Springer Science & Business Media: Berlin/Heidelberg, Germany, 2011.
50. Maleki, A.; Pourfayaz, F. Optimal sizing of autonomous hybrid photovoltaic/wind/battery power system with LPSP technology by using evolutionary algorithms. *Sol. Energy* **2015**, *115*, 471–483. [CrossRef]
51. INEGI (National Institute of Statistics). *Anuario Estadìstico y Geogràfico de Michoacàn de Ocampo*; INEGI (National Institute of Statistics): Aguascalientes, Mexico, 2017.
52. IEA (International Energy Agency) National Solar Radiation Database. 2021. Available online: https://nsrdb.nrel.gov/ (accessed on 13 November 2022).
53. Sengupta, M.; Xie, Y.; Lopez, A.; Habte, A.; Maclaurin, G.; Shelby, J. The national solar radiation data base (NSRDB). *Renew. Sustain. Energy Rev.* **2018**, *89*, 51–60. [CrossRef]
54. INEGI (Instituto Nacional de Estadí y Stica Geografía). *Anuario Estadístico y Geográfico de Michoacán de Ocampo 2017*; Instituto Nacional de Estadí y Stica Geografía: Aguascalientes, Mexico, 2018; Volume 10, p. 132. . [CrossRef]
55. IRENA (International Renewable Energy Agency). *Battery Storage for Renewables: Market Status and Technology Outlook*; IRENA (International Renewable Energy Agency): Abu Dhabi, United Arab Emirates, 2015.
56. IRENA (International Renewable Energy Agency). *Renewable Power Generation Costs in 2017 Report*; International Renewable Energy Agency: Abu Dhabi, United Arab Emirates, 2018.
57. Gbémou, S.; Eynard, J.; Thil, S.; Guillot, E.; Grieu, S. A Comparative Study of Machine Learning-Based Methods for Global Horizontal Irradiance Forecasting. *Energies* **2021**, *14*, 3192. [CrossRef]

58. Medina-Santana, A.A.; Hewamalage, H.; Cárdenas-Barrón, L.E. Deep Learning Approaches for Long-Term Global Horizontal Irradiance Forecasting for Microgrids Planning. *Designs* **2022**, *6*, 83. [CrossRef]
59. Medina-Santana, A.A.; Flores-Tlacuahuac, A.; Cárdenas-Barrón, L.E.; Fuentes-Cortés, L.F. Optimal design of the water-energy-food nexus for rural communities. *Comput. Chem. Eng.* **2020**, *143*, 107120. [CrossRef]
60. Hyndman, R.J.; Athanasopoulos, G. *Forecasting: Principles and Practice*; Monash University: Clayton, Australia , 2018.
61. Taieb, S.B.; Bontempi, G.; Atiya, A.F.; Sorjamaa, A. A review and comparison of strategies for multi-step ahead time series forecasting based on the NN5 forecasting competition. *Expert Syst. Appl.* **2012**, *39*, 7067–7083. [CrossRef]
62. Tesorflow Documentation. Available online: https://www.tensorflow.org/api_docs/python/tf/keras/layers/LSTM/ (accessed on 3 January 2022.
63. Dunning, I.; Huchette, J.; Lubin, M. JuMP: A Modeling Language for Mathematical Optimization. *SIAM Rev.* **2017**, *59*, 295–320. [CrossRef]
64. Bukar, A.L.; Tan, C.W.; Lau, K.Y. Optimal sizing of an autonomous photovoltaic/wind/battery/diesel generator microgrid using grasshopper optimization algorithm. *Sol. Energy* **2019**, *188*, 685–696. [CrossRef]

Article

Dynamic Thermal Transport Characteristics of a Real-Time Simulation Model for a 50 MW Solar Power Tower Plant

Haoyu Huang [1], Ershu Xu [1,2,*], Lengge Si [1], Qiang Zhang [3] and Qiang Huang [4]

[1] School of Energy Power and Mechanical Engineering, North China Electric Power University, Beijing 102206, China
[2] Key Laboratory of Power Station Energy Transfer Conversion and System, North China Electric Power University, Ministry of Education, Beijing 102206, China
[3] School of Electrical, Energy and Power Engineering, Yangzhou University, Yangzhou 225127, China
[4] CGN New Energy Holding Co., Ltd., Beijing 100070, China
* Correspondence: xuershu@ncepu.edu.cn; Tel.: +86-(10)-6177-3918

Abstract: A dynamic simulation model of a heliostat field and molten salt receiver system are developed on the STAR-90 simulation platform. In addition, a real-time simulation model coupling the above two models is built to study the photothermal conversion process of Delingha's 50 MW solar power tower plant. The nonuniform and time-varying characteristics of the energy flux density on the receiver surface and the dynamic characteristics under different operating conditions are studied. The operational process of the receiver of a typical day is simulated. It was found that there was a strong positive correlation between the energy flux and DNI, and the maximum energy flux density on the surface of the heat absorbing tube panel moved from the first tube panel to the fourth in sequence from 12:00 to 18:00. At the same time, the energy flux density of the last four panels decreased gradually along the arrangement order of the panels. DNI, molten salt mass flow rate and inlet temperature step disturbance simulations are carried out, and the response curves of the molten salt outlet temperature and tube wall temperature are obtained. The conclusion of this paper has important guiding significance for the establishment of an operational strategy for photothermal coupling in a molten salt solar power tower plant.

Keywords: solar power tower plant; receiver; heliostat; model; dynamic characteristics

Citation: Huang, H.; Xu, E.; Si, L.; Zhang, Q.; Huang, Q. Dynamic Thermal Transport Characteristics of a Real-Time Simulation Model for a 50 MW Solar Power Tower Plant. *Energies* **2023**, *16*, 1946. https://doi.org/10.3390/en16041946

Academic Editors: Sharul Sham Dol and Anang Hudaya Muhamad Amin

Received: 12 December 2022
Revised: 5 January 2023
Accepted: 17 January 2023
Published: 15 February 2023

1. Introduction

With the rapid development of the world economy in recent years, the problems of energy consumption and shortage are becoming increasingly serious. According to world energy statistics, China, for 18 consecutive years, ranked first in global energy consumption growth, while industrial consumption accounted for more than 70% [1]. How to reduce the consumption of fossil fuels has once again been put on the agenda, and the power and heat industries have been greatly impacted. Since then, China has focused on making renewable energy the mainstay of its new power system, controlling fossil energy consumption and reducing greenhouse gases.

However, there are many problems in the development of renewable energy power generation. Energy absorption is a key problem to be solved, but the concept of the "integration of wind, solar and thermal storage" may solve the power grid coordination problem [2]. Renewable energy generation is usually characterized by low energy density, intermittency and geographical constraints, which greatly restrict its efficient and large-scale development [3,4]. The development of an energy storage system can mitigate the impact of intermittency and help to realize uninterrupted power generation. The volatility of renewable energy makes it necessary to cooperate with energy storage technology to achieve peak-filling purposes. Energy storage technologies will serve as the cornerstone of renewable energy integration [5].

In the field of energy and electricity, solar energy is favored for its cleanness, long duration and great radiant energy. Molten salt solar power tower (SPT) technology with good dispatchability is equipped with thermal storage systems to help it achieve continuous power generation. This renewable energy generation technology will be widely used in the future in new power systems. At present, there are still many technical problems in the operation of an SPT plant, which need to be constantly explored. The energy flux density on the receiver surface is very uneven, which will seriously affect the safe operation of the receiver to a certain extent [6]. In order to reduce the influence of nonuniform energy fluxes on the operation of the receiver, most researchers focus on simulating the distribution of a solar flux. Qiu [7] developed a model to simulate the operation of the receiver in the Dahan power station and obtained the solar flux distribution on the heat absorbing tube in the receiver. Cruz et al. [8] developed an optimization algorithm to equalize the energy flux density on the endothermic surface of a receiver. Xu et al. [9] established a three-dimensional transient model and calculated the temperature field of the receiver under the time-varying flux to simulate the transient characteristics of the receiver. Sanchez et al. [10] studied the nonuniform energy flux distribution on an external receiver. Ho [11] used the photographic flux (PHLUX) mapping method to determine the distribution of a solar flux from a concentrating system. Besarati et al. [12] proposed a new optimization algorithm and combined it with the HFLCAL model for verification, finding that the new algorithm can effectively reduce the maximum energy flux density. In previous studies, more attention has been paid to the energy flux distribution of the cavity receiver, but the research on the external receiver cannot be neglected. The energy flux on the surface of the receiver is usually measured by infrared thermography, but the direct normal irradiance (DNI) is a real-time variable, which will make the measurement result inaccurate. Therefore, it is necessary to develop a real-time simulation model for an SPT plant to predict the change in the energy flux, which can also effectively improve the system security.

In addition, improving the control strategy of a heliostat field to make the energy flux distribution uniform is also a focus of research. Yu [13] developed a multi focus aiming model and proposed a grouping method for the focus distribution of a heliostat field of a cavity receiver. Sun [14] proposed a method for calculating the reference deviation of the two rotating axes of a heliostat, which is used to reduce the tracking error of heliostat. Hu [15] studied a three-reflection heliostat and found that the flux density of it was more uniform than that of a single-reflection heliostat. Victor [16] designed a heliostat field with an equilibrium mirror density according to the mirror density. During the operation of an SPT plant, the light side changes quickly, while the heat side responds slowly. The two sides are difficult to synchronize, which is one of the reasons for the receiver's overtemperature. For each study on photothermal coupling models, the influence of the variation in the energy flux with time on the molten salt temperature was not considered. Furthermore, an effective control strategy of the heliostat field should also be coordinated to ensure the safe operation of the power plant to a greater extent.

The variable working environment (cloud cover, windy weather, etc.) is a great challenge to the safety of a receiver's operation. It is helpful to study the dynamic characteristics of the receiver through simulation to avoid problems that may occur in an actual operation. In this paper, heliostat field model was built using the Monte Carlo ray tracing method. A model of the molten salt receiver was established using the modular modeling method. A dynamic simulation model of a 50 MW photothermal coupling system was established by coupling the two models. The thermal transport characteristics and dynamic characteristics of the molten salt receiver under a nonuniform and time-varying energy flux were studied by simulating the operational process of a real power station. The results are helpful for ensuring the safe operation and formulation of a control strategy for a molten salt SPT plant.

2. Model Description

2.1. Model of the Heliostat

The mathematical model of a two-axis heliostat with azimuth-elevation tracking was established using the Monte Carlo ray tracing method. Ray tracing is a statistical method by which a certain amount of incident rays are reflected to the surface of a receiver by a heliostat. In addition, the amount of reflected rays reaching the receiver determines the efficiency. A schematic diagram of the heliostat modeling is shown in Figure 1. A ground coordinate system (X_g, Y_g, Z_g) was established to calibrate the position of the heliostats and receiver. The heliostat coordinate system (X_h, Y_h, Z_h) was used to calculate the positional relationship between the incident ray and the reflected ray. The receiver coordinate system (X_r, Y_r, Z_r) was applied to calculate the intersection of the reflected ray and the surface of the receiver. The reflecting mirror was divided into several regions, and the reflecting point was the vertex of the boundary of each region. The incident ray was reflected to the surface of the receiver after a parallel incident, and the reflecting path of the reflected ray was traced. The heliostat model made the following assumptions, taking into account the engineering practice and computational speed:

(1) Ignoring the influence of the small spacing of the mirror on the whole focusing process, the heliostat was regarded as a complete sphere.

(2) All rays of the sun in the solar cone carry the same amount of energy.

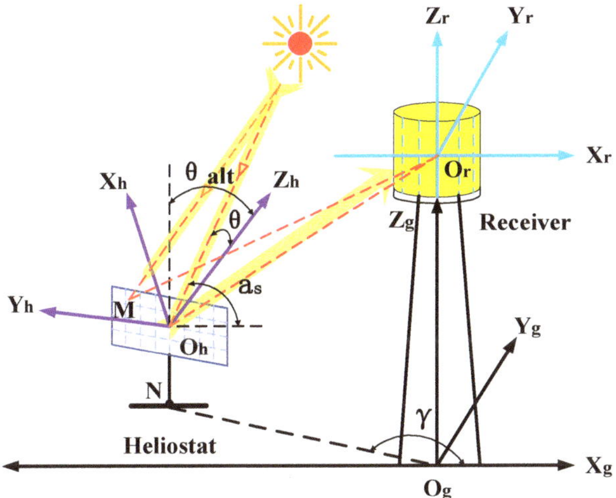

Figure 1. A schematic diagram of the heliostat modeling.

The azimuth of an arbitrary number of heliostats in the field can be expressed as

$$
\gamma = \begin{cases} \arccos \dfrac{X_{Ng}}{X_{Ng}{}^2 + Y_{Ng}{}^2}, Y_{Ng} \geq 0 \\ \arctan \dfrac{Y_{Ng}}{X_{Ng}} + 2\pi, X_{Ng} \geq 0, Y_{Ng} < 0 \\ \arctan \dfrac{Y_{Ng}}{X_{Ng}} + \pi, X_{Ng} < 0, Y_{Ng} < 0 \end{cases}
\tag{1}
$$

where $N(X_{Ng}, Y_{Ng}, Z_{Ng})$ is the coordinate of the heliostat position in the ground coordinate system.

The vectorized representation of the incident ray ($\vec{\beta}_{ing}$), reflected ray ($\vec{\beta}_{refg}$), and the central normal of the heliostat ($\vec{\beta}_{norg}$) in the ground coordinate system

$$
\begin{cases}
\vec{\beta}_{ing} = [\cos(\alpha_s)\cos(\beta_s) \quad \cos(\alpha_s)\sin(\beta_s) \quad \sin(\alpha_s)] \\
\vec{\beta}_{refg} = [-\cos(\gamma)\sin(\phi) \quad -\sin(\gamma)\sin(\phi) \quad \cos(\phi)] \\
\vec{\beta}_{norg} = [\sin(\theta_{alt})\cos(\theta_{azi}) \quad \sin(\theta_{alt})\sin(\theta_{azi}) \quad \cos(\theta_{alt})]
\end{cases}
\tag{2}
$$

where α_s is the solar altitude angle, β_s is the solar azimuth angle, θ_{azi} represents the rotational angle of the axis, which is perpendicular to the ground, and θ_{alt} represents the pitch angle formed by pitching on the another axis.

$$
\vec{\beta}_{norg} = \frac{\vec{\beta}_{ing} + \vec{\beta}_{refg}}{2\cos\theta}
\tag{3}
$$

$$
a_{norg}^2(x) + a_{norg}^2(y) + a_{norg}^2(z) = 1
\tag{4}
$$

The sun incidence angle (θ) and the tracking rotation angle of the two axes of the heliostat can be obtained by

$$
\theta = \arccos(\frac{\sqrt{2}}{2}[\sin(\alpha_s)\cos(\phi) - \cos(\gamma - \beta_s)\cos(\alpha_s)\sin(\phi) + 1]^{\frac{1}{2}})
\tag{5}
$$

$$
\theta_{azi} = \arctan(\frac{\cos(\alpha_s)\sin(\beta_s) - \sin(\gamma)\sin(\phi)}{\cos(\alpha_s)\cos(\beta_s) - \cos(\gamma)\sin(\phi)})
\tag{6}
$$

$$
\theta_{alt} = \arccos(\frac{\cos(\phi) + \sin(\alpha_s)}{2\cos\theta})
\tag{7}
$$

The vector of the solar incident ray in the heliostat coordinate system can be expressed as

$$
\vec{\beta}_{inh} =
\begin{bmatrix}
1 & 0 & 0 \\
0 & \cos(\theta_{alt}) & \sin(\theta_{alt}) \\
0 & -\sin(\theta_{alt}) & \cos(\theta_{alt})
\end{bmatrix}
\times
\begin{bmatrix}
\cos(\theta_{azi} + \frac{\pi}{2}) & \sin(\theta_{azi} + \frac{\pi}{2}) & 0 \\
-\sin(\theta_{azi} + \frac{\pi}{2}) & \cos(\theta_{azi} + \frac{\pi}{2}) & 0 \\
0 & 0 & 1
\end{bmatrix}
\times
\begin{bmatrix}
\vec{\beta}_{ing}(x) \\
\vec{\beta}_{ing}(y) \\
\vec{\beta}_{ing}(z)
\end{bmatrix}
\tag{8}
$$

The vector of the reflected ray in the heliostat coordinate system

$$
\begin{bmatrix}
\vec{\beta}_{refh}(x) \\
\vec{\beta}_{refh}(y) \\
\vec{\beta}_{refh}(z)
\end{bmatrix}
=
\left\{
\vec{e}(3) - 2 \times
\begin{bmatrix}
\vec{n}_{Mh}(x) \\
\vec{n}_{Mh}(y) \\
\vec{n}_{Mh}(z)
\end{bmatrix}
\times
\begin{bmatrix}
\vec{n}_{Mh}(x) & \vec{n}_{Mh}(y) & \vec{n}_{Mh}(z)
\end{bmatrix}
\right\}
\times
\begin{bmatrix}
\vec{\beta}_{inh}(x) \\
\vec{\beta}_{inh}(y) \\
\vec{\beta}_{inh}(z)
\end{bmatrix}
\tag{9}
$$

where $\begin{bmatrix} \vec{n}_{Mh}(x) & \vec{n}_{Mh}(y) & \vec{n}_{Mh}(z) \end{bmatrix}$ is the normal vector of any point $M(X_{Mh}, Y_{Mh}, Z_{Mh})$ in the heliostat coordinate system.

The vector of the reflected ray and point M in the receiver coordinate system can be obtained by

$$
\vec{\beta}_{refr} =
\begin{bmatrix}
1 & 0 & 0 \\
0 & \cos(-\frac{\pi}{2}) & -\sin(-\frac{\pi}{2}) \\
0 & \sin(-\frac{\pi}{2}) & \cos(-\frac{\pi}{2})
\end{bmatrix}
\times
\begin{bmatrix}
\cos(-\pi) & -\sin(-\pi) & 0 \\
\sin(-\pi) & \cos(-\pi) & 0 \\
0 & 0 & 1
\end{bmatrix}
\times
\begin{bmatrix}
\cos(\theta_{azi} + \frac{\pi}{2}) & -\sin(\theta_{azi} + \frac{\pi}{2}) & 0 \\
\sin(\theta_{azi} + \frac{\pi}{2}) & \cos(\theta_{azi} + \frac{\pi}{2}) & 0 \\
0 & 0 & 1
\end{bmatrix}
\times
\begin{bmatrix}
1 & 0 & 0 \\
0 & \cos(\theta_{alt}) & -\sin(\theta_{alt}) \\
0 & \sin(\theta_{alt}) & \cos(\theta_{alt})
\end{bmatrix}
\times
\begin{bmatrix}
\vec{\beta}_{refh}(x) \\
\vec{\beta}_{refh}(y) \\
\vec{\beta}_{refh}(z)
\end{bmatrix}
\tag{10}
$$

$$\begin{bmatrix} M_r(x) \\ M_r(y) \\ M_r(z) \end{bmatrix} = \begin{bmatrix} 1 & 0 & 0 \\ 0 & \cos\left(-\frac{\pi}{2}\right) & -\sin\left(-\frac{\pi}{2}\right) \\ 0 & \sin\left(-\frac{\pi}{2}\right) & \cos\left(-\frac{\pi}{2}\right) \end{bmatrix} \times \begin{bmatrix} \cos(-\pi) & -\sin(-\pi) & 0 \\ \sin(-\pi) & \cos(-\pi) & 0 \\ 0 & 0 & 1 \end{bmatrix} \times \left\{ \begin{bmatrix} X_{Ng} \\ Y_{Ng} \\ Z_{Ng} - Hor \end{bmatrix} + \begin{bmatrix} \cos\left(\theta_{azi} + \frac{\pi}{2}\right) & -\sin\left(\theta_{azi} + \frac{\pi}{2}\right) & 0 \\ \sin\left(\theta_{azi} + \frac{\pi}{2}\right) & \cos\left(\theta_{azi} + \frac{\pi}{2}\right) & 0 \\ 0 & 0 & 1 \end{bmatrix} \times \begin{bmatrix} 1 & 0 & 0 \\ 0 & \cos(\theta_{alt}) & -\sin(\theta_{alt}) \\ 0 & \sin(\theta_{alt}) & \cos(\theta_{alt}) \end{bmatrix} * \begin{bmatrix} M_h(x) \\ M_h(y) \\ M_h(z) \end{bmatrix} \right\} \tag{11}$$

Equation (12) is constructed from the reflection point and the direction vector of the reflected ray. Combined with Equation (13), which is the surface equation of the receiver, the coordinates of the intersection point can be solved.

$$\begin{cases} x = M_r(x) + \vec{\beta}_{refr}(x) \cdot t \\ y = M_r(y) + \vec{\beta}_{refr}(y) \cdot t \\ z = M_r(z) + \vec{\beta}_{refr}(z) \cdot t \end{cases} \tag{12}$$

$$x^2 + y^2 = R^2, -\frac{H_r}{2} < z < \frac{H_r}{2} \tag{13}$$

The energy flux density can be obtained by counting the number of intersections in a certain region

$$\varphi = \frac{DNI(t) \cdot \eta_h \cdot N_{jd} \cdot S_h}{N_{in} \cdot S_q} \tag{14}$$

where N_{jd} is the number of intersections in an area, N_{in} is the number of incident rays, η_h is the efficiency of the heliostat, S_h is the area of the heliostat, and S_q is the area of the reflection.

The heliostat field of a molten salt SPT plant is usually composed of thousands of the heliostats arranged in a circular shape, but there are no public data concerning the arrangement of a heliostat field in a power station at present. In order to achieve better simulation results, a heliostat field was optimized using SAM software in this paper. The arrangement of the heliostat field is shown in Figure 2. There were 5273 heliostats of 100 m^2. The weather parameters, the heliostat coordinates and the structural parameters of the heliostat were taken as input parameters, and the actual working state of the heliostat was simulated.

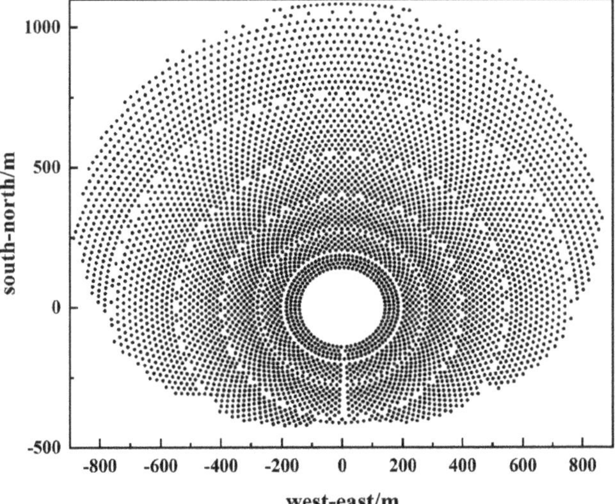

Figure 2. The layout of the heliostat field.

2.2. Model of the Receiver

The external molten salt receiver was composed of 16 tube panels with two flow paths, and solar salt (60% $NaNO_3$ and 40% KNO_3) was used as the heat-absorbing working fluid. The molten salt enters the receiver from the north side of the tower, flows through the four tube panels and crosses the two paths to exchange heat, making the temperature uniform, and then flows into the remaining tube panels. The flow path of the molten salt and the heat transfer process of the tube panel are shown in Figure 3.

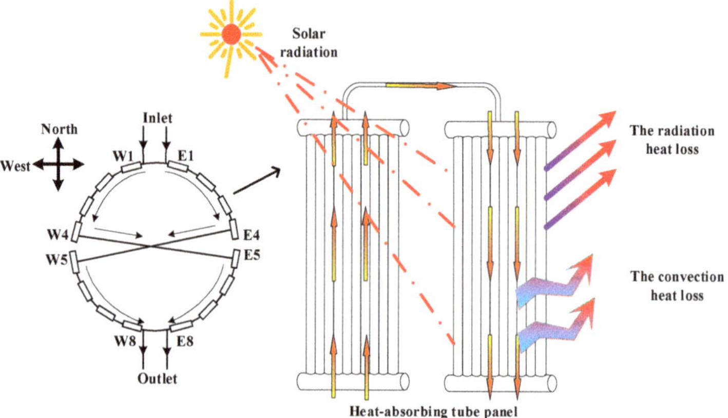

Figure 3. The flow path of the molten salt and the heat transfer process of the tube panel.

The structural parameters of the external molten salt receiver were taken from the 50 MW SPT plant in Delingha, China, as shown in Table 1. A dynamic mathematical model for the flow and heat transfer of a single tube panel of the receiver was established using the modular modeling method. The assumptions and simplifications of the model were as follows:

(1)   The flow in the tube was one-dimensional and unidirectional.
(2)   The variation in the potential energy and kinetic energy of the molten salt flow in the tube was ignored.

Table 1. Parameters of the molten salt receiver and collector field.

Parameter	Value	Unit
Heat load	251	MW
Diameter of the receiver	13	m
Height of the receiver	15	m
Effective heat absorbing area	615	m^2
Number of tube panels	16	
Number of heat absorbing tubes	1216	
Material of the receiver	316H	
Length of the heat absorbing tube	15	m
Outer diameter of the heat absorbing tube	31.75	mm
Inner diameter of the heat absorbing tube	29.55	mm
Surface absorptivity	$\geq$95%	
Installation height	190	m
The average heat flux density	460	kW/m^2
The maximum heat flux density	510	kW/m^2
Designed DNI	900	W/m^2

The radiant energy of the sun received by the surface of the receiver

$$Q_{sun} = \varphi \cdot A \tag{15}$$

Energy conservation of the tube panel

$$Q_{sun} - Q_{sm} - Q_{lc} - Q_{lr} = M_m \cdot C_m \cdot \frac{dt_m}{d\tau} \tag{16}$$

Energy conservation of the molten salt

$$M_s \cdot C_s \cdot \frac{dt_{ous}}{d\tau} = G_{in} C_s t_{ins} - G_{ou} C_s t_{ous} + Q_{sm} \tag{17}$$

Mass conservation of the molten salt

$$V_s \cdot \frac{d\rho_{ous}}{d\tau} = G_{in} - G_{ou} \tag{18}$$

The heat absorption of the molten salt

$$Q_{sm} = h_{sm} A (t_m - t_s) \tag{19}$$

$$h_{sm} = \frac{Nu_s \cdot \lambda}{d} \tag{20}$$

The Nusselt number is based on an empirical formula [17]

$$Nu_s = \frac{f_D}{8} \cdot \mathrm{Pr}_s \cdot \frac{\mathrm{Re}_s - 1000}{1 + 12.7 \cdot \sqrt{\frac{f_D}{8}} \cdot (\mathrm{Pr}_s^{\frac{2}{3}} - 1)} \cdot \left[1 + \left(\frac{d}{H_r}\right)^{\frac{2}{3}} \right] \cdot \left(\frac{\mathrm{Pr}_s}{\mathrm{Pr}_m}\right)^{0.11} \tag{21}$$

where f_D is the Darcy friction coefficient, which can be calculated with the Filonenko formula [18]

$$f_D = (1.82 \cdot \log_{10}(\mathrm{Re}_s) - 1.64)^{-2} \tag{22}$$

The Reynolds and Prandtl numbers of the molten salt can be expressed as

$$\begin{cases} \mathrm{Re}_s = \frac{u \times d}{\nu} \\ \nu = \frac{\mu}{\rho_{ous}} \\ \mathrm{Pr}_s = \frac{\nu}{a} \end{cases} \tag{23}$$

The convection heat loss

$$Q_{lc} = h_{com} A (t_m - t_{amb}) \tag{24}$$

Comprehensive convective heat transfer coefficient [19]

$$h_{com} = \sqrt[3.2]{h_{nc}^{3.2} + h_{fc}^{3.2}} \tag{25}$$

The natural convection heat transfer coefficient between the heat-absorbing tube panel and the surrounding air [20]

$$h_{nc} = \frac{\lambda_a \cdot Nu_{anc}}{H_r} \tag{26}$$

$$Nu_{anc} = 0.088 \cdot Gr_a^{\frac{1}{3}} \cdot \left(\frac{t_m + 273.15}{t_{amb} + 273.15}\right)^{0.18} \tag{27}$$

$$Gr_a = \frac{g \cdot \alpha_v \cdot \Delta t \cdot H_r^3}{\nu^2} \tag{28}$$

Considering the installation height of external molten salt receiver is large, there is no shelter in the high altitude, the wind level is higher, air and tube panel forced convection occurs. The forced convection heat transfer coefficient is calculated according to the forced convection heat transfer of the air flow across the cylinder [21],

$$h_{fc} = \frac{\lambda_a \cdot Nu_{afc}}{d} \tag{29}$$

$$Nu_{afc} = 0.0266 \cdot \text{Re}_a^{0.8} \cdot \text{Pr}_a^{\frac{1}{3}} \tag{30}$$

The radiation heat loss.

According to the Stefan–Boltzmann law, the radiant heat transfer between the tube panel and the environment is calculated as

$$Q_{lr} = \varepsilon \sigma_b A \left[(t_m + 273.15)^4 - (t_{amb} + 273.15)^4 \right] \tag{31}$$

where ε is the emissivity of the metal surface. The surface coating was Pyromark, and 0.95 was chosen here [22]. σ_b is the Boltzmann constant, 5.67×10^{-8} W/(m^2·K).

The mathematical model was solved using the Euler algorithm. The molten salt outlet temperature and tube wall temperature can be calculated with Equation (32).

$$y_{n+1} = y_n + \frac{dy_n}{dx_n} \Delta \tau \tag{32}$$

where n is the calculation step, and $\Delta \tau$ is the integral time step size.

2.3. Model Validation

The model's parameters were set in accordance with the parameters of the Solar Two power plant, and the test data on 5 March 1999 were compared [23]. The data comparisons are shown in Table 2. The simulation results of the molten salt outlet temperature were for 558.1 °C and the operational data for 565.0 °C when the flow rate was 79 kg/s; thus, the relative error was 1.22%. The relative error of the molten salt outlet temperature was 0.53% when the flow rate was reduced to 36 kg/s. The error of the thermal efficiency was lower than 1.6%; therefore, the reliability of the model was validated.

Table 2. The comparison of the simulation results and test data.

Parameter	Operation	Simulation	Operation	Simulation
Flow rate, kg·s^{-1}	79		36	
Inlet temperature, °C	308.0		308.5	
Ambient temperature, °C	16		16	
Wind speed, m·s^{-1}	3.0		3.0	
Outlet temperature, °C	565.0	558.1	564.0	561.0
The error of the outlet temperature, %	1.22		0.53	
Thermal efficiency, %	86.6	85.8	78.3	77.1
The error of the thermal efficiency, %	0.92		1.53	

3. Results with Analysis

3.1. The Dynamic Characteristics of the Energy Flux Density

The dynamic simulation model of the integrated heat-collecting and heat-absorbing by the 50 MW molten salt SPT plant was run on the STAR-90 simulation platform. The 2021 equinox day (i.e., March 20) was selected as a typical day and the operational conditions on this date were simulated. The observed meteorological data for the station location were taken as the input parameters. The meteorological data were from Delingha, China. The change in the DNI data on the vernal equinox is displayed in Figure 4. A three-dimensional diagram of the energy flux distribution on the surface of the receiver at each typical moment is shown in Figure 5. The plane of the x-axis and y-axis in the diagram represents the eight tube panels on the west side of the molten salt receiver.

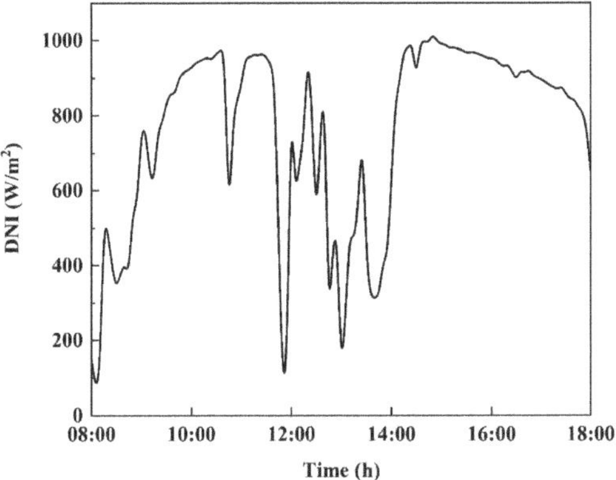

Figure 4. DNI on the vernal equinox.

The figures indicate that the energy flux distribution on the surface of the tube panel was very uneven. The energy flux of a single tube panel was high in the middle and low on both sides; that is, the position of the entrance and exit of the tube panel was low and the position of the middle was high. The energy flux density on a single tube panel was in a Gaussian distribution. This distribution mode was influenced by the central point focusing mode adopted in the heliostat field. The variation of the DNI had a direct effect on the total energy absorbed by the receiver, and the energy flux on the surface of the receiver had a positive correlation with the DNI. Moreover, the maximum energy flux at each specific time appeared in sequence on the W1 to W4 panels. Over time, the highest point moved toward the tube panel behind the flow path. The energy flux on the surface of the last four tube panels (from W5 to W8) decreased in turn.

According to the distribution of the energy flux density, the wall temperature of the tube panel and molten salt temperature were calculated, and the temperature distribution at a specific time is shown in Figure 6. It can be seen that the distribution curve of the wall temperature on a single tube panel was similar to that of the energy flux along the axial direction, both rising at first and then decreasing. This shows that the wall temperature at the inlet and outlet of the tube panel was lower, but the wall temperature at the middle was higher. It also demonstrates the consistency of the temperature and energy flux changes. The nonuniform distribution of the energy flux made the highest wall temperature appear in the middle of the panel, which was also the reason for the tube wall temperature fluctuation. The temperature of the molten salt increased gradually with the flow process in the receiver. The temperature of the molten salt at the outlet of the receiver was the highest because of the cascade arrangement of each panel and the superposition effect of the temperature. The difference between the molten salt temperature and the tube wall temperature decreased gradually along the flow direction. The maximum temperature difference was at the first panel, and the minimum temperature difference was at the eighth panel. At 16:00, the DNI was the largest of the four time points; thus, the wall temperature and molten salt outlet temperature were also the highest. At this time, the maximum temperature difference at the first panel was 66.8 °C in the middle of the panel, and the minimum temperature difference at the eighth panel was 25.3 °C. It is shown that the first tube panel was easily affected by a large temperature difference, while the eighth panel was easily affected by a high temperature. More attention should be paid in both of these locations in the operation of the receiver. At 14:00 and 18:00, the DNI decreased seriously, and the molten salt outlet temperature was 383.1 and 388.2 °C, respectively, which was much lower than the rated outlet temperature of 565 °C. At this time, the thermal storage system will play a role to ensure the operation of the power plant system. Overall, the energy flux density varied with the change in the DNI and had a certain rule with the change in the time.

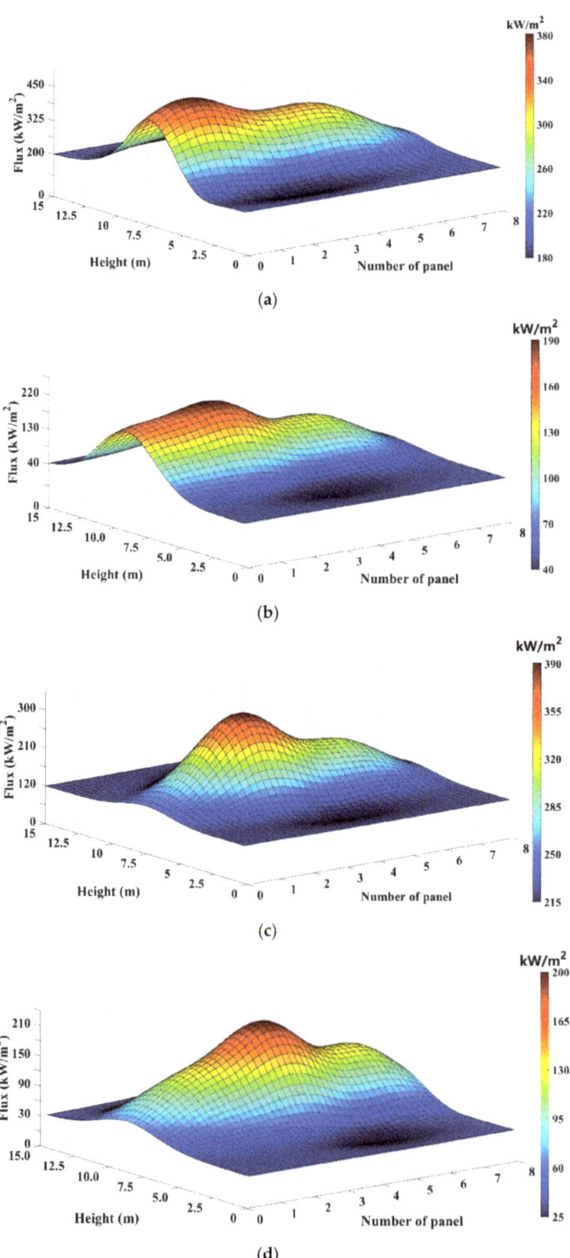

Figure 5. Energy flux distribution on the surface of the receiver at each typical moment: (**a**) energy flux distribution at 12:00; (**b**) energy flux distribution at 14:00; (**c**) energy flux distribution at 16:00; (**d**) energy flux distribution at 18:00.

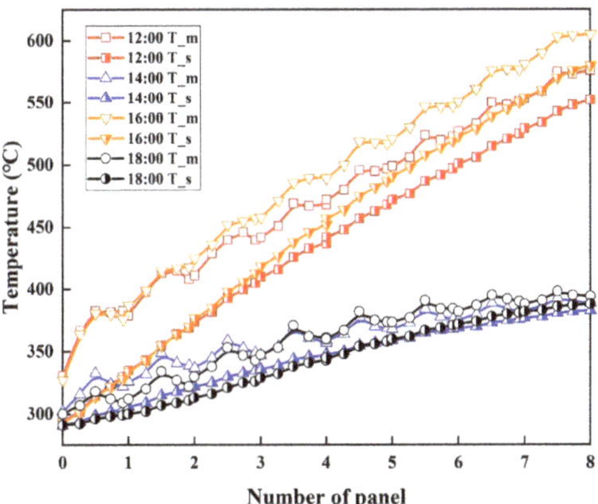

Figure 6. The temperature distribution at a specific time.

3.2. The Dynamic Characteristics of the Receiver under Variable Operating Conditions

The operation of the external molten salt receiver was affected by many factors, including the DNI, flow rate and molten salt inlet temperature. Based on the operation of the 50 MW photothermal coupled dynamic model under a 100% load condition, the three factors mentioned above were taken as step disturbance variables, the dynamic characteristics of the molten salt receiver under disturbance were studied and the outlet temperature changes of the molten salt and tube wall were observed. The main parameters of the receiver under full load conditions are shown in Table 3.

Table 3. The main parameters of the receiver under full load conditions.

Parameter	Value
Molten salt inlet temperature, °C	290
Molten salt outlet temperature, °C	565
Outlet temperature of tube panel wall, °C	596.5
Flow rate, t/h	680
Ambient temperature, °C	20
DNI, W/m^2	900
Wind speed, m/s	5

3.2.1. DNI Disturbance

In the actual operation of the SPT plant, the solar radiation reaching the heliostat field will change sharply due to the weather, thus causing a large fluctuation in the molten salt outlet temperature of the receiver. The rapid change in temperature seriously affects its safe operation. In order to improve the operational safety, the dynamic response characteristics of the receiver under variable operating conditions were investigated, and a DNI step disturbance simulation was carried out.

The receiver operated under the rated operating condition, and the first disturbance was applied after 15 s, which makes the DNI step decreased by 30%, from 900 to 630 W/m^2. The other quantities remained constant, and the second disturbance was applied after 60 s and the DNI increased by 10% to 720 W/m^2; the changes in the molten salt outlet temperature and tube wall temperature are shown in Figure 7. When the temperature was stable, the temperature of the tube wall decreased by 28.1 °C, and the molten salt outlet temperature decreased by 25.3 °C; this shows that the tube wall temperature was more susceptible to disturbance. The temperature response time was less than 10 s when the disturbance occurs.

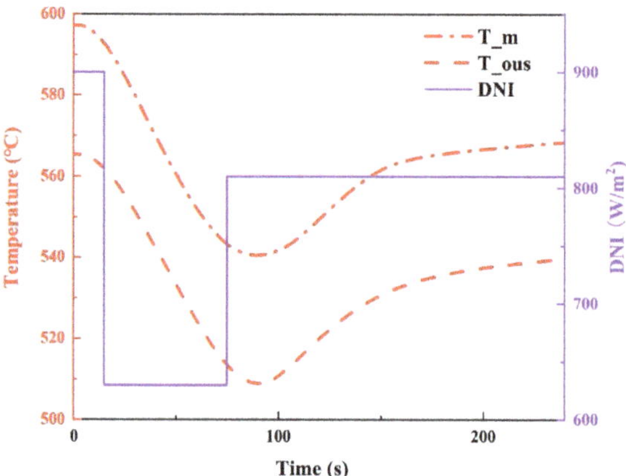

Figure 7. The results of the DNI disturbance.

3.2.2. The Flow Rate Disturbance

For the sake of keeping the molten salt outlet temperature constant, the flow rate of the molten salt entering the receiver was adjusted in real time, according to the variation of the solar radiation. Therefore, as the main regulation of the receiver, it is very important to master the rule of the molten salt flow rate, which affects the temperature of the receiver. Therefore, a step disturbance simulation of the molten salt flow rate was carried out.

The receiver operated under 100% operating conditions, and the molten salt flow rate was reduced by 15% from 680 to 578 t/h at 15 s. After the disturbance, which lasted for 60 s, the flow rate increased by 5% to 612 t/h and kept constant until the temperature was stable. The results of the flow rate disturbance are shown in Figure 8. It can be seen that there was a negative correlation between the temperature and flow rate, because at a constant input energy, the flow rate decreased, while the unit volume of the molten salt absorbed more heat. Before the second disturbance was applied, the molten salt outlet temperature reached 598.1 °C, which seriously overheated and affected the safe operation of the receiver. The temperature was also less than 10 s to complete the response to the flow rate disturbance.

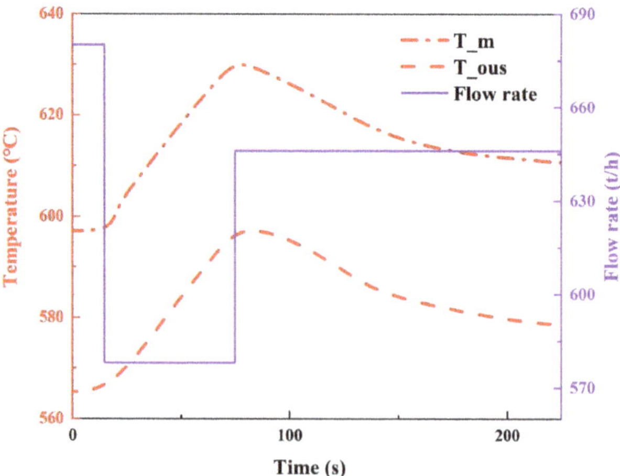

Figure 8. The results of the flow rate disturbance.

3.2.3. Inlet Temperature Disturbance

Salt tanks are equipped in large-scale SPT plants to ensure continuous operation. However, a large volume salt tank results in the uneven distribution of the molten salt temperature. The molten salt inlet temperature is difficult to maintain at a constant value. Therefore, it is necessary to study the dynamic characteristics of the receiver by a step disturbance simulation of the molten salt inlet temperature.

The molten salt inlet temperature increased from 290 to 320 °C at 15 s and then decreases to 300 °C at 150 s. The results of the disturbance test are shown in Figure 9. It can be seen from the diagram that the response of the molten salt outlet temperature to the step disturbance of the molten salt inlet temperature was obviously delayed, and the temperature changed approximately 35 s after the first disturbance was applied. At the end of the disturbance, the molten salt inlet temperature increased by 10 °C, while the molten salt outlet temperature increased by 5.5 °C. It was obvious that the change in the outlet temperature was less than the inlet temperature. In addition, the change in the wall temperature was only 4.2 °C. This shows that the effect of the molten salt inlet temperature on the molten salt outlet temperature was mainly reflected in the longer response time.

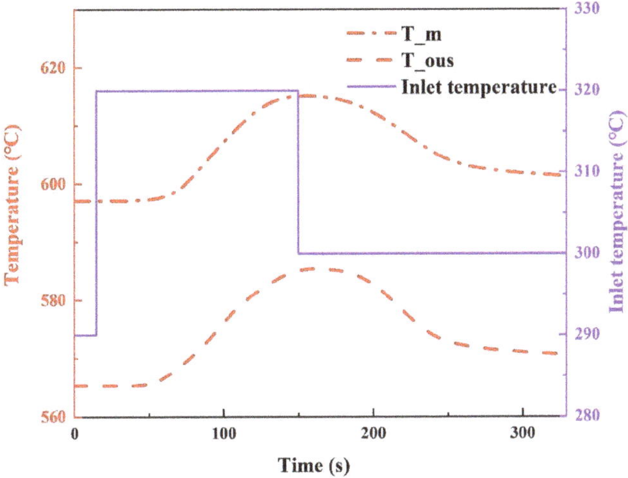

Figure 9. The results of the inlet temperature disturbance.

The results of the step disturbance simulations show that the changes in the DNI, molten salt mass flow rate and molten salt inlet temperature had different effects on the molten salt outlet temperature and tube wall temperature. The variation in the DNI and flow rate directly affected the molten salt outlet temperature and caused it to fluctuate in a large range, while the influence of the molten salt inlet temperature was weaker. The response time of the molten salt outlet temperature to the inlet temperature was significantly longer, and the variation degree of the outlet temperature was less than the fluctuation range of the inlet temperature. Therefore, the wide range of the changes in the DNI should be closely watched, as well as the accurate control of the flow rate changes in the process of operation. The long response time of the molten salt outlet temperature to the inlet temperature should also be considered in the formulation of the control strategy.

4. Conclusions

In this paper, a heliostat field model and an external molten salt receiver model were established, and a 50 MW integrated photothermal dynamic model was developed using the design parameters of a practical SPT plant. The nonuniform and time-varying characteristics of the surface energy flux of the receiver were studied by simulating the operation of an actual power plant on the vernal equinox day. The dynamic characteristics of the molten salt receiver under different operating conditions were investigated using the step disturbance simulations of the DNI, molten salt flow rate and molten salt inlet temperature. The obtained results are helpful for the safe operation, as well as the formulation of a control strategy, of an SPT plant. The conclusions are as follows:

(1) There was a strong positive correlation between the overall change trend in the nonuniform energy flux density on the surface of the external receiver and the change in the DNI. From 12:00 to 18:00, the maximum energy flux density on the surface of the tube panel moved from the first panel on the west side gradually to the fourth, and the energy flux density of the last four panels decreased in sequence.

(2) The energy flux distribution on a single tube panel was high in the middle and low at the inlet and outlet. In addition, the temperature distribution on the tube wall was basically consistent with the energy flux distribution along the axial direction, which increased first and then decreases. The temperature difference between the inner and outer walls of the first tube panel was the largest, and the molten salt temperature of the eighth panel was the highest.

(3) The step disturbances in the DNI and molten salt flow rate had a more direct influence on the change in the molten salt outlet temperature and tube wall temperature than that of the molten salt inlet temperature disturbance. However, the response time of the molten salt inlet temperature disturbance was longer than that of the former two. Therefore, DNI monitoring and accurate control of the molten salt flow rate should be paid more attention during the operation of an SPT plant. At the same time, the influence of the long response time characteristics of the molten salt inlet temperature on the formulation of the control strategy should be considered.

Author Contributions: Conceptualization, E.X.; investigation, H.H.; data curation, H.H.; writing—original draft preparation, H.H.; writing—review and editing, L.S. and Q.Z.; supervision, Q.H.; funding acquisition, E.X. All authors have read and agreed to the published version of the manuscript.

Funding: This work was funded by the National Natural Science Foundation of China (Grant No. 51976058).

Institutional Review Board Statement: Not applicable.

Informed Consent Statement: Not applicable.

Data Availability Statement: Not applicable.

Conflicts of Interest: The authors declare that they have no known competing financial interests or personal relationships that may affect the work reported in this paper.

Nomenclature

a	thermal diffusivity, m^2/s
A	solar collecting area, m^2
C	specific heat capacity, $kJ/(kg\cdot k)$
d	inner diameter of the receiver, m
f_D	friction coefficient
Gr	Grashof number
h	heat transfer coefficient, $kW/m^2\cdot K$
H_r	height of the receiver, m
H_{or}	installation height, m
M	mass, kg
Nu	Nusselt number
Pr	Prandtl number calculated based on the fluid temperature
Q	heat, kW

R	radius of the heat receiver, m
Re	Reynolds number
t	temperature, °C
Δt	temperature difference between the air and tube wall, K
T_m	tube wall temperature, K
T_{ous}	molten salt outlet temperature, K
u	velocity, m/s
V	volume, m^3
G	mass flow rate, kg/s
α_v	expansion coefficient of air, K^{-1}
$\vec{\beta}$	direction vector
γ	azimuth of the heliostat, rad
ε	emissivity
θ	sun incidence angle, rad
θ_{azi}	azimuth angle of the rotating shaft, rad
θ_{alt}	pitching angle of the rotating shaft, rad
λ	thermal conductivity, kW/m·K
μ	dynamic viscosity, N·s/m^2
v	kinematic viscosity, m^2/s
ρ	density, kg/m^3
σ_b	Stefan–Boltzmann constant
τ	time, s
φ	energy flux density, kW/m^2
a	air
abs	energy absorbed
amb	ambient
fc	forced convection
g	ground coordinate system
h	heliostat coordinate system
in	inlet/incident
lc	convective heat loss
lr	radiative heat loss
m	metal tube wall
nc	natural convection
nor	normal
ou	outlet
r	receiver coordinate system
ref	reflected ray
s	molten salt
DNI	direct normal irradiation
SPT	solar power tower

References

1. Ma, X.-J.; Chen, R.-M.; Su, H. Driving Factors and Decoupling Analysis on Industrial Energy Consumption in China. *Stat. Inf. Forum* **2021**, *36*, 70–81.
2. Li, L.; Lin, J.; Wu, N.; Xie, S.; Meng, C.; Zheng, Y.; Wang, X.; Zhao, Y. Review and Outlook on the International Renewable Energy Development. *Energy Built Environ.* **2020**, *3*, 139–157. [CrossRef]
3. Mekhilef, S.; Saidur, R.; Safari, A. A review on solar energy use in industries. *Renew. Sustain. Energy Rev.* **2011**, *15*, 1777–1790. [CrossRef]
4. He, Y.-L.; Qiu, Y.; Wang, K.; Yuan, F.; Wang, W.-Q.; Li, M.-J.; Guo, J.-Q. Perspective of concentrating solar power. *Energy* **2020**, *198*, 117373. [CrossRef]
5. Yu, H.; Duan, J.; Du, W.; Xue, S.; Sun, J. China's energy storage industry: Develop status, existing problems and countermeasures. *Renew. Sustain. Energy Rev.* **2017**, *71*, 767–784. [CrossRef]
6. Du, B.-C.; He, Y.-L.; Zheng, Z.-J.; Cheng, Z.-D. Analysis of thermal stress and fatigue fracture for the solar tower molten salt receiver. *Appl. Therm. Eng.* **2016**, *99*, 741–750. [CrossRef]
7. Qiu, Y.; He, Y.-L.; Li, P.-W. A comprehensive model for analysis of real-time optical performance of a solar power tower with a multi-tube cavity receiver. *Appl. Energy* **2017**, *185*, 589–603. [CrossRef]

8. Cruz, N.C.; Redondo, J.L.; Álvarez, J.D.; Berenguel, M.; Ortigosa, P.M. A parallel Teaching–Learning-Based Optimization procedure for automatic heliostat aiming. *J. Supercomput.* **2016**, *73*, 591–606. [CrossRef]

9. Xu, L.; Stein, W.; Kim, J.-S.; Too, Y.C.S.; Guo, M.; Wang, Z. Transient numerical model for the thermal performance of the solar receiver. *Appl. Therm. Eng.* **2018**, *141*, 1035–1047. [CrossRef]

10. Sánchez-González, A.; Santana, D. Solar flux distribution on central receivers: A projection method from analytic function. *Renew. Energy* **2015**, *74*, 576–587. [CrossRef]

11. Ho, C.K.; Khalsa, S.S. A Photographic Flux Mapping Method for Concentrating Solar Collectors and Receivers. *J. Sol. Energy Eng.* **2012**, *134*, 041004. [CrossRef]

12. Besarati, S.M.; Yogi Goswami, D.; Stefanakos, E.K. Optimal heliostat aiming strategy for uniform distribution of heat flux on the receiver of a solar power tower plant. *Energy Convers. Manag.* **2014**, *84*, 234–243. [CrossRef]

13. Yu, Q.; Wang, Z.; Xu, E. Analysis and improvement of solar flux distribution inside a cavity receiver based on multi-focal points of heliostat field. *Appl. Energy* **2014**, *136*, 417–430. [CrossRef]

14. Sun, F.-H.; Wang, Z.-F.; Guo, M.-H.; Liang, W.-F. An automatic alignment system of heliostat based on the method of track axis reference dislocation. *J. Sol. Energy* **2016**, *37*, 877–883.

15. Hu, Y.; Xu, Z.; Zhou, C.; Du, J.; Yao, Y. Design and performance analysis of a multi-reflection heliostat field in solar power tower system. *Renew. Energy* **2020**, *160*, 498–512. [CrossRef]

16. Victor, G.; Kypros, M.; Clotilde, C. Heliostat fields with a balanced mirror density. *Sol. Energy* **2022**, *243*, 336–347.

17. Gnielinski, V. New Equations For Heat And Mass-Transfer In Turbulent Pipe And Channel Flow. *Int. Chem. Eng.* **1976**, *16*, 359–368.

18. Colebrook, C. Turbulent flow in pipes with particular reference to the transition region between smooth and rough pipe laws. *Ice* **1939**, *11*, 133–156. [CrossRef]

19. Winter, C.-J.; Sizmann, R.L.; Vant-Hull, L.L. *Solar Power Plants: Fundamentals, Technology, Systems, Economics*; Springer: Berlin, Germany, 1991.

20. Siebers, D.L.; Kraabel, K.J. *Estimating Convective Energy Losses from Solar Central Receivers*; Sandia National Laboratories: Albuquerque, NM, USA, 1984.

21. Wagner, M. *Simulation and Predictive Performance Modeling of Utility-Scale Central Receiver System Power Plants*; University of Wisconsin-Madison: Madison, WI, USA, 2008.

22. Zhang, Q.; Cao, D.; Ge, Z.; Du, X. Response characteristics of external receiver for concentrated solar power to disturbance during operation. *Appl. Energy* **2020**, *278*, 115709. [CrossRef]

23. Bradshaw, R.W.; Dawson, D.B.; Rosa, D.L.; Gilbert, R.; Goods, S.H.; Hale, M.J.; Jacobs, P.; Jones, S.A.; Kolb, G.J.; Pacheco, J.E.; et al. *Final Test and Evaluation Results from the Solar Two Project*; National Renewable Energy Laboratory (NREL): Golden, CO, USA, 2002.

Article

Biological Hydrogen Energy Production by Novel Strains *Bacillus paramycoides* and *Cereibacter azotoformans* through Dark and Photo Fermentation

Eldon Chung Han Chua [1], Siaw Khur Wee [1], Jibrail Kansedo [1], Sie Yon Lau [1], King Hann Lim [2], Sharul Sham Dol [3,4,*] and Anuj Nishanth Lipton [5]

[1] Department of Chemical Engineering, Faculty of Engineering and Science, Curtin University Malaysia, CDT 250, Miri 98009, Sarawak, Malaysia; eldonchuach@postgrad.curtin.edu.my (E.C.H.C.); wee.siaw.khur@curtin.edu.my (S.K.W.); jibrail.k@curtin.edu.my (J.K.)

[2] Department of Electrical & Computer Engineering, Faculty of Engineering and Science, Curtin University Malaysia, CDT 250, Miri 98009, Sarawak, Malaysia; glkhann@curtin.edu.my

[3] Mechanical Engineering Department, Abu Dhabi University, Abu Dhabi P.O. Box 59911, United Arab Emirates

[4] Center for Renewable Energy and Energy Efficiency, Abu Dhabi University, Abu Dhabi P.O. Box 59911, United Arab Emirates

[5] Singapore Institute of Food and Biotechnology Innovation, 31 Biopolis Way, #01-02, Nanos, Singapore 138669, Singapore; anujnish99@gmail.com

* Correspondence: sharulshambin.dol@adu.ac.ae

Abstract: In daily life, energy plays a critical role. Hydrogen energy is widely recognized as one of the cleanest energy carriers available today. However, hydrogen must be produced as it does not exist freely in nature. Various methods are available for hydrogen production, including electrolysis, thermochemical technology, and biological methods. This study explores the production of biological hydrogen through the degradation of organic substrates by anaerobic microorganisms. *Bacillus paramycoides* and *Cereibacter azotoformans* strains were selected as they have not yet been studied for biological hydrogen fermentation. This study investigates the ability of these microorganisms to produce biological hydrogen. Initially, the cells were identified using cell morphology study, gram staining procedure, and 16S ribosomal RNA (rRNA) gene polymerase chain reaction. The cells were revealed as *Bacillus paramycoides* (MCCC 1A04098) and *Cereibacter azotoformans* (JCM 9340). Moreover, the growth behaviour and biological hydrogen production of the dark and photo fermentative cells were studied. The inoculum concentrations experimented with were 1% and 10% inoculum size. This study found that *Bacillus paramycoides* and *Cereibacter azotoformans* are promising strains for hydrogen production, but further optimization processes should be performed to obtain the highest hydrogen yield.

Keywords: hydrogen energy; dark fermentation; photo fermentation; biological hydrogen; biodegradation

Citation: Chung Han Chua, E.; Wee, S.K.; Kansedo, J.; Lau, S.Y.; Lim, K.H.; Dol, S.S.; Lipton, A.N. Biological Hydrogen Energy Production by Novel Strains *Bacillus paramycoides* and *Cereibacter azotoformans* through Dark and Photo Fermentation. *Energies* **2023**, *16*, 3807. https://doi.org/10.3390/en16093807

Academic Editor: Marcin Dębowski

Received: 31 March 2023
Revised: 23 April 2023
Accepted: 25 April 2023
Published: 28 April 2023

1. Introduction

Energy is vital in this burgeoning epoch of science and technology. The oil crisis in the 1970s indicated that the recent energy management system is not sustainable over a long period of time [1]. Global reliance on a myriad quantity of fossil fuels has led to excessive greenhouse gas (GHG) emissions. GHG created by the combustion of fossil fuels have caused climate change and global warming which has led to increasing sea levels [2,3]. The use of alternative technologies for energy conversion is necessary to reduce global dependence on non-renewable energy sources. Renewable energy, particularly hydrogen energy, has been proposed due to its high calorific value (140 kJ/g), carbon-neutral nature, and environmentally friendliness [4,5]. The development of hydrogen energy will play a significant role in the future of energy management and contribute to the reduction in environmental pollution.

Hydrogen, although the most abundant element in the universe, cannot be found in pure form on Earth and requires further isolation from various molecules, such as hydrocarbons, water, hydrides, and acids, to obtain it in its gaseous state [6]. The production of hydrogen energy can be classified into three types: grey, blue, and green hydrogen, depending on the feedstocks and production methods used (Figure 1) [7]. Grey hydrogen is produced from hydrocarbons through high-energy-requiring thermochemical conversion technology. Blue hydrogen is produced in the same way but with carbon capture and storage [8]. On the other hand, green hydrogen is obtained from renewable resources such as biomass and water, and metabolic engineering is one of the production technologies for green hydrogen generation. This method is adapted to create a biorefinery sustainable pathway for biomass waste conversion into valuable biochemicals by-products, or namely biohydrogen fuel [9].

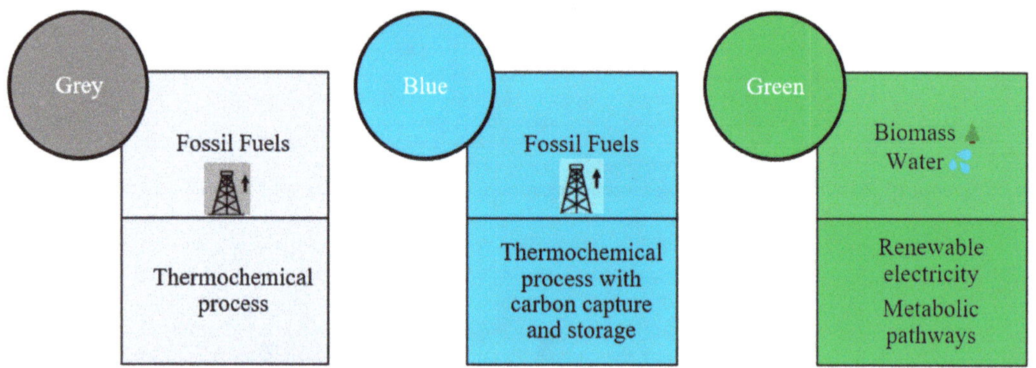

Figure 1. Differentiation of hydrogen production pathways into grey, blue, and green colour code.

Biological hydrogen production through bacteria biodegradation offers several environmental and economic benefits if cost-effective biomass waste is utilized as the feedstock [10]. Lignocellulosic and biomass wastes, such as palm oil mill effluent, winery wastewater, paper waste, post-harvest agricultural waste, and woody biomass, have been shown to significantly contribute to the economic viability of sustainable biopathway hydrogen formation [11–13]. Despite the many efforts made to produce biological hydrogen through fermentation technology, the main limitation remains the low hydrogen yield from fermentative bacteria. Furthermore, different microorganism strains may possess varying potentials in hydrogen production. Therefore, to ensure sustainable biofuel generation, the development of biological hydrogen generation through fermentation necessitates further exploration of new hydrogen-producing strains [14].

In recent years, the use of novel microbial strains for biological hydrogen production has gained significant attention. These novel strains have the potential to improve the efficiency and yield of hydrogen production and to overcome the limitations of traditional hydrogen production methods. For instance, some of these strains can tolerate high concentrations of organic matter and produce hydrogen gas through a biodegradation process. A research study conducted by Pu et al. (2019) investigated the effect of substrate concentration on biological hydrogen production by anaerobic seed sludge. The study revealed that the anaerobic seed sludge was able to produce biological hydrogen even at high concentrations of volatile solid substrate, with a hydrogen yield of 5.3 mL per gram of volatile solid substrate. This clarifies that the high substrate concentration did not prevent the production of biological hydrogen from anaerobic degradation [15]. Palafox-Félix et al. (2022) found that the concentration of a carbon source can influence the formation of metabolites in biochemical processes, as indicated by proteomic analysis. Specifically, when glucose is not present, the bacterium shifts its carbon flow towards the production of antifungal substances by utilizing alternative carbon and nitrogen sources. Conversely, when glucose is available, the bacterium prioritizes energy generation and cell growth [16]. Therefore, it is important to explore the metabolic behaviour of this novel strain under different operational conditions, particularly for the generation of biological hydrogen.

Furthermore, some strains can produce hydrogen gas over a wide range of pH and temperature conditions, making the process more flexible and adaptable. For example, Tang et al. (2022) studied the effect of various initial pH levels on hydrogen fermentation by *Clostridium sensu stricto 12* sp. The study found that the maximum cumulative hydrogen production occurred at an initial pH of 5, with 70.94 mL of hydrogen gas produced per gram of volatile solid substrate. Additionally, even at a low initial pH of 4, the dark fermentation process still yielded 24.93 mL of hydrogen per gram of volatile solid substrate [17]. Another fermentation study investigated the effect of temperature on anaerobic mixed microflora to produce biological hydrogen. The study examined three temperature conditions: mesophilic (37 °C), thermophilic (55 °C), and hyper-thermophilic (80 °C). The experimental results showed that the thermophilic condition yielded the highest biological hydrogen, with a production rate of 12.28 mmol/g cellulose. At the hyper-thermophilic temperature of 80 °C, a hydrogen yield of 9.72 mmol/g cellulose was obtained [18]. Another study examined the impact of metabolic heat within a biofilm on biological hydrogen production. The researchers employed a fiber Bragg grating (FBG) sensor to measure the temperature of *Rhodopseudomonas palustris* CQK-01 biofilm and investigate its effect on hydrogen generation. The study found that both the biofilm thickness and the temperature had a significant influence on the efficiency of biological hydrogen production via fermentation [19].

In this regard, this research work is a preliminary trial that explores the potential of two novel strains, the dark fermentative *Bacillus paramycoides* and the photo fermentative *Cereibacter azotoformans*, for biological hydrogen production. *Bacillus paramycoides* and *Cereibacter azotoformans* are facultative anaerobes, allowing them to produce hydrogen under anaerobic and mild aerobic conditions, providing flexibility in the production process [20,21]. Additionally, *Bacillus* sp. is widely available and *Cereibacter* sp. has the potential to reduce the cost of hydrogen production as it requires less nutrition. Thus, they were selected for the experimental study. Moreover, these novel strains have not yet been investigated for biological hydrogen production. According to the literature, *Cereibacter* sp. could produce additional hydrogen by using the by-products from *Bacillus* sp. [22]. Therefore, this research explores the potential of the two novel strains for hydrogen production, followed by single strain parametric study and co-culture system for future research and development. In this study, the cells were identified and their growth behaviour study was performed. The preliminary trial showed that *Bacillus paramycoides* and *Cereibacter azotoformans* can produce biological hydrogen energy through dark and photo fermentation, respectively. Additionally, these strains demonstrated a fast onset of hydrogen evolution from media inoculation, indicating their potential for efficient biological hydrogen production. Nevertheless, this study is just a basic exploration of cells and hydrogen production, and further optimization processes will be required to maximize the potential of these strains for biological hydrogen production. The results of these optimization processes will be presented as future work, building upon the initial findings presented in this research.

2. Materials and Methods

2.1. Microorganisms and Culture Medium

The freeze-dried cells (DSM14 and DSM5864) were obtained from DSMZ and activated in Pyrex borosilicate conical flasks for 48 h before the experimental study. The photo fermentative microbe was cultivated in a nutrient broth solution consisting of 8 g nutrient broth in 1 L of Milli-Q water, with an initial pH of 6.61. On the other hand, the dark fermentative microbe was grown in a culture media comprising 10 g of glucose, 3 g of peptone, 1 g of yeast extract, 2.8 g of K_2HPO_4, 3.9 g of KH_2PO_4, 0.2 g of $MgSO_47H_2O$, 0.1 g of NaCl, 0.01 g of $CaCl_26H_2O$, 0.05 g of $FeSO_47H_2O$, 0.2 g of L-cysteine, and 1 mL of microelements in 1 L of solution [22]. The initial pH of the activation broth for the dark fermentative microbe was 6.68. The microelements solution (1 L) contained 0.07 g of $ZnCl_2$,

0.1 g of $MnCl_2 4H_2O$, 0.06 g of H_3BO_3, 0.2 g of $CoCl_2 6H_2O$, 0.02 g of $CuCl_2 2H_2O$, 0.02 g of $NiCl_2 6H_2O$, and 0.04 g of $NaMoO_4 2H_2O$. Prior to every experimental study, the conical flasks and cultivating media were autoclaved (HV-110 Hirayama) at 121 °C for 20 min. Carbon sources were autoclaved separately at 110 °C for 20 min before being added to the cultivating media.

2.2. Experimental Setup and Analytical Method

The growth behaviour study was performed based on bacterial concentrations of 1% and 10% in 250 mL Pyrex conical flasks, with a working volume of 200 mL of medium. To generate biological hydrogen, both dark and light fermentative bacteria require anaerobic conditions. To achieve this condition during the experimental study, the conical flask with fermentation medium was first flushed with oxygen-free argon gas for 15 min to remove any remaining oxygen. The flask was then closed with a rubber stopper. Bacteria culturing was performed in a closed desiccator with lit candles to remove any excess oxygen from the environment. Additionally, for the light fermentation process, the setup was constantly supplied with illumination by an Osram 300 W Ultra-Vitalux lamp, with light intensity of 15 klx. For the dark fermentation process, it was performed in a closed incubator to avoid light. Fermentation temperature for both dark and photo fermentation was fixed at 33 °C. A detailed illustration of the experimental setup is presented in Figure 2:

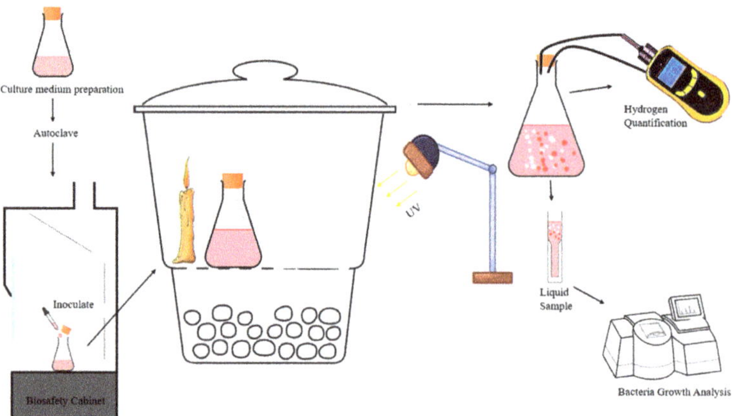

Figure 2. Experimental setup for biological hydrogen fermentation.

The analytical method is being used to monitor the growth behaviour of bacteria and measure the concentration of biological hydrogen produced by dark and photo fermentation processes. For the growth behaviour study, 2 mL samples of the solution were collected periodically every 4 h and inserted into a 2.5 mL cuvette (Kartell) for analysis. The liquid sample was analysed using a UV-VIS spectrometer (Perkin Elmer) with an optical density (OD) of 600 nm for both dark and photo fermentative cells [23,24]. This method of analysis is known as spectrophotometry and involves measuring the amount of light absorbed by the sample at a specific wavelength [25]. The absorption of light is directly proportional to the concentration of the sample, so by measuring the amount of light absorbed, the concentration of bacteria in the solution can be determined. Additionally, to measure the concentration of biological hydrogen produced by both dark and photo fermentation processes, a portable hydrogen gas detector (ATO) equipped with an electrochemical detector was used. To measure the gas produced from the fermentation process, the inlet and outlet of the hydrogen detector were plugged into the conical flask, creating a loop to detect the real-time concentration in the conical flask. The constant pump rate of 75 mL/min from the detector was utilized. After recording the hydrogen reading from the detector, the outlet was unplugged, and the detector continued to pump to flush away the hydrogen gas present in the conical flask, thus resetting the hydrogen gas concentration to zero. The procedure was repeated every 4 h to measure the hydrogen gas produced periodically. Furthermore, the inoculation, the liquid sample collection, and the hydrogen quantification procedures were performed in a class II biological safety cabinet (Esco Scientific) to prevent contamination.

2.3. Gram Staining Procedure

The gram staining procedure was performed to classify the cells as either gram-positive or gram-negative microorganisms. First, the cells were air-dried and heat-fixed onto a microscope glass slide. Next, the heat-fixed cells were flooded with a crystal violet staining reagent for 60 s. In the following step, the glass slide was washed with a gentle stream of tap water for 2 s, followed by flooding with gram's iodine for 60 s. Furthermore, the glass slide was washed with ethyl alcohol for 5 to 10 s to remove the iodine reagent. Lastly, safranin was used as a counterstain, which was flooded onto the glass slide for roughly 45 s, followed by flushing with an indirect stream of tap water until no colour appeared in the effluent [26].

3. Results and Discussion

3.1. Morphology and Gram Staining

The morphology of dark and light fermentative cells was observed using a scanning electron microscope (SEM), and the results are presented in Figure 3. The dark fermentative cell, as shown in Figure 3A, was identified as a rod-shaped bacterium with a cell length ranging from 1.8–2.2 µm and a cell width ranging from 0.8–1.2 µm. On the other hand, the photo fermentative cell, as shown in Figure 3B, was identified as an ovoid bacterium with a cell length ranging from 1.5–2 µm and a cell width ranging from 1.2–1.5 µm.

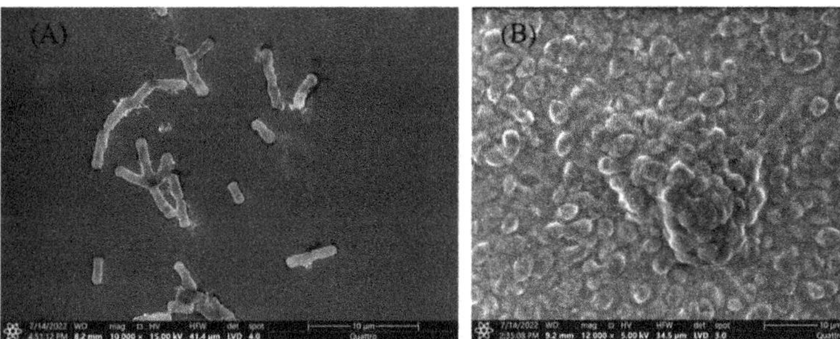

Figure 3. (**A**) Dark fermentative cell and (**B**) photo fermentative cell under SEM.

The distinction between a gram-positive and a gram-negative bacterium lies in the thickness of the peptidoglycan layer in the cell walls. Gram-positive bacteria have thick layers of peptidoglycan in their cell walls, while gram-negative bacteria have thin layers of peptidoglycan in their cell walls [27]. The gram staining procedure was employed to differentiate between gram-positive and gram-negative bacteria. Gram-positive organisms retain the primary colour and appear purple under a microscope, whereas gram-negative organisms appear red under a microscope (Figure 4) [26].

According to the gram staining procedure, the cells were classified as gram-positive and gram-negative organisms, displaying purple and red staining under the microscope. This finding suggests that the identity of these cells could be from the *Corynebacterium*, *Clostridium*, or *Bacillus* family for purple staining cells [28–30]. Conversely, the red staining observed under the microscope confirmed the cell identity as gram-negative bacteria. These cells could possibly belong to the *Neisseria*, *Pseudomonas*, or *Cereibacter* family [31–33]. To further confirm the strains of the microbes, DNA sequencing was performed for the respective strains.

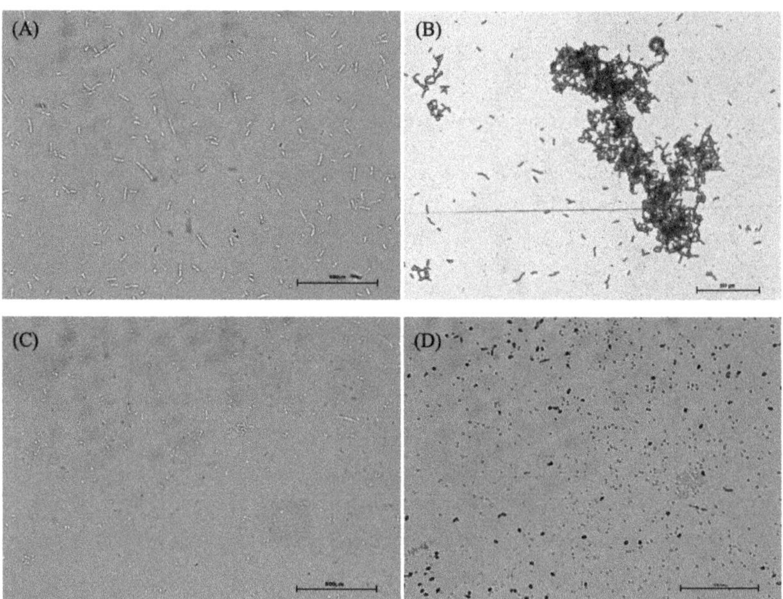

Figure 4. (**A**) *Bacillus* sp. before and (**B**) after gram staining, and (**C**) *Cereibacter* sp. before and (**D**) after gram staining.

3.2. DNA Sequencing

The DNA was isolated and identified under the 16S ribosomal RNA (rRNA) gene polymerase chain reaction (PCR). The reaction for PCR is shown in Figure 5. From the total DNA, the 16S rRNA gene from the gDNA of bacterial isolates were PCR amplified using the primer 785F and 907R. The amplified product was run on to a 0.6 agarose. The agarose gel was documented and the PCR amplified product weight showed prominent DNA bands with approximate sizes of 1500 base pairs. PCR amplified products were run on 1% agarose gel. Lane M indicates the DNA ladder (DNA Ladder Mix 250 to 10,000 base pair, catalogue number BIO-5140). Markers with high intensity were indicated by their size. Lanes 1 and 2 indicate the PCR amplified 16S rRNA gene of the respective bacterial isolate. This analysis indicates specific amplification of the 16S rRNA gene. Sequence analysis was performed using both forward and reverse primers and results were edited and assembled into one full-length sequence.

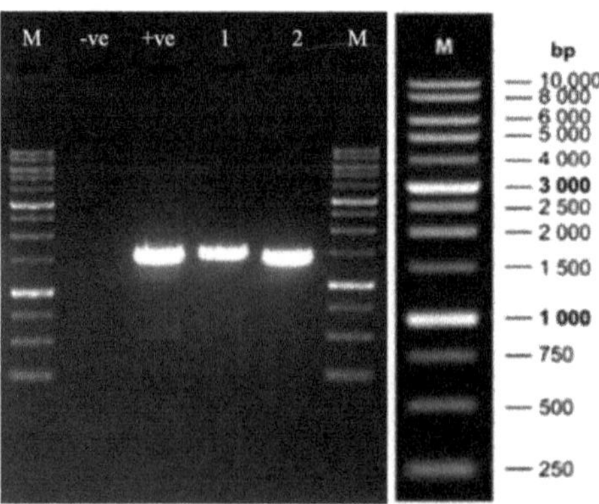

Figure 5. Agarose gel electrophoresis analysis of 16S rRNA genes amplified from two bacterial isolates.

The sizes of the 16S rRNA sequences obtained for each of the bacterial isolates are presented in Table 1. The 16S rRNA sequence of isolate 1 shows 99.93% identity to *Bacillus paramycoides*. For isolate 2, the 16S rRNA sequence is showing 100% identity to *Cereibacter azotoformans*.

Table 1. Bacteria isolated identified.

Sample	16S rRNA Sequenced Gene Size (Base Pair)	GenBank Accession Number	% Identity	% Query Cover	Scientific Name
1	1509	NR_157734.1	99.93	100	*Bacillus paramycoides*
2	1418	NR_113300.1	100	99	*Cereibacter azotoformans*

Additionally, the isolates were subjected to phylogenetic analysis based on their 16S rRNA gene sequences, which were compared to the 16S rRNA gene sequences of hit species. This comparison highlighted the differential alignment of bacterial isolates with various species. The resulting phylogenetic tree (see Figure 6) classifies the dark and photo fermentative bacteria as *Bacillus paramycoides* (MCCC 1A04098) and *Cereibacter azotoformans* (JCM 9340).

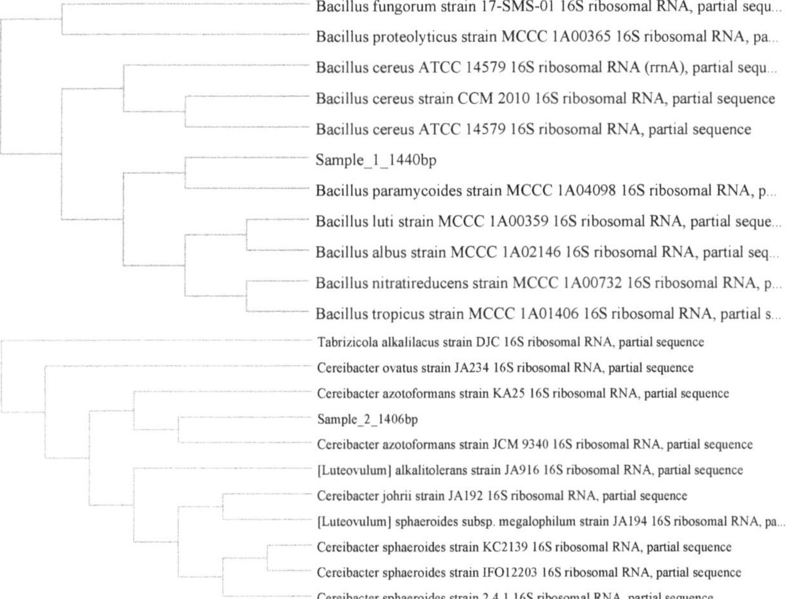

Figure 6. Phylogenetic tree for both isolate.

Once the morphology, the gram staining characteristics, and the identity of both bacteria were determined, a growth behaviour study was conducted. The results of this study will be presented in the following section.

3.3. Growth Behaviour Study and Biological Hydrogen Production with 1% Inoculum

During the growth behaviour study for both strains, simultaneous analysis of biological hydrogen production was performed. Prior to the inoculation process, the initial pH for the *Cereibacter* sp. culture broth was measured as 6.61, while the initial pH for the *Bacillus* sp. broth was recorded as 6.68. Figure 7 displays the growth curves for both bacterial strains along with the biological hydrogen production. The biological hydrogen generation and liquid samples were collected and analysed periodically every 4 h for a duration of 4 days.

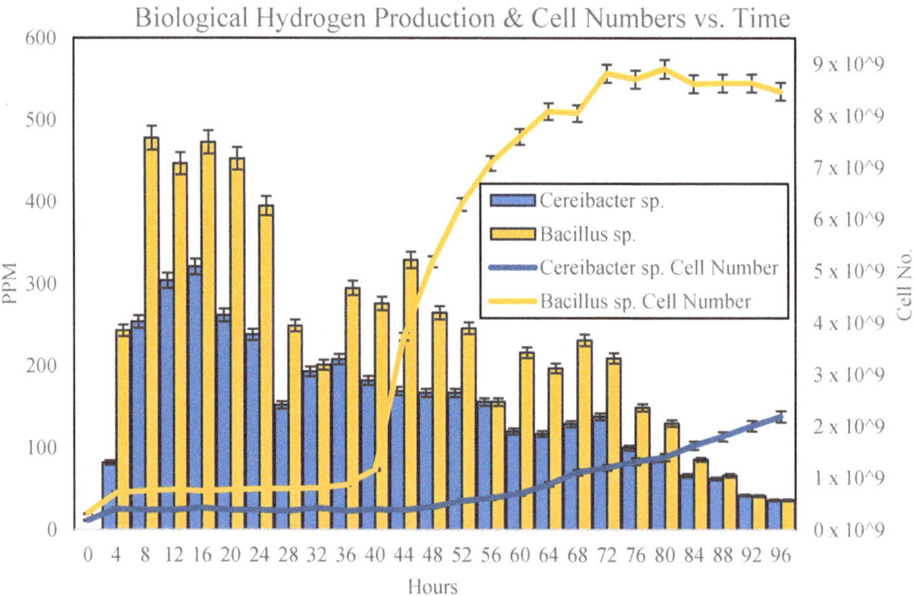

Figure 7. Growth curve and simultaneous biological hydrogen production by *Cereibacter azotoformans* and *Bacillus paramycoides* (1% inoculum).

Figure 7 illustrates that rapid biological hydrogen production occurred during the first 24 h of dark and light biodegradation, and that the hydrogen production rate began to decrease and achieve a steady state from 48 h onwards. The cumulative biological hydrogen production was 5739 ppm and 3654 ppm for dark fermentative *Bacillus* sp. and light fermentative *Cereibacter* sp., respectively. Additionally, two growth curves for *Cereibacter azotoformans* and *Bacillus paramycoides* were shown based on a 1% inoculum concentration. The biological hydrogen was produced rapidly during the initial phase for both dark and photo fermentative bacteria. Once the bacteria began to grow and increase in cell numbers, the hydrogen production decreased significantly. The growth behaviour of *Bacillus* sp. and *Cereibacter* sp. was studied for 96 hours. During 0 to 38 hours, *Bacillus* sp. experienced a lag phase due to strain adaptation to a new environment. After the lag phase, the exponential phase lasted for 28 hours, from 38 to 66 hours of fermentation time, where maximum growth rate appeared. Lastly, from 66 to 96 hours of growth behaviour study, a stationary phase occurred with no net increase in cell numbers. For *Cereibacter* sp., the lag phase occurred from 0 to 50 hours fermentation time, after which the exponential phase appeared from 52 to 96 hours of fermentation time. The doubling mechanism of *Cereibacter* sp. occurred during the log phase, illustrating a proportional growth curve in Figure 7. Nonetheless, the growth curve study for *Cereibacter* sp. did not reach the stationary phase, so a growth study for both dark and light fermentative bacteria based on a 10% inoculum concentration will be conducted to obtain a full growth curve study and biological hydrogen production analysis. In addition, the biological hydrogen yield can also be compared between small (1%) and large (10%) inoculum sizes.

3.4. Growth Behaviour Study and Biological Hydrogen Production with 10% Inoculum

During the study on growth behaviour at a 1% inoculum concentration, it can be observed that *Bacillus* sp. underwent lag, log, and stationary phases. However, *Cereibacter* sp. was only able to achieve lag and part of the exponential phase. As a result, another growth curve study was at a 10% inoculum concentration, which is illustrated in Figure 8.

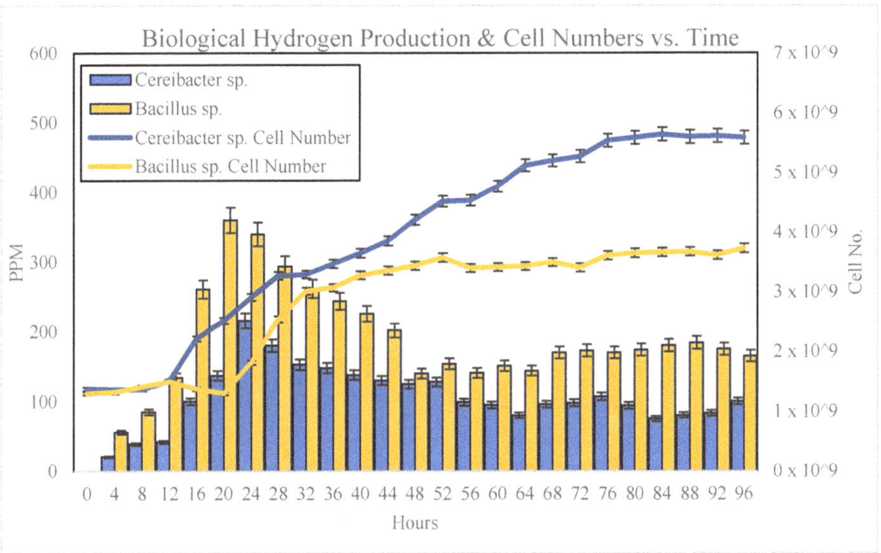

Figure 8. Growth curve and simultaneous biological hydrogen production by *Cereibacter azotoformans* and *Bacillus paramycoides* (10% inoculum).

At a 10% inoculum concentration, *Bacillus* sp. and *Cereibacter* sp. were inoculated to study their growth behaviour and biological hydrogen production for 96 h. During this period, simultaneous analysis of biological hydrogen production was carried out for both dark and light fermentation. The results showed that rapid biological hydrogen production occurred during the first 20 h of both dark and light fermentation. However, the biological hydrogen production rate decreased significantly from the 20th to the 44th hour of fermentation time. After the 44th hour, a steady state biological hydrogen production was achieved until the end of the fermentation period. Dark fermentative *Bacillus* sp. accumulated 4668 ppm of cumulative biological hydrogen production, while photo fermentative *Cereibacter* sp. produced 2564 ppm after 96 h of fermentation. Both bacterial strains achieved up to the stationary phase in their growth behaviour. The growth curve for *Bacillus* sp. in Figure 8 demonstrates that the lag phase was from 0 to 20 h, and the log phase occurred from the 20th to 44th hours of fermentation time. The stationary phase for *Bacillus* sp. began at the 48th hour and lasted until the end of the fermentation period. In contrast, for *Cereibacter* sp., the lag phase occurred from 0 to 12 h, followed by the exponential phase that lasted up to the 76th hour. The steady phase then appeared from the 76th to the 96th hour of the fermentation period.

In both the 1% and 10% inoculum growth behaviour studies, the lag phase was observed as the initial period in the life of a bacterial population where cells adapted to a new environment. During the lag phase, cells prepared to generate proteins and cellular enzymes, increasing in size but not in cell numbers. The duration of the lag phase could be affected by the inoculum size, physiochemical environment of both origin and new fermentation broth, and physiological history of the bacteria [34]. In both the 1% and 10% inoculum growth studies, rapid biological hydrogen production occurred during the initial lag phase of bacterial growth. Additionally, the hydrogen yield from 1% inoculum was higher than that of the 10% inoculum size. Comparatively, 1% inoculum of dark and photo fermentation showed 22.9% and 42.5% hydrogen yield enhancements, respectively, compared to 10% inoculum size. The phenomenon of a high hydrogen yield resulting from a lower inoculum size could be explained by the density-dependent communication system known as quorum sensing [35]. In the quorum sensing communication system, bacterial cell-to-cell communication involves production, detection, and response to extracellular signaling molecules called autoinducers. During the lag phase of bacterial growth, the cells could send autoinducer signals between the inter-species community to improve the microbial concentration. Furthermore, quorum sensing is involved in regulating enzyme production for microbial growth purposes [36]. The quorum sensing system between bacteria could improve hydrogenase or nitrogenase enzyme activity, regulating a higher hydrogen production rate. Therefore, when the bacterial community achieved significant numbers, or namely the stationary phase, biological hydrogen production decreased and became stable, which could be due to the decreasing rate of quorum sensing. A similar result was obtained by Ulhiza et al. (2018) in an experimental research work to investigate the biological hydrogen yield from different yeast concentrations. They reported that a 5% inoculum resulted in a higher hydrogen yield compared to a 9% inoculum size. The biological hydrogen yields were 52 µmol and 32 µmol for 5% and 9% inoculum sizes, respectively [37].

Palafox-Félix et al. (2022) utilized a label-free quantitative proteomic method to investigate the metabolic control of *Amycolatopsis* sp. BX17's metabolism under different glucose concentrations. The findings of their research showed that when glucose was absent from the culture medium, a metabolic shift occurred that favoured the utilization of alternative carbon sources, resulting in a decrease in the bacterial growth rate and an increase in the production of secondary metabolites [16]. In this study, a lower inoculum may have led to higher biological hydrogen production, possibly due to the availability of sufficient glucose for hydrogen metabolism. Under a high inoculum concentration and glucose depletion, the metabolism of microorganisms may shift towards the utilization of alternative carbon sources, resulting in an increased production of secondary metabolites and an inhibition of hydrogen production. Another study enhanced biological hydrogen production by optimizing biofilm growth in a photobioreactor. The study found that a thick biofilm can reduce its porosity, which can result in increased resistance to mass transfer. Additionally, the substrate at the bottom of the biofilm layer may become limited, while the outer layer of the biofilm may experience a lack of light. Consequently, this can cause a decrease in hydrogen production by bacteria [38]. As a result, high inoculum concentration may lead to lower biological hydrogen production due to biofilm inhibition. In summary, biological hydrogen production is significantly affected by the community of inter-bacteria species.

After the lag phase, bacterial growth enters the exponential phase where cell doubling occurs through binary fission, resulting in a periodic doubling of the population [39]. However, during this phase, the rate of biological hydrogen production decreases with increasing cell numbers. This may be due to a reduction in the rate of quorum sensing as the cell population grows, leading to a decrease in enzyme activity responsible for hydrogen production. Furthermore, accumulation of toxic metabolites during cell doubling may inhibit enzyme activity, further decreasing the biological hydrogen production rate. The pH changes observed during the experiment, with a final pH of 6.89 for *Cereibacter* sp. and 4.81 for *Bacillus* sp., indicate that increasing cell numbers can alter the fermentation environment, potentially leading to changes in the biological hydrogen metabolism. As a result, higher cell numbers can lead to a lower hydrogen production rate. Nevertheless, with a 1% inoculum concentration of *Cereibacter* sp., only a portion of the log phase is achieved due to the low cell concentration, resulting in a lower cell division rate. A complete growth curve for *Cereibacter* sp. can be achieved with a 10% inoculum, as shown in Figure 8. During the steady state, the rate of dividing cells equals the rate of dying cells, resulting in no net increase in the number of viable cells. Moreover, bacteria enter the stationary phase due to various reasons, including limited nutrients, stress factors, such as changes in osmolarity, pH or temperature, and an accumulation of toxic metabolites [40,41]. It is essential to consider all these changing parameters in order to optimize hydrogen production.

4. Conclusions

This journal paper reports the results of a DNA identification study which identified two strains of bacteria as *Bacillus paramycoides* (MCCC 1A04098) and *Cereibacter azotoformans* (JCM 9340). This study on the growth behaviour of these strains indicated that both dark and photo fermentative phases, including lag, log, and steady phases, were achieved using a 10% inoculum size. Furthermore, this study aimed to investigate the potential of the novel strains, *Bacillus paramycoides* and *Cereibacter azotoformans*, to produce biological hydrogen, making this study a preliminary trial. The experimental results revealed that both strains could carry out hydrogen fermentation and produced the highest yield when a lower bacteria inoculum size was used. This study also highlighted the significant impact of quorum sensing on hydrogen fermentation. Overall, *Bacillus paramycoides* and *Cereibacter azotoformans* are considered promising strains for hydrogen fermentation due to their fast onset of hydrogen evolution from media inoculation. Due to their ability to thrive under facultative anaerobic conditions, these strains have shown promise for use in hydrogen fermentation. This paper suggests that future studies should focus on exploring various operating conditions, co-culture symbiotic fermentation, and genetic engineering modifications to enhance biological hydrogen yield through fermentation. This article provides new evidence for the potential of hydrogen production through fermentation and makes a significant contribution towards the development of green energy.

Author Contributions: Methodology, E.C.H.C.; Validation, K.H.L.; Formal analysis, E.C.H.C. and J.K.; Investigation, E.C.H.C., S.K.W., S.Y.L. and A.N.L.; Resources, J.K. and S.S.D.; Writing—original draft, E.C.H.C.; Writing—review & editing, S.K.W., S.Y.L. and S.S.D.; Supervision, S.K.W., J.K., S.Y.L., K.H.L. and S.S.D. All authors have read and agreed to the published version of the manuscript.

Funding: This research was conducted at Curtin University Malaysia and was funded by the Ministry of Higher Education Malaysia (MoHE) Fundamental Research Grant Scheme (FRGS/1/2019/TK10/CURTIN/02/1). The authors gratefully acknowledge the financial support provided by MoHE for this research project.

Data Availability Statement: The data presented in this study are available on request from the corresponding author. The data are not publicly available due to confidentiality concerns.

Conflicts of Interest: The authors declare no conflict of interest.

References

1. Adeosun, O.A.; Tabash, M.I.; Anagreh, S. Oil price and economic performance: Additional evidence from advanced economies. *Resour. Policy* **2022**, *77*, 102666. [CrossRef]
2. Karmaker, A.K.; Rahman, M.M.; Hossain, M.A.; Ahmed, M.R. Exploration and corrective measures of greenhouse gas emission from fossil fuel power stations for Bangladesh. *J. Clean. Prod.* **2020**, *244*, 118645. [CrossRef]
3. Wang, W.-K.; Hu, Y.-H.; Liao, G.-Z.; Zeng, W.-L.; Wu, S.-Y. Hydrogen fermentation by photosynthetic bacteria mixed culture with silicone immobilization and metagenomic analysis. *Int. J. Hydrogen Energy* **2022**, *47*, 40590–40602. [CrossRef]
4. Lee, D.-H. Biohydrogen yield efficiency and the benefits of dark, photo and dark-photo fermentative production technology in circular Asian economies. *Int. J. Hydrogen Energy* **2020**, *46*, 13908–13922. [CrossRef]
5. Li, X.; Sui, K.; Zhang, J.; Liu, X.; Xu, Q.; Wang, D.; Yang, Q. Revealing the mechanisms of rhamnolipid enhanced hydrogen production from dark fermentation of waste activated sludge. *Sci. Total. Environ.* **2021**, *806*, 150347. [CrossRef] [PubMed]
6. Abad, A.V.; Dodds, P. Production of Hydrogen. In *Encyclopedia of Sustainable Technologies*; Elsevier: Berlin/Heidelberg, Germany, 2017; pp. 293–304. [CrossRef]
7. Hermesmann, M.; Müller, T. Green, Turquoise, Blue, or Grey? Environmentally friendly Hydrogen Production in Transforming Energy Systems. *Prog. Energy Combust. Sci.* **2022**, *90*, 100996. [CrossRef]
8. Srivastava, N.; Srivastava, M.; Malhotra, B.D.; Gupta, V.K.; Ramteke, P.; Silva, R.N.; Shukla, P.; Dubey, K.K.; Mishra, P. Nanoengineered cellulosic biohydrogen production via dark fermentation: A novel approach. *Biotechnol. Adv.* **2019**, *37*, 107384. [CrossRef]

9. Vernès, L.; Li, Y.; Chemat, F.; Abert-Vian, M. Biorefinery Concept as a Key for Sustainable Future to Green Chemistry—The Case of Microalgae. In *Plant Based "Green Chemistry 2.0"*; Springer: Singapore, 2019; pp. 15–50. [CrossRef]
10. Jung, J.-H.; Sim, Y.-B.; Ko, J.; Park, S.Y.; Kim, G.-B.; Kim, S.-H. Biohydrogen and biomethane production from food waste using a two-stage dynamic membrane bioreactor (DMBR) system. *Bioresour. Technol.* **2022**, *352*, 127094. [CrossRef]
11. Mejía-Saucedo, C.; Buitrón, G.; León-Galván, M.F.; Carrillo-Reyes, J. Biomass purge strategies to control the bacterial community and reactor stability for biohydrogen production from winery wastewater. *Int. J. Hydrogen Energy* **2021**, *47*, 5891–5900. [CrossRef]
12. Zhang, L.; Wang, Y.-Z.; Zhao, T.; Xu, T. Hydrogen production from simultaneous saccharification and fermentation of lignocellulosic materials in a dual-chamber microbial electrolysis cell. *Int. J. Hydrogen Energy* **2019**, *44*, 30024–30030. [CrossRef]
13. Rorke, D.C.S.; Lekha, P.; Kana, G.E.B.; Sithole, B.B. Effect of pharmaceutical wastewater as nitrogen source on the optimization of simultaneous saccharification and fermentation hydrogen production from paper mill sludge. *Sustain. Chem. Pharm.* **2022**, *25*, 100619. [CrossRef]
14. Wang, J.; Yin, Y. Progress in microbiology for fermentative hydrogen production from organic wastes. *Crit. Rev. Environ. Sci. Technol.* **2019**, *49*, 825–865. [CrossRef]
15. Pu, Y.; Tang, J.; Wang, X.C.; Hu, Y.; Huang, J.; Zeng, Y.; Ngo, H.H.; Li, Y. Hydrogen production from acidogenic food waste fermentation using untreated inoculum: Effect of substrate concentrations. *Int. J. Hydrogen Energy* **2019**, *44*, 27272–27284. [CrossRef]
16. Palafox-Félix, M.; Huerta-Ocampo, J.; Hernández-Ortíz, M.; Encarnación-Guevara, S.; Vázquez-Moreno, L.; Guzmán-Partida, A.M.; Cabrera, R. Proteomic analysis reveals the metabolic versatility of Amycolatopsis sp. BX17: A strain native from milpa agroecosystem soil. *J. Proteom.* **2021**, *253*, 104461. [CrossRef]
17. Tang, T.; Chen, Y.; Liu, M.; Du, Y.; Tan, Y. Effect of pH on the performance of hydrogen production by dark fermentation coupled denitrification. *Environ. Res.* **2022**, *208*, 112663. [CrossRef] [PubMed]
18. Gadow, S.I.; Jiang, H.; Watanabe, R.; Li, Y.-Y. Effect of temperature and temperature shock on the stability of continuous cellulosic-hydrogen fermentation. *Bioresour. Technol.* **2013**, *142*, 304–311. [CrossRef] [PubMed]
19. Chen, M.; Xin, X.; Liu, H.; Wu, Y.; Zhong, N.; Chang, H. Monitoring Biohydrogen Production and Metabolic Heat in Biofilms by Fiber Bragg Grating Sensors. *Anal. Chem.* **2019**, *91*, 7842–7849. [CrossRef]
20. Yamamura, S.; Ike, M.; Fujita, M. Dissimilatory arsenate reduction by a facultative anaerobe, Bacillus sp. strain SF-1. *J. Biosci. Bioeng.* **2003**, *96*, 454–460. [CrossRef]
21. Arai, H.; Roh, J.H.; Kaplan, S. Transcriptome Dynamics during the Transition from Anaerobic Photosynthesis to Aerobic Respiration in *Rhodobacter sphaeroides* 2.4.1. *J. Bacteriol.* **2008**, *190*, 286–299. [CrossRef]
22. Zagrodnik, R.; Laniecki, M. The role of pH control on biohydrogen production by single stage hybrid dark- and photofermentation. *Bioresour. Technol.* **2015**, *194*, 187–195. [CrossRef]
23. Orsi, E.; Folch, P.L.; Monje-López, V.T.; Fernhout, B.M.; Turcato, A.; Kengen, S.W.M.; Eggink, G.; Weusthuis, R.A. Characterization of heterotrophic growth and sesquiterpene production by *Rhodobacter sphaeroides* on a defined medium. *J. Ind. Microbiol. Biotechnol.* **2019**, *46*, 1179–1190. [CrossRef]
24. Jang, W.J.; Lee, K.-B.; Jeon, M.-H.; Lee, S.-J.; Hur, S.W.; Lee, S.; Lee, B.-J.; Lee, J.M.; Kim, K.-W.; Lee, E.-W. Characteristics and biological control functions of Bacillus sp. PM8313 as a host-associated probiotic in red sea bream (*Pagrus major*) aquaculture. *Anim. Nutr.* **2023**, *12*, 20–31. [CrossRef]
25. Eyring, M.; Martin, P. Spectroscopy in Forensic Science. In *Reference Module in Chemistry, Molecular Sciences and Chemical Engineering*; Elsevier: Amsterdam, The Netherlands, 2013. [CrossRef]
26. O'Toole, G.A. Classic Spotlight: How the Gram Stain Works. *J. Bacteriol.* **2016**, *198*, 3128. [CrossRef] [PubMed]
27. Ruhal, R.; Kataria, R. Biofilm patterns in gram-positive and gram-negative bacteria. *Microbiol. Res.* **2021**, *251*, 126829. [CrossRef]
28. Sizar, O.; Unakal, C.G. Gram Positive Bacteria. In *Management of Antimicrobials in Infectious Diseases*; Humana Press: Totowa, NJ, USA, 2022; pp. 29–41. [CrossRef]
29. Carter, G.R. Corynebacterium. In *Diagnostic Procedure in Veterinary Bacteriology and Mycology*; Academic Press: Cambridge, MA, USA, 1990; pp. 263–270. [CrossRef]
30. Figueiredo, G.G.O.; Lopes, V.R.; Romano, T.; Camara, M.C. Clostridium. In *Beneficial Microbes in Agro-Ecology: Bacteria and Fungi*; Academic Press: Cambridge, MA, USA, 2020; pp. 477–491. [CrossRef]
31. Kollipara, A.; Lee, D.; Darville, T. Sexually Transmitted Infections and the Urgent Need for Vaccines: A Review of Four Major Bacterial STI Pathogens. In *Mucosal Vaccines: Innovation for Preventing Infectious Diseases*; Academic Press: Cambridge, MA, USA, 2020; pp. 625–647. [CrossRef]
32. Wu, M.; Li, X. Klebsiella pneumoniae and Pseudomonas aeruginosa. In *Molecular Medical Microbiology*; Academic Press: Cambridge, MA, USA, 2020; pp. 1547–1564. [CrossRef]
33. Do, Y.S.; Schmidt, T.M.; Zahn, J.A.; Boyd, E.S.; de la Mora, A.; DiSpirito, A.A. Role of *Rhodobacter* sp. Strain PS9, a Purple Non-Sulfur Photosynthetic Bacterium Isolated from an Anaerobic Swine Waste Lagoon, in Odor Remediation. *Appl. Environ. Microbiol.* **2003**, *69*, 1710–1720. [CrossRef] [PubMed]
34. Swinnen, I.A.M.; Bernaerts, K.; Dens, E.J.J.; Geeraerd, A.H.; Van Impe, J.F. Predictive modelling of the microbial lag phase: A review. *Int. J. Food Microbiol.* **2004**, *94*, 137–159. [CrossRef] [PubMed]
35. Raju, D.V.; Nagarajan, A.; Pandit, S.; Nag, M.; Lahiri, D.; Upadhye, V. Effect of bacterial quorum sensing and mechanism of antimicrobial resistance. *Biocatal. Agric. Biotechnol.* **2022**, *43*, 102409. [CrossRef]

36. Urvoy, M.; Lami, R.; Dreanno, C.; Delmas, D.; L'helguen, S.; Labry, C. Quorum Sensing Regulates the Hydrolytic Enzyme Production and Community Composition of Heterotrophic Bacteria in Coastal Waters. *Front. Microbiol.* **2021**, *12*, 3831. [CrossRef] [PubMed]
37. Ulhiza, T.A.; Puad, N.I.M.; Azmi, A.S. Preliminary optimization of process conditions for biohydrogen production from sago wastewater. In Proceedings of the 2017 IEEE 3rd International Conference on Engineering Technologies and Social Sciences, ICETSS, Bangkok, Thailand, 7–8 August 2017; Volume 2018, pp. 1–4. [CrossRef]
38. Liao, Q.; Zhong, N.; Zhu, X.; Huang, Y.; Chen, R. Enhancement of hydrogen production by optimization of biofilm growth in a photobioreactor. *Int. J. Hydrogen Energy* **2015**, *40*, 4741–4751. [CrossRef]
39. Barer, M.R. Bacterial growth, physiology and death. In *Medical Microbiology*, 18th ed.; Churchill Livingstone: London, UK, 2012; pp. 39–53. [CrossRef]
40. Preiss, L.; Hicks, D.B.; Suzuki, S.; Meier, T.; Krulwich, T.A. Alkaliphilic bacteria with impact on industrial applications, concepts of early life forms, and bioenergetics of ATP synthesis. *Front. Bioeng. Biotechnol.* **2015**, *3*, 75. [CrossRef]
41. Jaishankar, J.; Srivastava, P. Molecular Basis of Stationary Phase Survival and Applications. *Front. Microbiol.* **2017**, *8*, 2000. [CrossRef] [PubMed]

Article

Performance Evaluation of Communication Infrastructure for Peer-to-Peer Energy Trading in Community Microgrids

Ali M. Eltamaly [1,2,3,*] and Mohamed A. Ahmed [4]

1   Sustainable Energy Technologies Center, King Saud University, Riyadh 11421, Saudi Arabia
2   K.A. CARE Energy Research and Innovation Center, Riyadh 11451, Saudi Arabia
3   Department of Electrical Engineering, Mansoura University, Mansoura 35516, Egypt
4   Department of Electronic Engineering, Universidad Técnica Federico Santa María, Valparaíso 2390123, Chile; mohamed.abdelhamid@usm.cl
*   Correspondence: eltamaly@ksu.edu.sa

Abstract: With the rapidly growing energy consumption and the rising number of prosumers, next-generation energy management systems are facing significant impacts by peer-to-peer (P2P) energy trading, which will enable prosumers to sell and purchase energy locally. Until now, the large-scale deployment of P2P energy trading has still posed many technical challenges for both physical and virtual layers. Although the communication infrastructure represents the cornerstone to enabling real-time monitoring and control, less attention has been given to the performance of different communication technologies to support P2P implementations. This work investigates the scalability and performance of the communication infrastructure that supports P2P energy trading on a community microgrid. Five levels make up the developed P2P architecture: the power grid, communication network, cloud management, blockchain, and application. Based on the IEC 61850 standard, we developed a communication network model for a smart consumer that comprised renewable energy sources and energy storage devices. Two different scenarios were investigated: a home area network for a smart prosumer and a neighborhood area network for a community-based P2P architecture. Through simulations, the suggested network models were assessed for their channel bandwidth and end-to-end latency utilizing different communication technologies.

Keywords: communication infrastructure; P2P energy trading; smart prosumer; microgrid

Citation: Eltamaly, A.M.; Ahmed, M.A. Performance Evaluation of Communication Infrastructure for Peer-to-Peer Energy Trading in Community Microgrids. *Energies* **2023**, *16*, 5116. https://doi.org/10.3390/en16135116

Academic Editors: Sharul Sham Dol and Anang Hudaya Muhamad Amin

Received: 27 May 2023
Revised: 29 June 2023
Accepted: 30 June 2023
Published: 2 July 2023

1. Introduction

Traditional power systems were developed decades ago based on centralized architectures, in which electricity is generated from central electric power plants, transmitted to the substation via transmission power lines, and then distributed to the consumers. In order to lower greenhouse gas emissions and boost energy efficiency, all nations are now creating new rules and regulations to encourage the integration of renewable energy sources. In Saudi Arabia, the plan of Saudi Vision 2030 for energy transition and sustainability aims to reach 50% renewable energy using solar and wind power in the overall energy mix by 2030 and Net Zero by 2060 [1]. In this direction, in the residential sector, Saudi Arabia's electricity and cogeneration regularity authority introduced in the August 2017 regulations allows households to generate their own electricity using solar energy [2]. As a result, in 2018, the number of installed solar panels increased in the majority of regions, especially in the Riyadh region. Since then, a significant number of customers have participated in connecting their generation systems with electric utilities, especially photovoltaic energy systems. More customers are predicted to participate in the use of the smart grid via demand-side management incorporation in the power system [3–6].

Peer-to-peer (P2P) energy trading is set to play an important role in direct energy sharing among prosumers in the future smart grid; however, issues relating to large-scale implementation still need to be addressed in order to enable reliable and secure operation [7]. In P2P energy trading, instead of relying on the main power grid, prosumers ate able to sell/buy from each other. However, such implementation of P2P energy trading poses many technical challenges in relation to electric power constraints, communication network requirements, and security and privacy issues [8].

In order to support the development of a future smart grid, there are different organizations and standardizations which have been working to define smart grid requirements and architectures from various aspects such as grid management, communication, networking, and security and privacy. Among the international organizations and standardizations are the International Electrotechnical Commission (IEC), the IEEE Standard Association, Internet Engineering Task Force (IETF), and International Telecommunication Union-Telecommunication (ITU-T) standardization sector [9]. Examples of widely accepted smart grid standards are the national institute of Standards and Technology (NIST), the IEEE 2030 standard, and the smart grid architecture model (SGAM). The SGAM architecture consists of five interoperability dimensions (component, communication, information, function, and business), five domains (generation, transmission, distribution, DER, and customer), and five zones (process, field station, operation, enterprise, and market) [10].

With the increased growth of distributed energy systems such as solar photovoltaic, battery storage systems, and electric vehicles, P2P energy trading has emerged as the next-generation form of energy management that is set to play an important role in enabling local energy trading for selling/buying among prosumers in the residential power grid. Such P2P energy trading could bring many benefits to all participants, including consumers, prosumers, and service providers. In such implementations, communication infrastructure will play an important role in enabling all market participants to communicate with one another in P2P energy trading. However, different communication architectures could be considered to support information exchange. Such architectures include, in general, centralized architecture, decentralized architecture, and hybrid-based architecture. The selection among different architectures should consider the performance requirements that need to be fulfilled, e.g., latency, throughput, and security. Another important aspect is the energy management system (EMS) that can enable the prosumer to have access to real-time demand and supply information, as this EMS influences the role of prosumers in deciding their participation as seller or buyer [11].

Communication infrastructure is the first step toward establishing reliable P2P energy trading among prosumers and consumers in microgrids. Such communication infrastructure would aim to enable reliable data delivery among different entities [12,13]. Most of the research in P2P energy trading has been focused on market operations, pricing mechanisms, blockchain technologies, and smart contracts, while less attention has been given to the performance of communication technologies to support P2P implementations. With a large number of participants, issues relating to communication and computation complexities must be addressed for robust and secure operations. To the best of our knowledge, there is no detailed communication network model for P2P energy trading in a community microgrid. To address this knowledge gap, this work investigated the performance of the communication infrastructure that supports P2P energy trading among smart prosumers in a community microgrid. The main contributions were:

- To develop a system architecture for local P2P energy trading which consists of five layers: power grid, communication network, cloud management, blockchain, and application.
- To develop a communication network model based on the IEC 61850 standard for smart prosumers integrated with renewable energy resources and energy storage systems.
- Evaluate the performance of HAN for a smart prosumer and NAN for a community-based P2P architecture using the OPNET modeler.
- Provide the performance analysis of different communication technologies with respect to bandwidth and end-to-end delay.

The rest of this paper is organized as follows: Section 2 presents the related work, while Section 3 presents the system architecture for P2P energy trading. Section 4 provides a detailed network model for smart prosumers in a community microgrid based on the IEC 61850 standard. The performance evaluation and simulation results are presented in Section 5. Section 6 illustrates the conclusion and future work.

2. Related Work

Many related works have discussed the main features and benefits of P2P energy trading from different aspects for consumers, prosumers, and distribution system operators in both physical and virtual layers [4–8,11–15]. For example, the survey for the socio–technical interaction and P2P energy trading mechanism was given by the authors in Ref. [7]. This work covered information and communication technology (ICT) for energy trading interactions, the energy services of multi-scale P2P energy trading, and the operational framework for the P2P marketplace. In Refs. [4–6], the authors provided a thorough examination of existing research and implementation initiatives for hybrid renewable energy systems and future research, while in Ref. [11], the authors examined the current state of P2P energy trading strategies, features, energy markets, and the benefits for both consumers and the grid. Furthermore, this work identified the research challenges in virtual and physical layers as well as the future research directions. Ref. [14] outlined the functions of blockchain and microgrids to enable peer-to-peer energy trading, while Ref. [15] offered an assessment of the literature on P2P kilowatts and megawatts in the distribution of power systems.

In [16], the authors developed P2P energy trading for microgrids using blockchain and the Internet of Things (IoT) technologies. This work provided basic information for the technical implementation of the P2P energy trading setup, including hardware (current sensor, Arduino board, LED, power supply, relay, and light bulbs) and software (Node-Red). The work considered a basic scenario between two peers only for the P2P energy trading model. In [17], the authors presented low-cost, open-source P2P energy trading for a remote community in Pakistan based on IoT and blockchain technologies. This system was designed and implemented based on six main components: field devices, ESP32, a relay, a user interface, a local WiFi network, and Ethereum private blockchain. In [18], the authors presented a review of P2P energy trading with respect to concepts, approaches, and control architectures in microgrids for local communities. This work provided a case study of the energy system in Nepal, highlighting constraints, challenges, and opportunities.

In [19], the authors designed a P2P energy trading method for optimal operation and cost-sharing when building microgrids based on a data envelopment analysis. Three cases were considered: Case 1 with all buildings holding prosumers (suitable to install PV modules), Case 2 with some buildings not suitable to install PV modules, and Case 3 with shared PV and energy storage systems. In [20], the authors presented centralized P2P energy trading for residential smart communities. A smart community with ten prosumers, an energy community manager, and an energy retailer was taken into account in the simulation scenario. Each prosumer had a solar PV and battery storage system. In [21], the authors presented the basic concept of a P2P energy trading platform using IoT and blockchain technologies. This work considered only two peers where the platform provided information that was related to metering, energy transfer, and money transfer.

In [22], the authors investigated the participation of sustainable users in P2P energy trading through social cooperation among prosumers. The feasibility of cost savings using a Canonical Coalition Game was investigated for five prosumers. In [23], the authors presented a P2P energy trading platform using the Internet of Things, Ethereum private blockchain, and a user interface based on Angular for remote communities. In [24], the authors proposed a P2P energy trading based on distributed ledger technology (DLT) for the Internet of Things application (IOTA) which provided low cost, low transaction time, and high scalability. In [25], the author introduced the 1 + 5 architectural view model for the design of cooperating IT systems for common business processes. Three case studies were given for the design of various solutions, including the electronic circulation of prescriptions, the communication between separate IT systems, and the design of a blockchain-based solution for renewable energy management.

Security and privacy are important factors to be considered in order to suit the requirements of different applications against any physical or cyberattacks in smart grids and microgrids. Cyberattacks can target the device level (sensors and measurement devices) as well as the communication network level. Physical tampering may include damage to physical devices such as smart meters, while manipulating control signals for circuit breakers or relays may result in the disruption of the power supply. Therefore, the main indicators to be considered include availability, integrity, and confidentiality. For peer-to-peer energy trading, cyberattacks with false data injection (FDI) for renewable energy, demand, and offered prices could affect energy costs to achieve financial gain, while a denial of service (DoS) could disable the availability of information such as available resources and prices. To mitigate such problems, authentication and access control are essential at the device level. For data transmission between the end devices and control systems, digital signature and encryption were essential elements to ensure data integrity [26,27].

3. System Architecture for P2P Energy Trading

Figure 1 shows the cyber-physical architecture of a group of prosumers in a residential grid. It consists of two main layers: the physical layer and the virtual layer. To enable P2P energy trading, the physical layer is responsible for providing the physical electric network for the transfer of electricity between sellers and buyers. Such a physical network could be the traditional distribution grid for power utility or a separate microgrid distribution grid in conjunction with the traditional grid. The main elements in the physical layer include the grid power connection, metering, and communication infrastructure. The virtual layer provides a secure connection among peers and gives equal access to all participants when deciding on the energy trading parameters for sell/buy orders. Other elements include market participants and regulations.

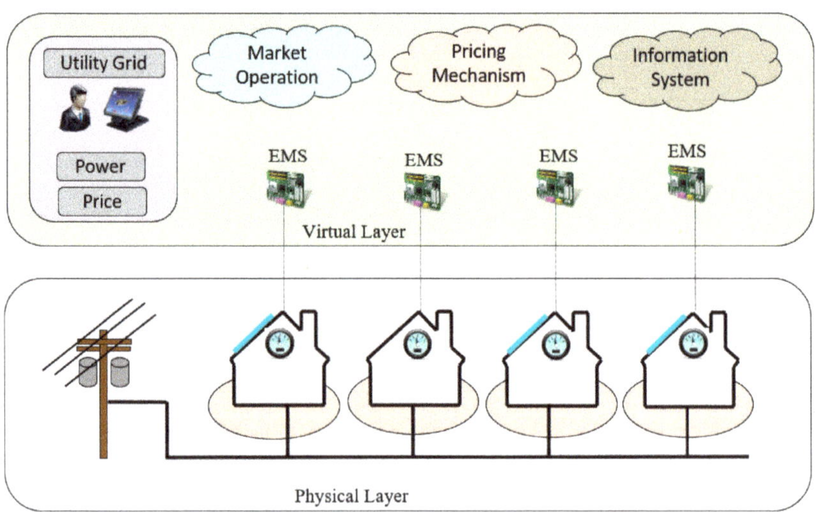

Figure 1. Schematic diagram for the physical layer and the cyber layer of a residential power grid.

To participate in P2P energy trading, a communication infrastructure was required to connect among peers. Such communication infrastructure could be implemented using different wired/wireless communication technologies. The following are the definitions of a peer, microgrid, and community microgrid.

- Peer: it refers to a single entity such as a smart home or a smart building.

- Microgrid: it consists of a collection of loads and distributed energy resources that can operate in a grid-connected mode or an island mode.
- Community Microgrid: it refers to a group of homes and/or buildings that are integrated with distributed energy resources and/or energy storage systems. Additionally, it could operate in both a grid-connected mode and an island mode.

Figure 2 depicts a system architecture diagram for local P2P energy trading in the distribution power system [28]. It consists of five layers: the power grid, communication network, cloud management, blockchain, and application.

- Power Grid Layer: It comprises the actual components of the electrical power system, including generation, storage, and consumption in households/buildings. Examples of the main appliances include HVAC, washing machines, ovens, and electric vehicles. Other elements in the distribution power system include transformers, power feeders, smart meters, etc. The data acquisition is performed using different sensor nodes, meters, measuring devices, RTUs, and IEDs.
- Communication Network Layer: it consists of communication network devices such as routers and switches that enable data transmission using wired/wireless communications. Data are collected from different devices such as smart meters, weather stations, and indoor environment sensors. Examples of wired technologies include Ethernet and power line communication, while wireless technologies include WiFi, ZigBee, mobile network, and LPWAN.
- Cloud Management Layer: this is a middleware layer that provides communication and data services for the upper layers. There are different solutions, such as VM-based, fog, and service-oriented. The cloud layer enables data storage and information retrieval.
- Blockchain Layer: it provides a decentralized distributed ledger that gives great benefits for automated energy trading processes, security, and transparency when sharing information among peers. Blockchain technology can enable the operation of smart contracts and distributed data storage among peers.
- Application Layer: there are different types of applications and services related to monitoring, management, control, and prediction, which can provide added values to the energy solution for all participants, including the utility, prosumer, and consumer.

The size and number of participants determine how energy trading is conducted in the P2P energy trading concept. Such architectures can be divided into three different types, as shown in Figure 3, including the full decentralized P2P architecture (Microgrid 1, Microgrid 3, and Microgrid 4), community-based P2P architecture (Microgrid 2), and hybrid P2P architecture [15,29].

- Full decentralized P2P architecture: each participant can directly negotiate with one another independently without any centralized unit to coordinate the trading among peers. Additionally, the peers have full autonomy over how they manage their own resources (generation/storage/load) with different preferences in energy trading, such as lower prices, serving the community, or using clean energy.
- Community-based P2P architecture: the members of the community share common goals. Each member of the community communicates their requirements for energy trading to a central unit (e.g., the community manager). In such cases, all measurements from the peers are communicated to a central unit (community manager), and such measurements are used to make decisions and take action.
- Hybrid P2P architecture: a combination of both fully decentralized P2P architecture and community-based P2P architectures. The role of the community manager can be used to manage the trading activities between the community members in the same microgrid or between members in different microgrids.

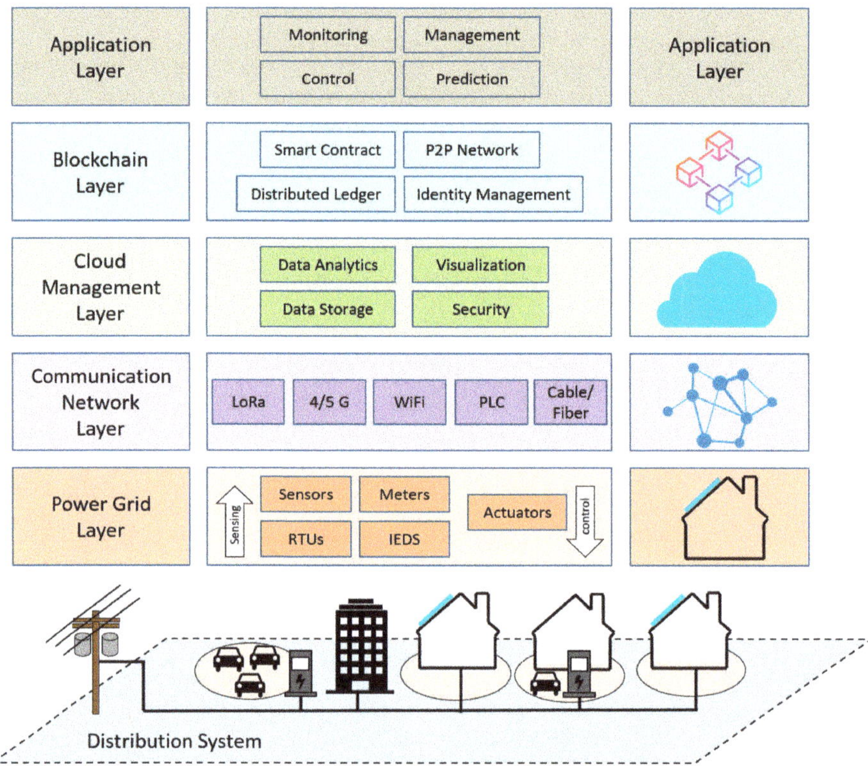

Figure 2. Schematic diagram for the system architecture of local P2P energy trading in distribution power system.

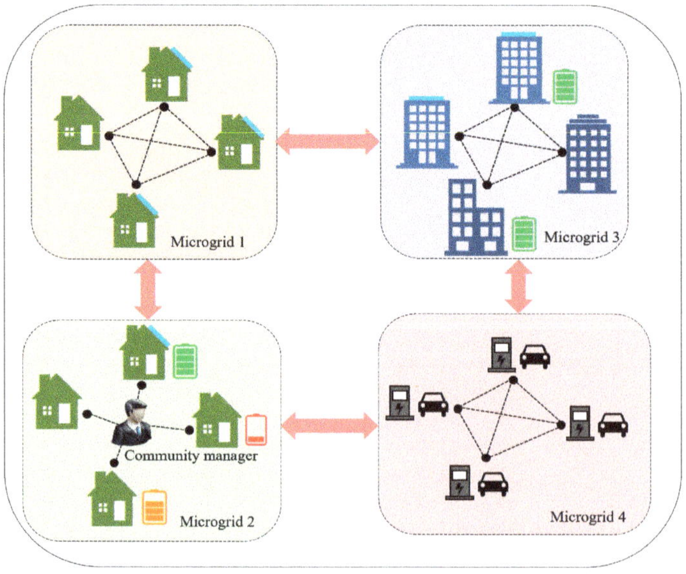

Figure 3. Schematic diagram for different P2P architectures in a community microgrid.

4. Modeling Smart Prosumers in a Community Microgrid

IEC 61850 is the international standard for communications in substations which defines a universal data model to facilitate the interoperability of different equipment and devices [30]. In this work, the IEC 61850-7-420 standard defined the information models which were used in the exchange of information with distributed energy resources (DER) and distribution automation (DA). DERs include connected generation systems, energy storage systems, and controllable loads. The DER management system includes an energy management system (EMS) in the microgrid, community, campus, and building. The DA equipment includes automated switches, fault indicators, and other management devices [31].

In this work, we developed a communication network architecture for smart prosumers based on the IEC 61850 standard's logical node concept. The suggested model was made up of several logical nodes. Each logical node carried several sorts of information, such as analog, status, and control data. Figure 4 shows the communication model for a smart prosumer. The developed communication model used in this work included a PV system, a battery energy storage system, loads, and the power grid. Detailed information for the sensing devices for each subsystem is given in Table 1 based on Ref. [32].

Table 1. Main elements of sensing devices for a smart prosumer.

Element	Sensing Devices
PV System	module temperature, current, voltage, power, tracker azimuth angle, and tracker tilt angle,
Battery System	characteristics of the rectifier, remote monitoring and control of battery charger, remote monitoring and control of battery system
Met. Mast	ambient air temperature, irradiance, wind direction, and wind speed,
Power Grid	current to/from grid, voltage_utility, power to/from grid

We considered that each smart prosumer included a home energy management system "local controller". This unit was responsible for the real-time monitoring and control of home appliances as well as optimizing local power generation, storage, and consumption. With respect to communication networks, the focus would be given to the commonly available technologies, including Ethernet and WiFi, for the feasibility of implementing this system in a real scenario for HAN [33] and NAN. The communication range of Ethernet and WiFi was suitable for HEMS to communicate with the sensor nodes and measurement devices in the community microgrid.

We considered that each smart prosumer included a small metrological station. Table 2 shows the measuring requirement for different sensors, including the temperature, irradiance, wind speed, and wind direction. Table 3 shows the configurations of IEDs such as P&C-IED, MU-IED, and CB-IED, while Table 4 shows the type of data and data size. The monitoring scope and control scope is given in Table 5.

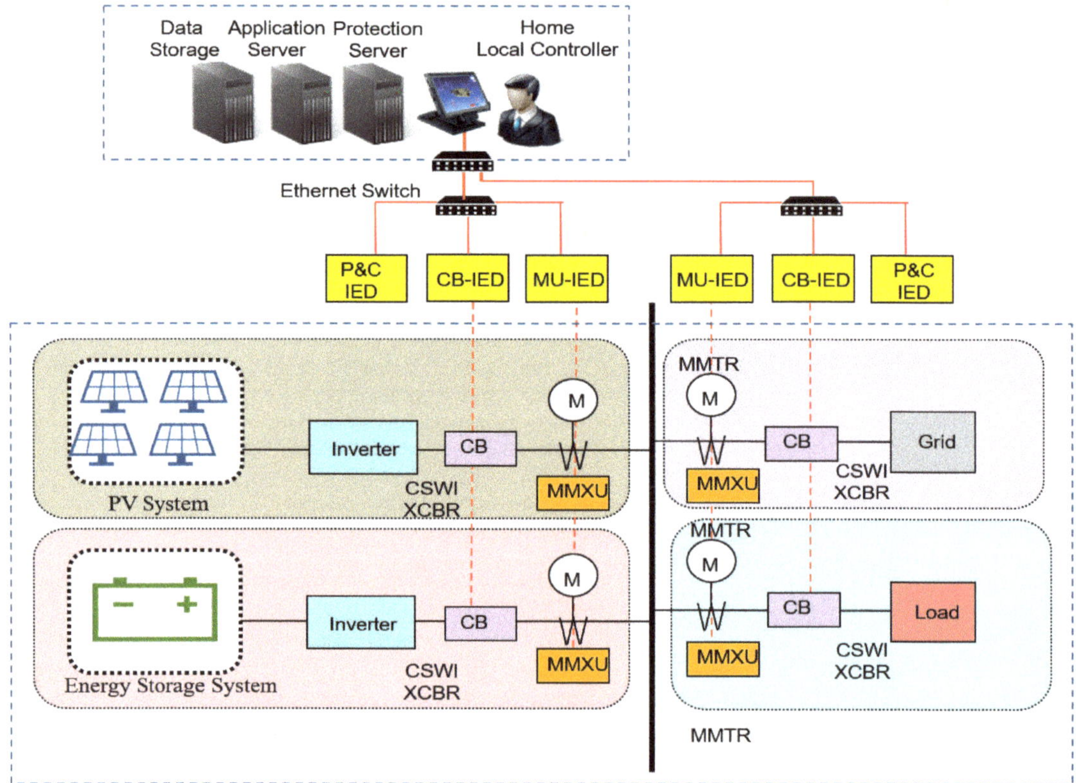

Figure 4. Communication model for a smart prosumer based on the IEC 61850 standard.

Table 2. Measuring requirements for the meteorological station.

Measurement	Ambient Temperature	Irradiance	Wind Speed	Wind Direction
Sampling Frequency	1 Hz	100 Hz	3 Hz	3 Hz
Number of Channels	1	1	1	1
Data Rate	2 bytes/s	200 bytes/s	6 bytes/s	6 bytes/s

Table 3. Configurations of IEDs.

	PV System	Energy Storage System	Load
CB-IED	1	1	1
MU-IED	1	1	1
P&C IED	1	1	1

Table 4. Data type and size for IEDs.

IED Type	CB-IED	MU-IED	P&C-IED
Data Type	Status of Breaker	Current and Voltage	Control information
Data Size	16-bytes	76,800-bytes	76,816-bytes

Table 5. Monitoring scope and control decisions.

	Generation	**Storage**	**Protection IEDs**	**Load**
Scope	Output Control	charge/discharge Control	Control and Protection	monitoring and control load
Connectivity	GC→HEMS	BESS→HEMS	IED→HEMS	LC→HEMS

The communication infrastructure is the heart of the P2P energy trading system. It enables each peer to access their own energy information locally, as well as share information with other peers based on the configured topology architecture. The communication infrastructure must be reliable and fulfill the requirements of the P2P energy trading system. With the increase in many services that request data from different microgrid components (prosumer, consumer, service provider), the communication infrastructure must be able to support data transfer and the capability of real-time monitoring and control [34,35]. Table 6 shows the communication requirements for the microgrid systems in view of the latency, throughput, and security to support different applications, including distributed energy resources and storage, such as smart meters, home energy management, and home automation.

Table 6. Microgrid time requirements for different applications.

Requirements	**DER and Storage**	**Smart Meter**	**HEMS**	**Home Automation**
Latency	15 s	Variable	300 ms–2 s	seconds
Throughput	96–56 kbps	10 kbps	9.6–56 kbps	4.8–48
Reliability	99–99.9%	>98%	>98%	>98%
Security	High	High	High	High

5. Simulation Results

There are different simulation tools that can be used to evaluate the network performance, such as NS-3, OMNet++, and OPNET. The OPNET Modeler is a network simulator that supports many library functions and different network protocols [36]. In this work, the OPNET Modeler was used for the modeling and simulation of the communication network. This software was user-friendly and included all network devices that could be used to build and configure complex network topologies. We considered a case study with 10 prosumers to assess the performance of the communication infrastructure (smart prosumers integrated with renewable energy resources). The OPNET Modeler was used to evaluate the communication network performance with different numbers of prosumers. The dimension of the site was based on the real dimension of the residential area in Riyadh City, Saudi Arabia.

The microgrid is a localized power grid that is integrated with distributed energy resources, an energy storage system, and loads which can be operated in a grid-connected mode with the main grid or a standalone mode that is independent of the grid. Based on the scale, a small microgrid can be a single home or a single building, while a large-scale microgrid can span a group of buildings, hospitals, or university campuses [37]. In this work, two scenarios were considered: the HAN of a standalone microgrid system for a "smart prosumer" and the NAN of a community-based P2P architecture with a group of prosumers.

5.1. Scenario 1: HAN Results for a Smart Prosumer

For the smart prosumer scenario, the communication links of the wired-based architecture were configured with a channel capacity of 100 Mbps and 1 Gbps. For the wireless-based solution, WiFi technology was configured using different data rates, including 54 Mbps, 24 Mbps, and 11 Mbps. Figure 5 shows the Ethernet-based architecture, while Figure 6 shows the network model using the WiFi-based architecture.

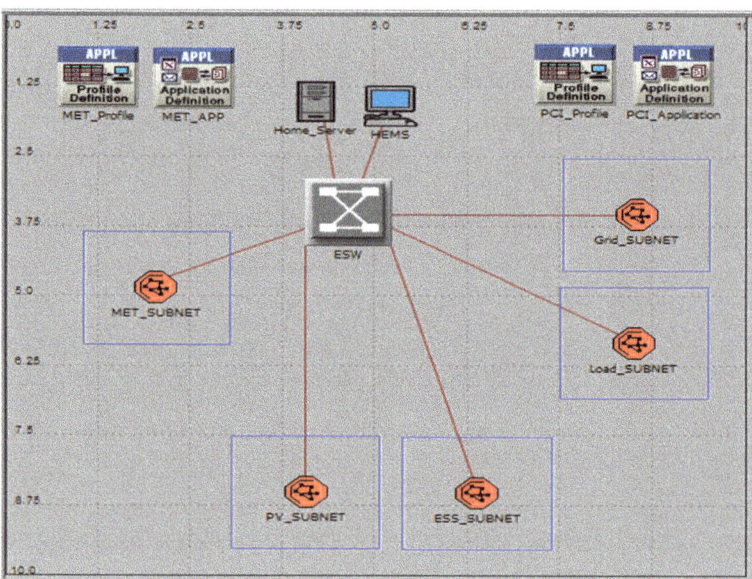

Figure 5. Ethernet-based communication network model for the smart prosumer.

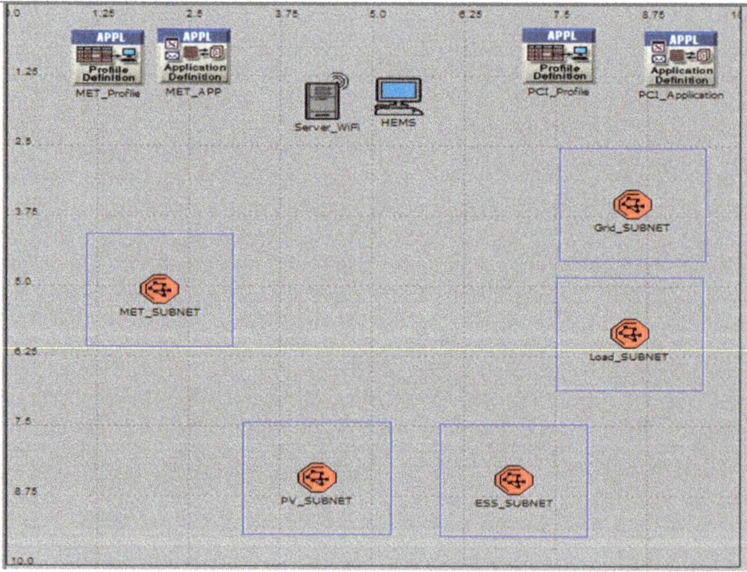

Figure 6. WiFi-based communication network model for the smart prosumer.

Figure 7 shows the simulation results of HAN for a smart prosumer, while Table 7 shows a summary of the average end-to-end delay in six different scenarios. For the smart prosumer, the average end-to-end delay was about 0.123 ms and 1.23 ms using a communication link with the Gigabit Ethernet and Fast Ethernet, respectively. In the case of the WiFi-based architecture, the average end-to-end delay was about 9.64 ms, 4.57 ms, and 2.69 ms, using the data rates of 11 Mbps, 24 Mbps, and 54 Mbps respectively.

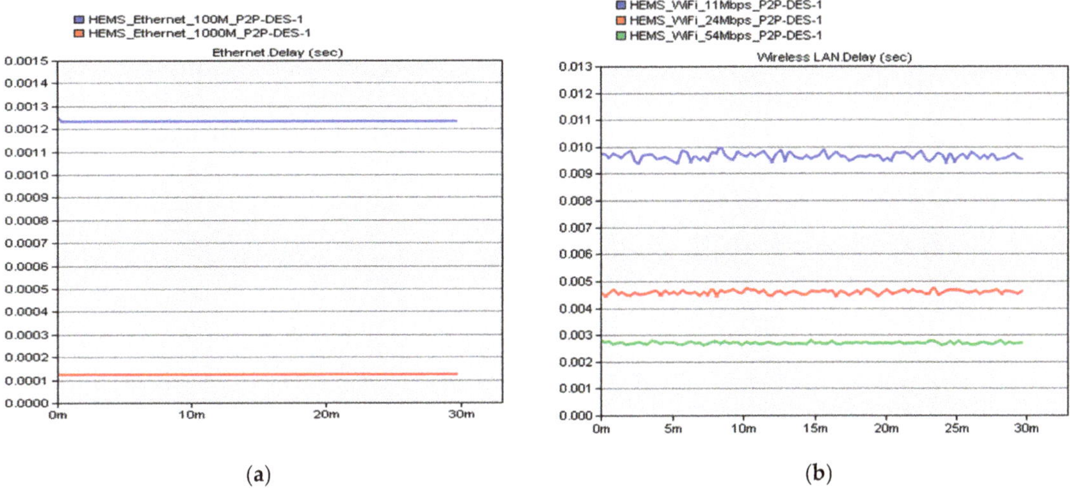

Figure 7. End-to-end delay for a smart prosumer scenario. (**a**) Ethernet with a channel capacity of 100 Mbps and 1 Gbps; (**b**) WiFi with data rate of 54 Mbps, 24 Mbps, and 11 Mbps.

Table 7. Average end-to-end for different scenarios for a smart prosumer.

	Ethernet	Average ETE Delay (s)	WiFi	Average ETE Delay (s)
Smart Home	HAN (100 Mbps)	0.00123	HAN (54 Mbps)	0.00269
	HAN (1 Gbps)	0.000123	HAN (24 Mbps)	0.00457
			HAN (11 Mbps)	0.00964

5.2. Scenario 2 NAN Results for Community-Based P2P Architecture

For NAN, the communication channels for the wired-based architecture were set up with channel capacities of 100 Mbps and 1 Gbps. For the community-based P2P architecture scenario, WiFi with different data speeds of 54 Mbps, 24 Mbps, and 11 Mbps was configured. The Ethernet-based design is depicted in Figure 8, while the WiFi-based network model is shown in Figure 9.

Table 8 shows the results of the end-to-end delay for the community-based P2P architecture. With 10 smart prosumers, the end-to-end delay was about 12.86 ms and 1.27 ms for the fast Ethernet and Gigabit Ethernet, respectively. The network performance satisfied the delay requirement for the different microgrid applications given in Table 6. However, for a WiFi-based architecture, stable performance was achieved with six prosumers (54 Mbps), five prosumers (24 Mbps), and four prosumers (11 Mbps), as shown in Table 9. The network performance was not stable in the case of 10 prosumers (54 Mbps), 7 prosumers (24 Mbps), and 6 prosumers (11 Mbps).

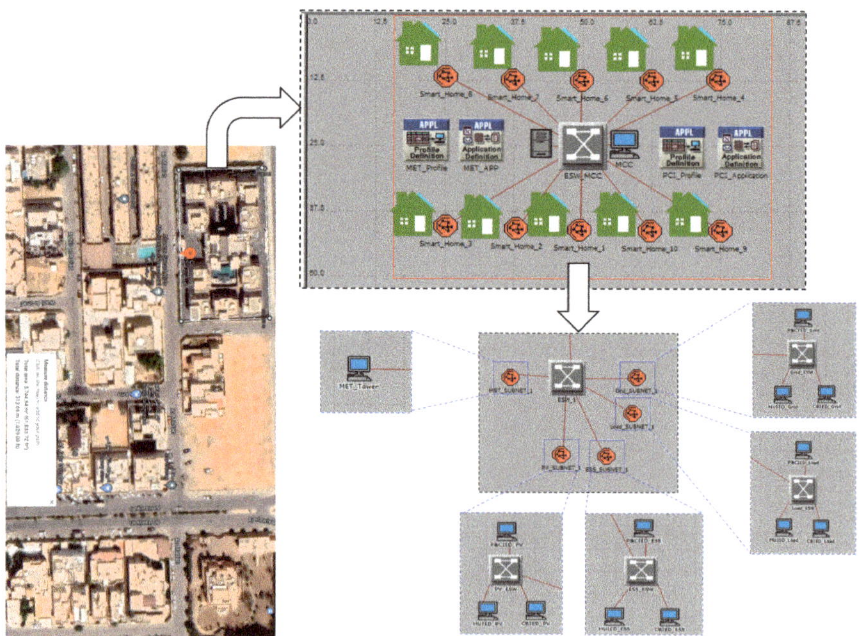

Figure 8. Ethernet-based communication network model for community-based P2P architecture with 10 smart prosumers.

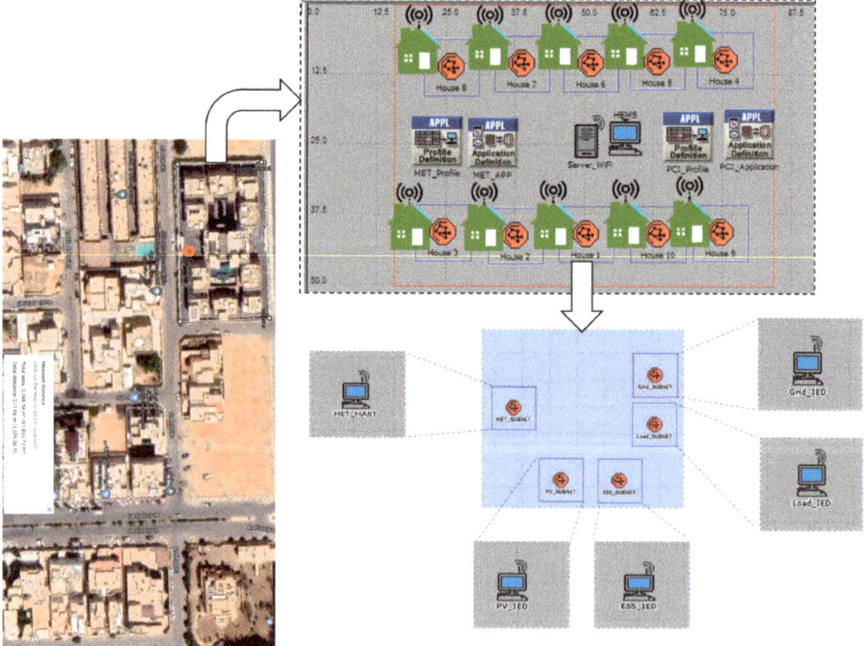

Figure 9. WiFi-based communication network model for community-based P2P architecture with 10 smart prosumers.

Table 8. End-to-end delay (second) for NAN based on Ethernet-based architecture for community-based P2P. Smart Prosumer (SP).

Scenario	Fast Ethernet (100 Mbps)	Gigabit Ethernet (1000 Mbps)
1 Prosumer	0.00123	0.000124
2 Prosumers	0.00247	0.000252
3 Prosumers	0.00371	0.000379
4 Prosumers	0.00510	0.000506
5 Prosumers	0.00621	0.000635
6 Prosumers	0.00753	0.000764
7 Prosumers	0.00878	0.000893
8 Prosumers	0.01021	0.001022
9 Prosumers	0.01149	0.001151
10 Prosumers	0.01286	0.001279

Table 9. Average end-to-end delay (second) for NAN based on WiFi-based architecture for community-based P2P.

Scenario	WiFi (54 Mbps)	WiFi (24 Mbps)	WiFi (11 Mbps)
1 Prosumer	0.00266	0.00452	0.00959
2 Prosumers	0.00451	0.00774	0.00774
3 Prosumers	0.00700	0.01213	0.01213
4 Prosumers	0.00919	0.01615	0.01615
5 Prosumers	0.01175	0.02620	0.44411
6 Prosumers	0.01507	0.20364	-
7 Prosumers	0.01791	-	-
8 Prosumers	0.02663	-	-
9 Prosumers	0.12120	-	-
10 Prosumers	-	-	-

5.3. Discussion and Limitations

The microgrid communication infrastructure is set to play an essential role in the real-time monitoring and control of peer-to-peer energy trading. To ensure its reliable and stable operation, various characteristics of communication technology, such as coverage, latency, and throughput, need to be evaluated. The results show that Ethernet-based communication architectures were sufficient for data sharing among peers at the HAN and NAN levels, with an end-to-end delay of 1.23 ms and 12.86 ms for a standalone prosumer and 10 prosumers, respectively. However, WiFi-based communication architectures were sufficient for a small number of participants. The end-to-end delay was about 2.66 ms and 121.2 ms for a standalone prosumer and 10 prosumers, respectively. To extend the current work, the main directions are given below:

- It is crucial to define the type and nature of peer-to-peer energy trading. There are different types of market mechanisms among peers, including centralized, decentralized, and distributed. Although this work is considered a centralized market mechanism based on the business model, other market mechanisms need to be investigated and compared.

- Microgrid subsystems cannot be monitored and controlled through a single communication network and/or technology. Although this work considered two different communication technologies, Ethernet and WiFi, further investigations are required for data transmission using heterogeneous communication technologies among different subsystems.

- Peer-to-peer energy trading is susceptible to different types of attacks which could compromise the microgrid infrastructures, including the physical and cyber layers. A thorough analysis of these types of cyberattacks, vulnerabilities, and countermeasures are very important elements to be investigated.

- Peer-to-peer energy trading is a complex cyber-physical system that is still in its early stage. Experimental work needs to be conducted to address the challenges of different physical and cyber layers, as well as to validate and compare the simulation results with real system implementations.

6. Conclusions

Peer-to-peer energy trading could provide new opportunities for future smart grids. In order to implement such systems, communication infrastructure is crucial in data transmission for monitoring, operation, and control. For a smart prosumer in a community microgrid, we developed a communication network model in this study. This model was on the IEC 61850 standard and included a PV system, a battery energy storage system, loads, and a grid connection. The smart prosumer model could be extended to include additional energy sources, such as wind turbines, electric vehicles, etc., owing to the modular design of the communication model. Two scenarios were considered: the HAN for a smart prosumer and NAN for a community-based P2P architecture. To evaluate the network performance, this work considered the commonly available technologies of Ethernet and WiFi for implementation. The simulation results showed that both Ethernet and WiFi technologies could satisfy the delay requirement for monitoring and control if the appropriate selection of channel capacity and data rate were selected. The end-to-end delay was about 12.86 ms and 1.27 ms for community-based P2P (10 prosumers) using the Fast Ethernet and Gigabit Ethernet, respectively. The average end-to-end delay was about 11.75 ms, 26.20 ms, and 444.11 ms for the WiFi-based architecture of a community-based P2P (5 prosumers) using different data rates of 54 Mbps, 24 Mbps, and 11 Mbps, respectively.

Although this work considered Ethernet and WiFi for the performance evaluation, other wired/wireless communication technologies need to be considered. Our current efforts are focused on putting the entire system into practice in order to verify these network performance findings. Furthermore, other communication technologies will be considered, such as BLE for HAN and LoRa for NAN, as well as a hybrid P2P architecture.

Author Contributions: Conceptualization, A.M.E. and M.A.A.; methodology, A.M.E. and M.A.A.; software, M.A.A.; validation, A.M.E. and M.A.A.; formal analysis, A.M.E. and M.A.A.; investigation, A.M.E. and M.A.A.; writing—original draft preparation, A.M.E. and M.A.A.; writing—review and editing, A.M.E. and M.A.A.; project administration, A.M.E.; funding acquisition, A.M.E. All authors have read and agreed to the published version of the manuscript.

Funding: This work was funded by the National Plan for Sciences and Technology program (Project No. 13-ENE2210-02) by King Saud University.

Data Availability Statement: Not applicable.

Acknowledgments: This work was supported by the National Plan for Sciences and Technology program (Project No. 13-ENE2210-02) by King Saud University for financial support to carry out the research work reported in this paper.

Conflicts of Interest: The authors declare no conflict of interest.

References

1. Saudi Vision 2030. Available online: https://www.vision2030.gov.sa/v2030/a-sustainable-saudi-vision/ (accessed on 13 April 2023).
2. Sadeem, A.; Alyamani, R. Data Insight: Saudi Arabia's Residential Solar Energy Use and Products. 02/13/2020. Riyadh: KAPSARC. 2020. Available online: https://www.kapsarc.org/research/publications/saudi-arabias-residential-solar-energy-use-and-prospects/ (accessed on 13 April 2023).
3. Alotaibi, M.; Eltamaly, A. A smart strategy for sizing of hybrid renewable energy system to supply remote loads in Saudi Arabia. *Energies* **2021**, *14*, 7069. [CrossRef]
4. Eltamaly, A.; Alotaibi, M. Novel Fuzzy-Swarm Optimization for Sizing of Hybrid Energy Systems Applying Smart Grid Concepts. *IEEE Access* **2021**, *9*, 93629–93650. [CrossRef]
5. Eltamaly, A.; Alotaibi, M.; Alolah, A.; Ahmed, M. A Novel Demand Response Strategy for Sizing of Hybrid Energy System with Smart Grid Concepts. *IEEE Access* **2021**, *9*, 20277–20294. [CrossRef]

6. Eltamaly, A.; Alotaibi, M.; Elsheikh, W.; Alolah, A.A.M. Novel Demand Side-Management Strategy for Smart Grid Concepts Applications in Hybrid Renewable Energy Systems. In Proceedings of the 2022 4th International Youth Conference on Radio Electronics, Electrical and Power Engineering (REEPE), Moscow, Russia, 17–19 March 2022; pp. 1–7. [CrossRef]
7. Wu, Y.; Wu, Y.; Cimen, H.; Vasquez, J.C.; Guerrero, J.M. P2P energy trading: Blockchain-enabled P2P energy society with multi-scale flexibility services. *Energy Rep.* **2022**, *8*, 3614–3628. [CrossRef]
8. Suthar, S.; Cherukuri, S.H.C.; Pindoriya, N.M. Peer-to-peer energy trading in smart grid: Frameworks, implementation methodologies, and demonstration projects. *Electr. Power Syst. Res.* **2023**, *214*, 108907. [CrossRef]
9. Lee, G.M.; Su, D.H. Standarization of smart grid in ITU-T. *IEEE Commun. Mag.* **2013**, *51*, 90–97. [CrossRef]
10. Panda, D.K.; Das, S. Smart grid architecture model for control, optimization and data analytics of future power networks with more renewable energy. *J. Clean. Prod.* **2021**, *301*, 126877. [CrossRef]
11. Tushar, W.; Saha, T.K.; Yuen, C.; Smith, D.; Poor, H.V. Peer-to-Peer Trading in Electricity Networks: An Overview. *IEEE Trans. Smart Grid* **2020**, *11*, 3185–3200. [CrossRef]
12. Wu, Y.; Wu, Y.; Guerrero, J.M.; Vasquez, J.C. Digitalization and decentralization driving transactive energy Internet: Key technologies and infrastructures. *Int. J. Electr. Power Energy Syst.* **2021**, *126*, 106593. [CrossRef]
13. Kabalci, Y.; Kabalci, E.; Padmanaban, S.; Holm-Nielsen, J.B.; Blaabjerg, F. Internet of Things Applications as Energy Internet in Smart Grids and Smart Environments. *Electronics* **2019**, *8*, 972. [CrossRef]
14. Wu, Y.; Wu, Y.; Cimen, H.; Vasquez, J.C.; Guerrero, J.M. Towards collective energy Community: Potential roles of microgrid and blockchain to go beyond P2P energy trading. *Appl. Energy* **2022**, *314*, 119003. [CrossRef]
15. Azim, M.I.; Tushar, W.; Saha, T.K.; Yuen, C.; Smith, D. Peer-to-peer kilowatt and negawatt trading: A review of challenges and recent advances in distribution networks. *Renew. Sustain. Energy Rev.* **2022**, *169*, 112908. [CrossRef]
16. Jabbar Aziz Baig, M.; Iqbal, M.T.; Jamil, M.; Khan, J. Peer-to-Peer Energy Trading in a Micro-grid Using Internet of Things and Blockchain. *Electron. ETF* **2021**, *25*, 39–49. [CrossRef]
17. Baig, M.J.A.; Iqbal, M.T.; Jamil, M.; Khan, J. A Low-Cost, Open-Source Peer-to-Peer Energy Trading System for a Remote Community Using the Internet-of-Things, Blockchain, and Hypertext Transfer Protocol. *Energies* **2022**, *15*, 4862. [CrossRef]
18. Shrestha, A.; Bishwokarma, R.; Chapagain, A.; Banjara, S.; Aryal, S.; Mali, B.; Thapa, R.; Bista, D.; Hayes, B.P.; Papadakis, A.; et al. Peer-to-Peer Energy Trading in Micro/Mini-Grids for Local Energy Communities: A Review and Case Study of Nepal. *IEEE Access* **2019**, *7*, 131911–131928. [CrossRef]
19. Liu, C.; Wang, Z.; Yu, M.; Gao, H.; Wang, W. Optimal peer-to-peer energy trading for buildings based on data envelopment analysis. *Energy Rep.* **2023**, *9*, 4604–4616. [CrossRef]
20. Gbadega, P.A.; Sun, Y. Centralized peer-to-peer transactive energy market approach in a prosumer-centric residential smart grid environment. *Energy Rep.* **2022**, *8*, 105–116. [CrossRef]
21. Baig, M.J.A.; Iqbal, M.T.; Jamil, M.; Khan, J. IoT and Blockchain Based Peer to Peer Energy Trading Pilot Platform. In Proceedings of the 2020 11th IEEE Annual Information Technology, Electronics and Mobile Communication Conference (IEMCON), Vancouver, BC, Canada, 4–7 November 2020; pp. 0402–0406. [CrossRef]
22. Tushar, W.; Saha, T.K.; Yuen, C.; Liddell, P.; Bean, R.; Poor, H.V. Peer-to-Peer Energy Trading With Sustainable User Participation: A Game Theoretic Approach. *IEEE Access* **2018**, *6*, 62932–62943. [CrossRef]
23. Baig, M.J.A.; Iqbal, M.T.; Jamil, M.; Khan, J. Blockchain-Based Peer-to-Peer Energy Trading System Using Open-Source Angular Framework and Hypertext Transfer Protocol. *Electronics* **2023**, *12*, 287. [CrossRef]
24. Mullaney, C.; Aijaz, A.; Sealey, N.; Holden, B. Peer-to-Peer Energy Trading meets IOTA: Toward a Scalable, Low-Cost, and Efficient Trading System. In Proceedings of the 2022 IEEE/ACM 15th International Conference on Utility and Cloud Computing (UCC), Vancouver, WA, USA, 6–9 December 2022; pp. 399–406. [CrossRef]
25. Górski, T. The 1+5 Architectural Views Model in Designing Blockchain and IT System Integration Solutions. *Symmetry* **2021**, *13*, 2000. [CrossRef]
26. Arbab-Zavar, B.; Golestan, S.; Vasquez, J.C.; Guerrero, J.M. Exploring Communication Architectures in Microgrids: Applications and Scenarios. *IEEE Ind. Electron. Mag.* **2023**, 2–15. [CrossRef]
27. Zhang, Y.; Krishnan, V.V.G.; Pi, J.; Kaur, K.; Srivastava, A.; Hahn, A.; Suresh, S. Cyber Physical Security Analytics for Transactive Energy Systems. *IEEE Trans. Smart Grid* **2020**, *11*, 931–941. [CrossRef]
28. Ahmed, M.A.; Chavez, S.A.; Eltamaly, A.M.; Garces, H.O.; Rojas, A.J.; Kim, Y.-C. Toward an Intelligent Campus: IoT Platform for Remote Monitoring and Control of Smart Buildings. *Sensors* **2022**, *22*, 9045. [CrossRef]
29. Liu, Y.; Wu, L.; Li, J. Peer-to-peer (P2P) electricity trading in distribution systems of the future. *Electr. J.* **2019**, *32*, 2–6. [CrossRef]
30. Lozano, J.C.; Koneru, K.; Ortiz, N.; Cardenas, A.A. Digital Substations and IEC 61850: A Primer. *IEEE Commun. Mag.* **2023**, *61*, 28–34. [CrossRef]
31. *IEC 61850-7-420:2021*; Communication Networks and Systems for Power Utility automation–Part 7-420: Basic Communication Structure–Distributed Energy Resources and Distribution Automation Logical Nodes. IEC Standards: Geneva, Switzerland, 2021. Available online: https://webstore.iec.ch/publication/34384#:~:text=IEC%2061850%2D%2D420%3A2021%20%7C%20IEC%20Webstore%20%7C%20LVDC (accessed on 27 May 2023).
32. Eltamaly, A.M.; Alotaibi, M.A.; Alolah, A.I.; Ahmed, M.A. IoT-Based Hybrid Renewable Energy System for Smart Campus. *Sustainability* **2021**, *13*, 8555. [CrossRef]

33. Condon, F.; Martínez, J.M.; Eltamaly, A.M.; Kim, Y.-C.; Ahmed, M.A. Design and Implementation of a Cloud-IoT-Based Home Energy Management System. *Sensors* **2023**, *23*, 176. [CrossRef]
34. Jogunola, O.; Ikpehai, A.; Anoh, K.; Adebisi, B.; Hammoudeh, M.; Gacanin, H.; Harris, G. Comparative Analysis of P2P Architectures for Energy Trading and Sharing. *Energies* **2018**, *11*, 62. [CrossRef]
35. Serban, I.; Cespedes, S.; Marinescu, C.; Azurdia-Meza, C.A.; Gomez, J.S.; Hueichapan, D.S. Communication Requirements in Microgrids: A Practical Survey. *IEEE Access* **2020**, *8*, 47694–47712. [CrossRef]
36. Zhuo, Z.; Huang, J.; Lu, W.; Lu, X. Research on Communication Stability of Inter-Cannonball Network Based on OPNET. *Appl. Sci.* **2023**, *13*, 4588. [CrossRef]
37. Kermani, M.; Adelmanesh, B.; Shirdare, E.; Sima, C.A.; Carnì, D.L.; Martirano, L. Intelligent energy management based on SCADA system in a real Microgrid for smart building applications. *Renew. Energy* **2021**, *171*, 1115–1127. [CrossRef]

Article

Passive Flow Control Application Using Single and Double Vortex Generator on S809 Wind Turbine Airfoil

Mustafa Özden [1,2], Mustafa Serdar Genç [1,3,4] and Kemal Koca [1,5,*]

1   Wind Engineering and Aerodynamic Research Laboratory, Department of Energy Systems Engineering, Erciyes University, 38039 Kayseri, Turkey; mustafaozden@erciyes.edu.tr (M.Ö.); musgenc@erciyes.edu.tr (M.S.G.)
2   Scientific Research Projects Unit, Erciyes University, 38039 Kayseri, Turkey
3   Energy Conversion Research and Application Center, Erciyes University, 38039 Kayseri, Turkey
4   MSG Teknoloji Ltd., Şti, Erciyes Teknopark Tekno-1 Binası, 61/20, 38039 Kayseri, Turkey
5   Department of Mechanical Engineering, Abdullah Gül University, 38080 Kayseri, Turkey
*   Correspondence: kemal.koca@agu.edu.tr

Abstract: The current study is aimed at investigating the influences of vortex generator (VG) applications mounted to the suction and pressure surfaces of the S809 wind turbine airfoil at low Reynolds number flow conditions. Both single and double VG applications were investigated to provide technological advancement in wind turbine blades by optimizing their exact positions on the surface of the airfoil. The results of the smoke-wire experiment for the uncontrolled case reveal that a laminar separation bubble formed near the trailing edge of the suction surface, and it was moved towards the leading edge as expected when the angle of attack was increased, resulting in bubble burst and leading-edge flow separation at $\alpha = 12°$. The u/U_∞, laminar kinetic energy and total fluctuation energy contours obtained from the numerical study clearly show that both the single and double VG applications produced small eddies, and those eddies in the double VG case led the flow to be reattached at the trailing edge of the suction surface and to gain more momentum by energizing. This situation was clearly supported by the results of aerodynamic force; the double VG application caused the lift coefficient to increase, resulting in an enhancement of the aerodynamic performance. A novel finding is that the VG at the pressure surface caused the flow at the wake region to gain more energy and momentum, resulting in a reattached and steadier flow condition.

Keywords: design optimization; wind energy; numerical simulation; experimental investigation; optimum VG applications; aerodynamic performance; airfoil

Citation: Özden, M.; Genç, M.S.; Koca, K. Passive Flow Control Application Using Single and Double Vortex Generator on S809 Wind Turbine Airfoil. *Energies* **2023**, *16*, 5339. https://doi.org/10.3390/en16145339

Academic Editor: Tek Tjing Lie

Received: 15 March 2023
Revised: 2 July 2023
Accepted: 7 July 2023
Published: 12 July 2023

1. Introduction

Due to the consuming of fossil energy resources such as natural gas, coal and petroleum alternatives, as well as global warming, sustainable progress in the energy sector has significantly grown in the past decades. Regarding the low carbon economy, it has started to become more crucial, especially among developed countries.

In recent years, wind energy has improved its reputation, and it has started to spearhead electricity production among renewables [1,2]. Regarding the increase in aerodynamic performance and the extraction of more power from wind energy [3], aerodynamic investigators have performed numerous studies so far [4]. They aimed to mitigate and suppress a few flow phenomena such as boundary layer separation, unsteady flow characteristics and laminar separation bubbles (LSB) formed over wind turbine blades [5] operating at low Reynolds number regimes [6–9]. Their experimental or numerical results indicate that these phenomena both negatively affected the aerodynamic performance of wind turbine blades and caused disturbing noise and vibration [10–12]. Therefore, a strong interest was developed among aerodynamic researchers in terms of flow control techniques [13–25].

As mentioned in the flow control study presented by Genç et al. [25], flow control techniques generally have two sub-branches, including laminar flow and turbulence manipulation. Also, turbulence manipulation methods can be divided into two essential branches, which are passive and active control techniques. Active control methods have not been as widely performed because both of their installation processes are more time-consuming and their experimental rigs are relatively expensive compared to passive control methods.

Among passive control techniques, vortex generators (VGs), which are the most famous in terms of the production of turbulence manipulation, have been extensively used in different applications such as all flight vehicles [26], solar chimneys [27], etc. The essential mission of VGs is to energize the flow in the boundary layer by transmitting momentum from the main flow, resulting in the boundary layer separation being hindered [28–30]. The pioneering study was studied by Taylor [31], and many aerodynamic engineers have so far studied the application of VGs in their studies, as shown in Table 1. Unlike the studies on VGs that are summarized in Table 1, this study aims to investigate the VGs mounted on both the suction surface and pressure surface of the S809 airfoil as a novelty. A detailed investigation of the function of VGs was conducted into the two main structures as follows: (i) the wind tunnel research, including the smoke-wire visualization and force measurement tests, as well as a comprehensive numerical simulation, were performed to observe the flow on both the suction and pressure surfaces of the uncontrolled S809 airfoil at different Reynolds numbers and angles of attack; (ii) after the suitable positions of VGs were found, the detailed numerical simulation was fulfilled for the controlled S809 airfoil.

Table 1. The relevant studies in the literature with regard to VG applications.

Test Specimen	Reynolds Number Range or Free-Stream Velocity	Study Type	Explanation	Citation
DU-97-W-300	3×10^6	CFD simulation	Using VGs resulted in both an increment in the maximum lift and increase in the stall angle.	[32,33]
NACA 23012C NACA $63_2$217	$0.7 \times 10^6 < \mathrm{Re_c} < 1.1 \times 10^6$ $0.27 \times 10^6 < \mathrm{Re_c} < 1.3 \times 10^6$	Wind tunnel research	For both airfoils, an increase in $C_{L,\,max}$ and delay in drag rise was observed. Furthermore, suppression of trailing-edge flow separation was observed.	[34]
LS(1)-0417GA(W)-1	0.8×10^5 and 1.6×10^5	Wind tunnel research	At higher Reynolds number, a rise in both the stall angle and maximum lift coefficient was observed.	[35]
DU-97-W-300	2×10^6	Wind tunnel research	Using VGs helped the stall to decrease. Furthermore, the load fluctuations were increased in the stall regime employing VGs.	[36]
NASA LS-0417	0.83×10^5 and 1.18×10^5	Wind tunnel research	The improvement in airfoil performance did not occur when implementing NASA LS-0417 at given Reynolds numbers. But at a larger Reynolds number, C_L/C_D increased by approximately 36%.	[37]
Simpler (2D) geometry	Between 1 m/s and 10 m/s	Wind tunnel research	Two different VG applications were investigated. The results show that VGs with counter-rotating types ensured better results compared to VGs with co-rotating configurations.	[38]

Table 1. Cont.

Test Specimen	Reynolds Number Range or Free-Stream Velocity	Study Type	Explanation	Citation
NREL S809_1	1×10^6	CFD simulation	Their results indicate that the installation of VGs at the 10% chord position caused the stall phenomenon to increase from $14°$ to $18°$. In addition, the output power was extracted around 96.48% when the double VG configuration was utilized.	[39]
DU91-W2-250	2×10^6	CFD simulation	The used computational models are compared with the experimental results presented in the literature. They exhibited great coherence between each other. Moreover, the vortex sheds coming from the VGs caused the stream flow to remain attached to the solid surface.	[40]
NACA 4415 S814	1.5×10^6	CFD simulation	Their results conclude that the stall was postponed, and the maximum lift increased.	[41]
NTUA-T18	0.87×10^6	Wind tunnel research	The stall cell was postponed for $\alpha = 5°$, and the lift rose to $\alpha = 15°$	[42]
RAF-19	1.93×10^5	CFD simulation	The positive effects of VGs exhibited a more dominant and crucial role in more cambered airfoils than less cambered airfoils.	[43]
NACA 4415	2×10^5	Wind tunnel research	In terms of controlling the boundary layer separation, the triangular VG type was best suited. Also, using the coupled VGs ensured the maximum lift coefficient increase by 21%, while the flow separation was postponed by $17°$.	[44]
NREL S809_2	1×10^6	CFD simulation	The vortex generators located at different x/c on the airfoil suction surface were investigated numerically. In addition to the location, a detailed study was conducted on the height of the vortex generator.	[45,46]

2. Investigation Methods

As mentioned earlier, the investigation methods of this study consist of both experimental investigations in the wind tunnel and detailed numerical simulations. These techniques are explained in detail in the sub-chapters of this paper.

2.1. Experimental Arrangements

2.1.1. Test Model and Wind Tunnel

A series of experiments were conducted in the low-speed suction-type wind tunnel belonging to the Wind Engineering and Aerodynamic Research Group (WEAR) at Erciyes University. The characteristic features of the wind tunnel were as follows: a 50 cm $\times$ 50 cm test chamber surrounded with Plexiglas material for visualization experiments; a low turbulence intensity of 0.3% at maximum free-stream velocity (~40 m/s), which was convenient to investigate flow characteristics forming, especially at low Reynolds number regimes [24,47,48]. Detailed technical specifications for the wind tunnel are illustrated in Table 2. As a test model, the S809 airfoil was utilized since it was frequently preferred for horizontal axis wind turbines (HAWTs) such as NREL Phase VI and other applications, which ensured that the most comprehensive data were obtained [49]. The test model was manufactured using a 3D printer. After manufacturing, all surfaces of the airfoil were cleaned using sandpaper, and its surfaces were then painted with a rapid-drying acrylic spray to provide a very polished surface. The Plexiglas was utilized at each tip of the airfoil in order to prevent the tip vortices effects. Its chord and span lengths were 200 mm and 200 mm, respectively. The air density is 1 kg/m3 in our laboratory. The blockage ratio of the model at $\alpha = 12°$ was under 7%, and there were no blockage corrections needed in the experimental results [48].

Table 2. Technical features of the wind tunnel [48].

Design	Suction Type and Low Speed
Length of tunnel	13 m
Test section	Length (4 m), Height (0.5 m), Width (0.5 m)
Motor	Type: DC motor; Power: 15 kW; Frequency: 50 Hz
Model	H4, 1000/15A
Capacity	45,000 m^3/h, 450 PA
Flow velocity	3 m/s < U < 40 m/s
Turbulence level	0.3% < Tu < ~0.9%
Nozzle	Contraction cone: 9:1

Related to the dimensions of the VG, they are illustrated as a sketched image in Figure 1. The height of the VG is symbolized by h, and its length is L. In this study, the height of the VG was determined to be 0.5 mm, and the length was determined to be 4 mm.

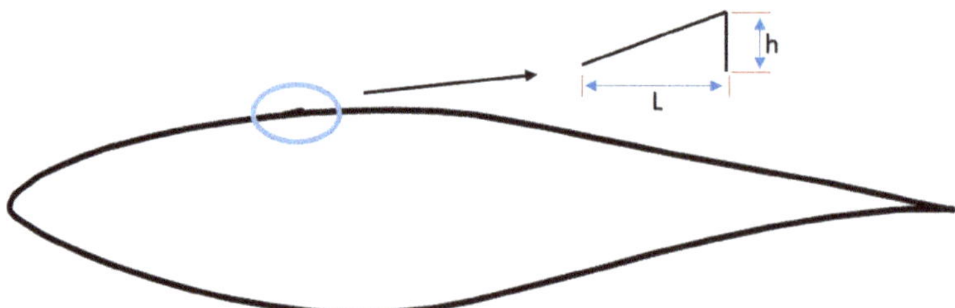

Figure 1. Schematic of the VG setting.

To obtain a better understanding of the position of the VGs, they were sketched and tabulated, as shown in Table 3. Furthermore, each configuration of the VGs is named from Case 1 to Case 8 hereinafter. It was clearly pointed out that a single application of VGs on the suction surfaces was determined between Case 1 and Case 5 [50], while a double application of VGs on both the suction and pressure surfaces was assigned between Case 6 and Case 8 according to their position chordwise.

Table 3. The configuration of VGs.

Configurations	Positions of VGs	Sketches
Case 1	Suction surface x/c = 0.1 Pressure surface Not applied	

Table 3. *Cont.*

Configurations	Positions of VGs	Sketches
Case 2	Suction surface x/c = 0.2 Pressure surface Not applied	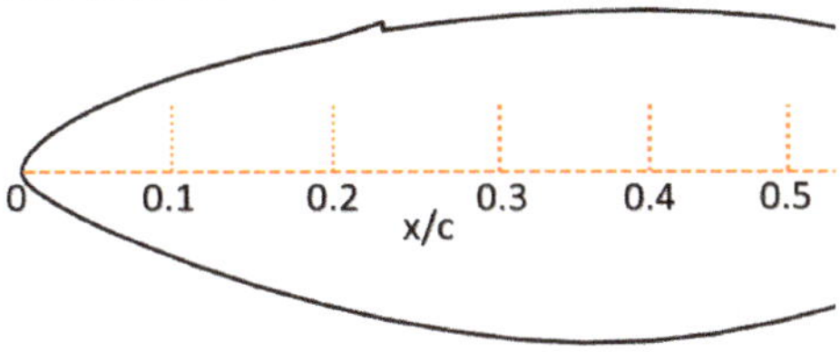
Case 3	Suction surface x/c = 0.3 Pressure surface Not applied	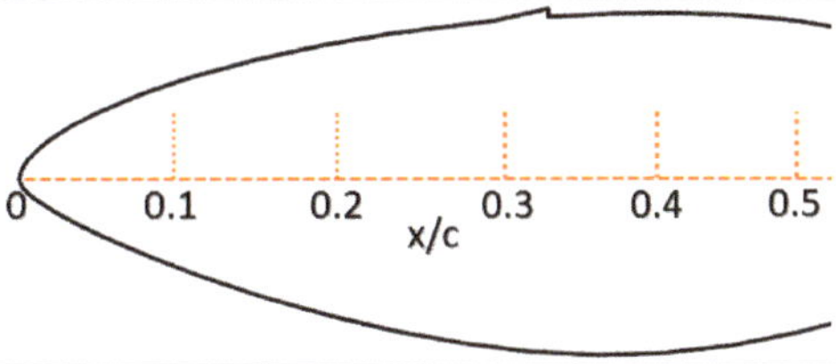
Case 4	Suction surface x/c = 0.4 Pressure surface Not applied	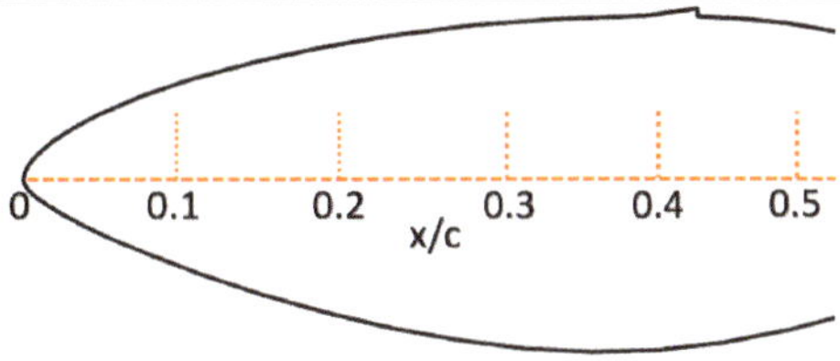
Case 5	Suction surface x/c = 0.5 Pressure surface Not applied	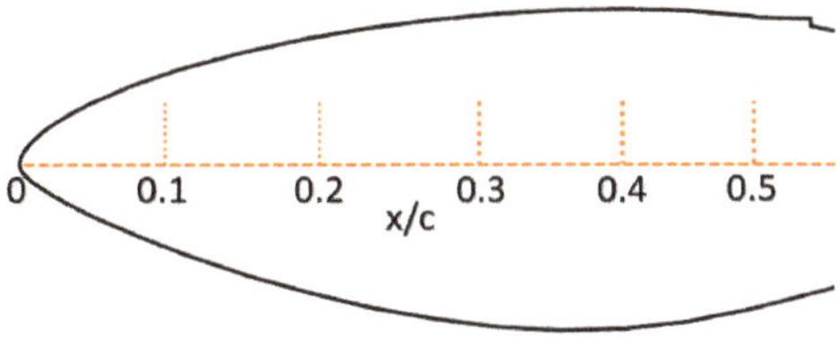
Case 6	Suction surface x/c = 0.3 Pressure surface x/c = 0.2	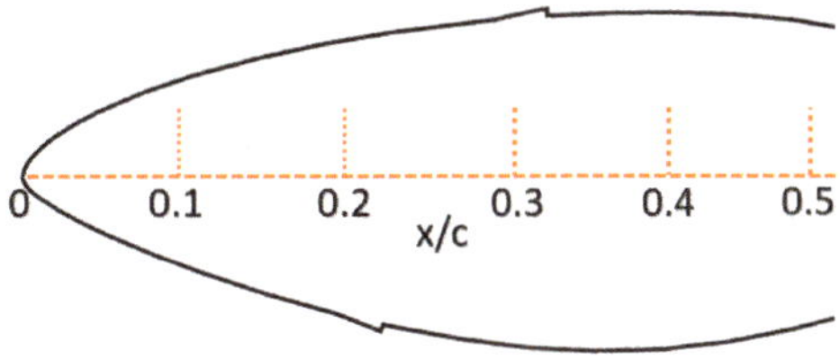
Case 7	Suction surface x/c = 0.3 Pressure surface x/c = 0.3	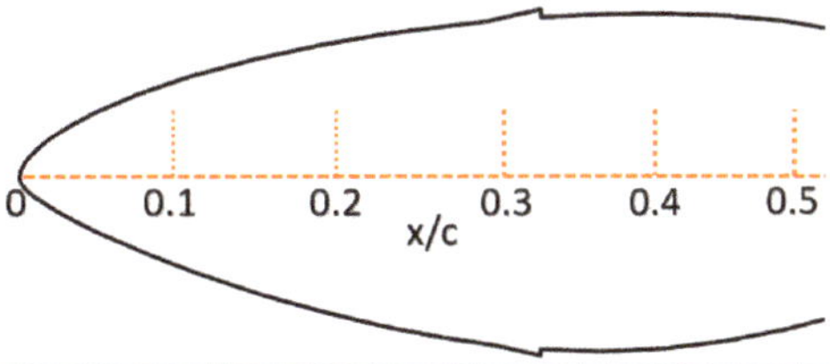

Table 3. *Cont.*

Configurations	Positions of VGs	Sketches
Case 8	Suction surface x/c = 0.3 Pressure surface x/c = 0.4	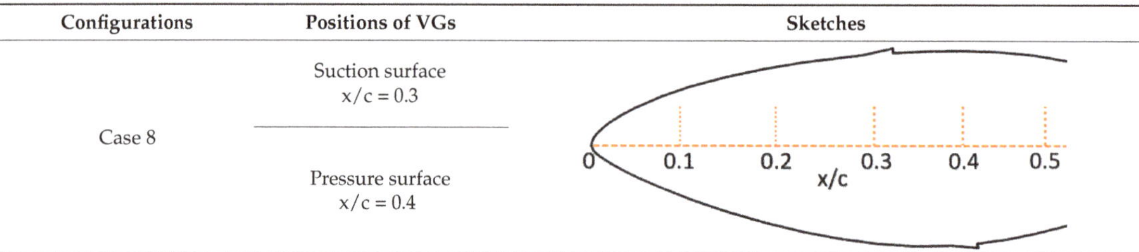

2.1.2. The Smoke-Wire Test

To visualize the flow phenomena on each surface of both the uncontrolled and the controlled S809 airfoils, the smoke-wire experiment was performed at different Reynolds numbers and angles of attack. The method was relatively easier than other flow visualization techniques. The equipment of the method included the following: (i) a thin copper wire in conjunction with its thickness of 0.03 mm, allowing us to form the suitable smoke sheets chordwise; (ii) an electric resistance heating system with a hand adjustment used for heating the wire and burning the oil droplets; (iii) oil; (iv) halogens to brighten and observe the flow phenomena in the test chamber; and (v) a compact camera with its tripod to record the flow phenomena during the tests. As seen in Figure 2, the thin copper wire was vertically positioned at 1.5c in front of the S809 airfoil. The draining process of the oil was manually performed on the wire. We waited for a while after the draining process so that homogeneous distribution was ensured along the wire. The electrically resistive heating was operated to burn a homogeneous oil droplet. The smoke sheets that formed in the test chamber after burning the wire were brightened by using halogens. A compact camera recorded videos of the smoke sheets. These videos were then divided into 1000 frames to obtain instantaneous images of the smoke sheets. Then, high-resolution instantaneous images were selected.

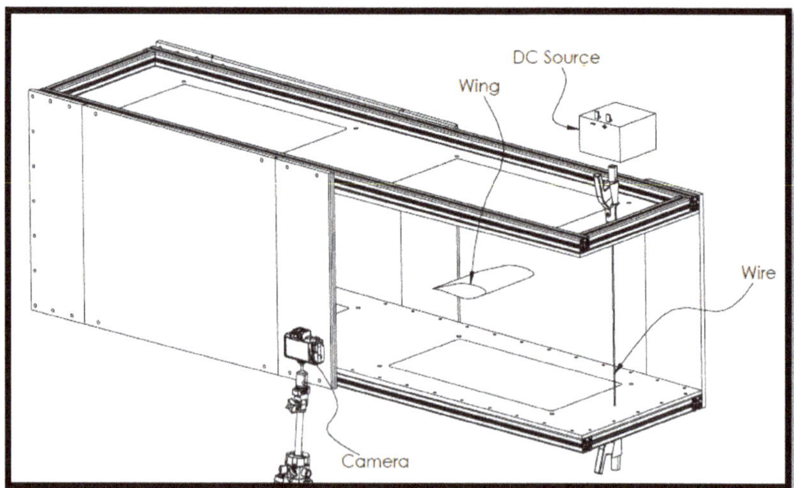

Figure 2. The sketch of the smoke-wire experiment at the test chamber.

2.1.3. Hotwire Experiment

The velocity profile at the wake region ensures good results in terms of the flow phenomena that form over the suction surface thanks to the shedding of the vortices. In this respect, the hotwire measurement was conducted, as seen in Figure 3. A one-dimensional

(55P16 type) hotwire probe was employed during the test in the wake of the S809 airfoil. The chord length of the model airfoil was 200 mm. It was positioned 0.5c away from the trailing edge, and it was run along the vertical axis (axis-y) at the wake region. In relation to the probe motion along the vertical axis, a traverse system that was able to move at both axis x and axis y was utilized. It was placed under the test chamber, and support rods were linked with the traverse system. Those support rods were used for the probe connection, as illustrated in the closed view. In relation to the technical specification of the hotwire measurement, the probe was run vertically at the wake region, and it gathered the data at 23 points with an interval of 5 mm (total measurement was carried out along 110 mm). Furthermore, it was run for 10 s at 2 kHz, resulting in 20,000 data samples obtained for each point. In terms of better accuracy, a probe calibration was carried out before every test using the reference velocity probe.

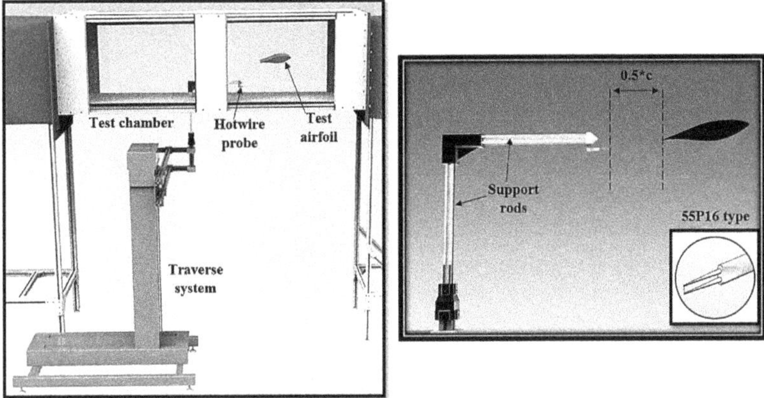

Figure 3. Scheme of the hotwire experiment.

2.2. Numerical Method

Apart from the experimental investigation, a numerical simulation was performed to determine the convenient position of VGs over the surfaces of the airfoil. RANS (Reynolds-Averaged Navier–Stokes) equations were solved using ANSYS-Fluent software to obtain the effects of VGs on the flow of the S809 airfoil. As seen in Figure 4, the domains of both the uncontrolled and the controlled S809 airfoil were formed by meshing. Regarding the mesh generation, two-dimensional (2D) cell types were used, resulting in a C-type structured quadrilateral mesh. To obtain a convenient and reliable number of nodes from mesh generation, the mesh independency study was fulfilled, as illustrated in Figure 5. It was then decided that the number of nodes of the mesh structures was about 1.5×10^6. Furthermore, the value of y^+ was about 1, allowing for the user to obtain more accurate results on the boundary layer of the airfoil. Regarding the residuals, the convergence criteria were preferred as 10^{-5}. The SIMPLEC solution algorithm was chosen, and the second-order upwind method was performed to obtain spatial discretization.

Regarding the detailed information on the C-type mesh as performed in the previous study [51], it consisted of inlet, outlet and far-field domains. The velocity inlet was applied to the inlet domain, while the pressure outlet was recognized in the outlet domain. Similarly to the authors' previous study [52], the airfoil was located 10c away from the far-field, and the velocity inlet was 20c away from the pressure outlet. The non-slip condition for the walls was assigned for the airfoil and VGs. For the 2D simulation, the transition k-k_L-ω model was performed. The k-k_L-ω model was a three-equation eddy-viscosity type, which includes transport equations for the turbulent kinetic energy (k_T), laminar kinetic energy (k_L), and the inverse turbulent time scale (ω). Moreover, the total fluctuation kinetic energy is the total of the turbulent kinetic energy (k_T) and laminar kinetic energy (k_L).

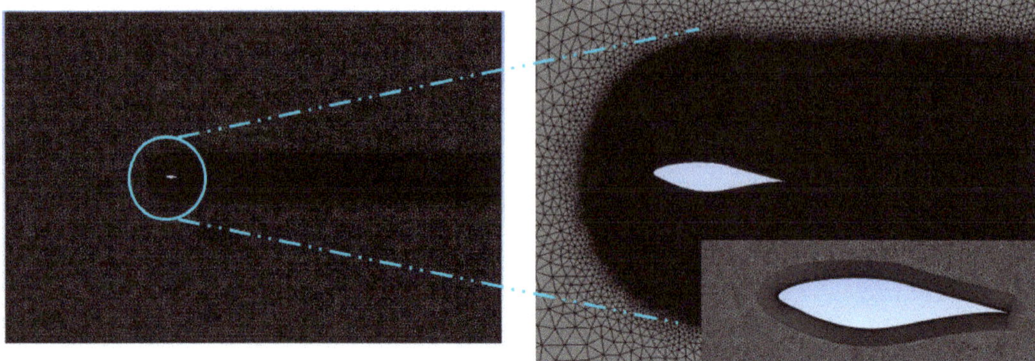

Figure 4. The close view of the flow domain.

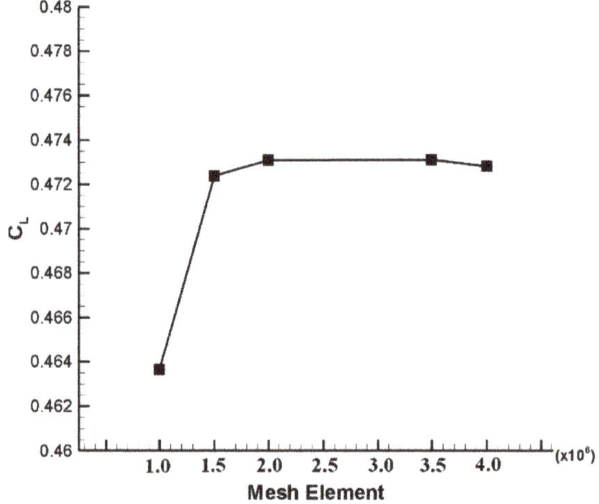

Figure 5. Mesh independency study at Re = 1.2×10^5.

3. Results and Discussion

The instantaneous flow images formed over the suction surface of the uncontrolled S809 airfoil are shown in Figure 6 by capturing them from the smoke-wire test at Re = 0.4×10^5 and 0.6×10^5. As the smoke-wire experiment did not allow for quality images to be obtained at higher speeds, the tests could be performed at these Reynolds numbers. For Re = 0.4×10^5, the LSB was seen at the trailing-edge part of the S809 airfoil at $\alpha = 0°$, and as expected, it started to move towards the leading edge when the angle of attack increased with the intervals of $4°$. After a while, the flow started to separate from the solid surface with the dominant adverse pressure gradients (APGs), causing the presence of the leading-edge flow separation with the LSB burst at $\alpha = 12°$. As mentioned in the detailed study presented by the current authors [11], vortex shedding, which gave information with regard to the progress of the LSB, was also illustrated, especially at larger angles of attack. It played a crucial role in terms of both the observation of the LSB characterization and how it affected the wake region of the airfoil. The size of the LSB was relatively minimized when the Reynolds number increased to Re = 0.6×10^5 as a result of the increasing inertial forces. This caused the APGs to overcome their effects, and there was no boundary layer separation, at least until $\alpha = 8°$. However, at $\alpha = 12°$, the flow separated nearly at $x/c = 0.3$, causing leading-edge separation.

Re= 0.4x10⁵   Re= 0.6x10⁵

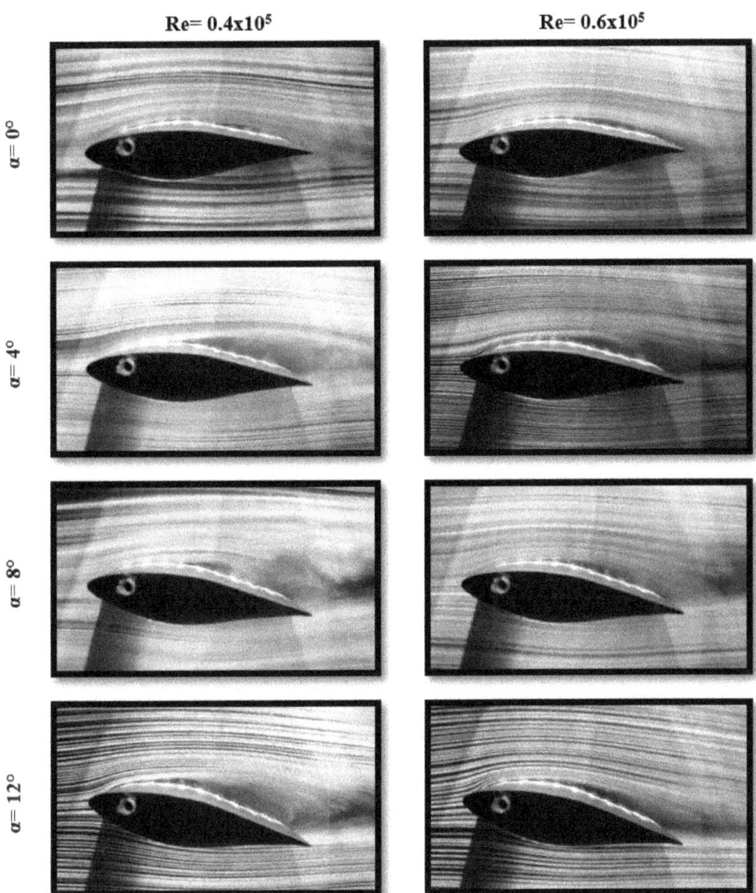

Figure 6. The instantaneous images of the uncontrolled S809 airfoil at different Reynolds numbers and angles of attack.

In this part, the numerical results at $\alpha = 8°$ for three different Reynolds numbers were compared with the experimental results [53] in Table 4 for validation. The experimental results at $\alpha = 8°$, which corresponded with our numerical results, were used for comparison. When all the results in Table 4 were examined, it was observed that our numerical results were close to the experimental results [53]. This shows that our numerical studies are reasonable and consistent. Thus, the VG studies were continued.

Table 4. The validation of the results of lift force coefficient for S809 airfoil at different Reynolds numbers for $\alpha = 8°$.

C_L at $\alpha = 8°$	Re = 0.4×10^5	Re = 0.8×10^5	Re = 1.2×10^5
Uncontrolled Experimental [53]	0.51	0.85	0.90
Uncontrolled Numerical	0.55	0.82	0.90

The aerodynamic forces as well as their ratios for $\alpha = 12°$ and Re = 0.4×10^5 were acquired by means of numerical simulation for the controlled S809 airfoil configurations, as illustrated in Table 5. The objective of Table 5 is to determine the S809 airfoil using VG configurations that exhibited the best aerodynamic performance. In this regard, the S809 airfoil named VGs_Case 3 with a C_L/C_D value of 13.27 as a single VGs application, and the S809 airfoil named VGs_Case 7 with a C_L/C_D value of 11.01 as a double VGs application were selected as the best performance configurations and are examined in detail in the rest of the paper.

Table 5. The aerodynamic force coefficients for S809 airfoil cases with different configurations of VGs at $\alpha = 12°$ and $Re = 0.4 \times 10^5$.

Case No.	C_L	C_D	C_L/C_D
VGs_Case 1	0.7203	0.1590	4.53
VGs_Case 2	1.0400	0.0787	13.21
VGs_Case 3	0.9800	0.0738	13.27
VGs_Case 4	0.7230	0.1560	4.63
VGs_Case 5	0.7310	0.1540	4.75
VGs_Case 6	0.9285	0.1761	5.27
VGs_Case 7	1.1688	0.1058	11.01
VGs_Case 8	1.0285	0.1489	6.90
Uncontrolled	0.9638	0.0868	11.10

In the numerical analysis, the vortex generators (single and double) were able to control the flow at low Reynolds number flows and at high angles of attack ($\alpha > 10°$). In Table 6, the numerical results of the aerodynamic force coefficient for the S809 airfoils without/with VGs are presented at $\alpha = 12°$ for different Reynolds numbers. As seen in the table, the increases in the lift forces were at low levels, and the drag force decreased due to the eddies. With the double vortex generator application, the lift increase was more significant at low Reynolds number flows, while the drag force also increased. This is due to the eddies that occur when a vortex generator is used on both surfaces, and as the lift is increased, the drag necessarily grows.

Table 6. The numerical results of aerodynamic force coefficients for S809 airfoils at $\alpha = 12°$.

$\alpha = 12°$	Uncontrolled		VGs_Case_3		VGs_Case_7	
Reynolds Number	C_L	C_D	C_L	C_D	C_L	C_D
$Re = 0.4 \times 10^5$	0.9638	0.0868	0.9800	0.0738	1.1688	0.1058
$Re = 0.6 \times 10^5$	1.1577	0.0781	1.0435	0.0701	1.3835	0.1689
$Re = 0.8 \times 10^5$	1.0120	0.0752	1.0225	0.0714	1.1450	0.0741
$Re = 1.2 \times 10^5$	1.1205	0.0685	1.1029	0.0682	1.1964	0.0681

The vortex shedding forms because of the presence of LSB, or flow separation. After formation, those flow phenomena move along the chord as denoted in Figure 7. The LSB-induced vortex shedding moved downwind on the suction surface, while the trailing-edge-induced vortex shedding moved downwind on the pressure surface. As mentioned in the vortex-shedding-based studies, both the LSB and trailing-edge flow separation induce vortex shedding mixed in the wake region, leading the velocity profile to decrease in the wake region (Figure 8). It is clearly seen that the value of u/U_∞ obtained from the experiments for the uncontrolled case decreased from 1.00 to 0.65, meaning that the velocity value decreased by 0.35. In addition to the experimental result, the value of u/U_∞ obtained from the numerical simulation for the uncontrolled case decreased from 0.90 to 0.77. The curve trends of those two results exhibited good coherence with each other. In relation to the results of the controlled cases, the value of u/U_∞ obtained from the numerical simulation for the VGs_Case 3 decreased from 0.97 to 0.80, while it decreased from 1.00 to 0.74 for the VGs_Case 7. The velocity value decreased by 17% and 26% for VGs_Case 3 and VGs_Case 7, respectively. It was clearly seen that the deficit in the velocity values for the uncontrolled case was more than those that occurred in the controlled cases. Utilizing VGs as a flow control technique ensured that the deficit in the velocity value was minimized. Technically speaking, the size of the LSB or effectiveness of the flow separation was minimized by means of VGs, as seen in the single VG application in Figure 9b, resulting in the presence of smaller and weaker vortex shedding, just as mentioned in the studies by the current authors [11] and by Ducoin et al. [54]. In Figure 9a, for the uncontrolled case, the LSBs at the leading edge and the trailing edge were observed, and the single VG case shortened the LSBs over the suction side, but not the LSB on the pressure side. Otherwise, the double VG application shortened the LSBs on the pressure side due to the vortex generator on the lower surface, the vortex formation gained more momentum due to the short LSB, and this merged with the vortex over the suction side at the trailing edge of the airfoil. This caused larger eddies to form, which created a vacuum effect that caused more flow to be formed from the upper surface of the airfoil. Thus, the velocity increase region over the suction side of the airfoil rose, as seen in the velocity contours for VGs_Case 7 in Figure 10a, which meant there was an increase in the pressure difference, and hence, the lift coefficient. However, as seen in Figure 10b,c, increases in laminar kinetic energy and total kinetic energy occurred, which pointed out an increase in fluctuations in the flow. All of this caused an increase in the drag force.

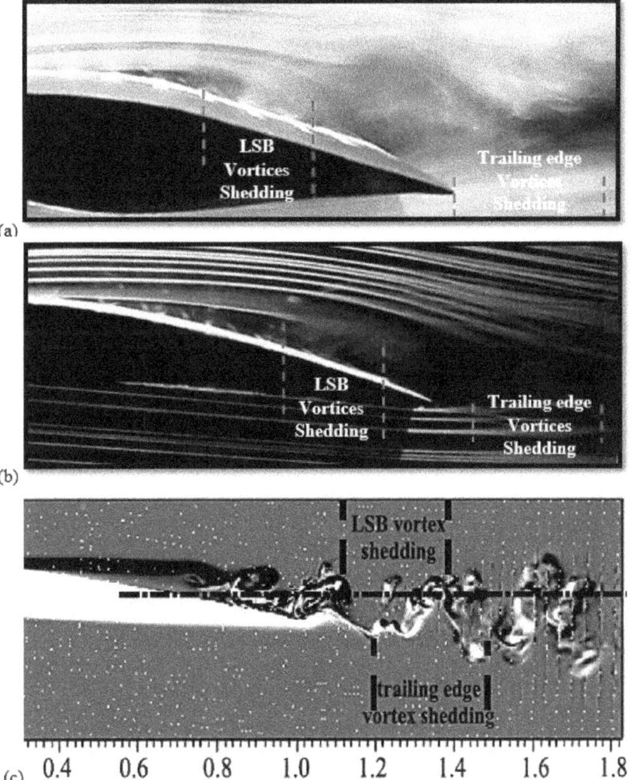

Figure 7. Visualization of LSB and trailing-edge vortex shedding: (**a**) current experimental result for the S809 airfoil at Reynolds number of 0.4×10^5; (**b**) experimental result from study performed by the current authors [11] for the NACA4412 airfoil at Reynolds number of 0.25×10^5; (**c**) numerical study performed by Ducoin et al. [54] for the SD7003 airfoil at Reynolds number of 0.2×10^5.

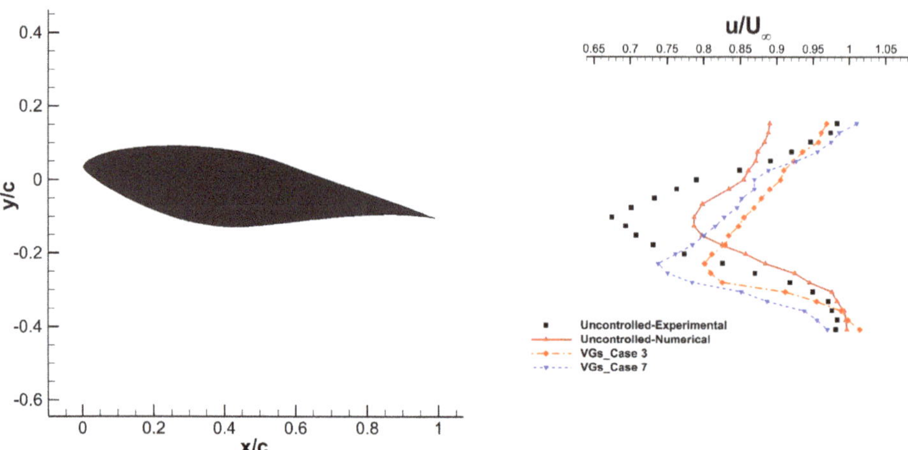

Figure 8. Values of u/U_∞ at the wake of the uncontrolled and controlled S809 airfoil at $\alpha = 8°$ and $Re = 0.6 \times 10^5$.

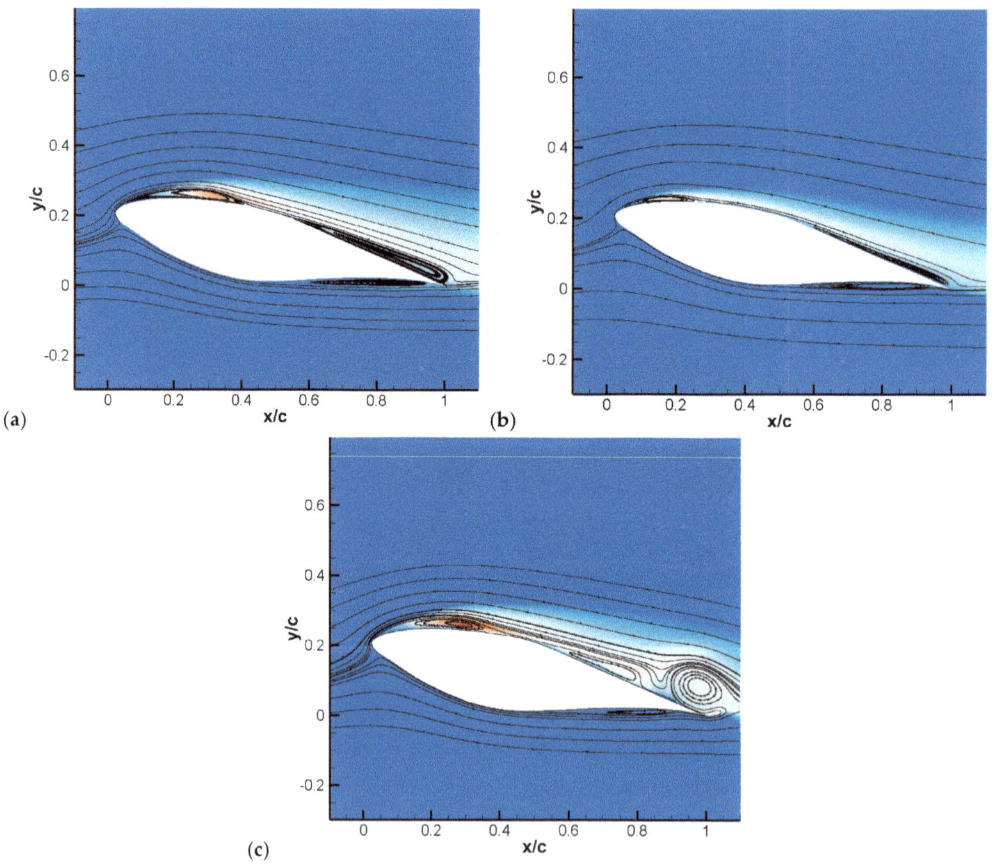

Figure 9. Streamlines with turbulent kinetic energy contours at $\alpha = 12°$ and $Re = 0.4 \times 10^5$ for (**a**) uncontrolled, (**b**) VGs_Case 3, (**c**) VGs_Case 7 cases.

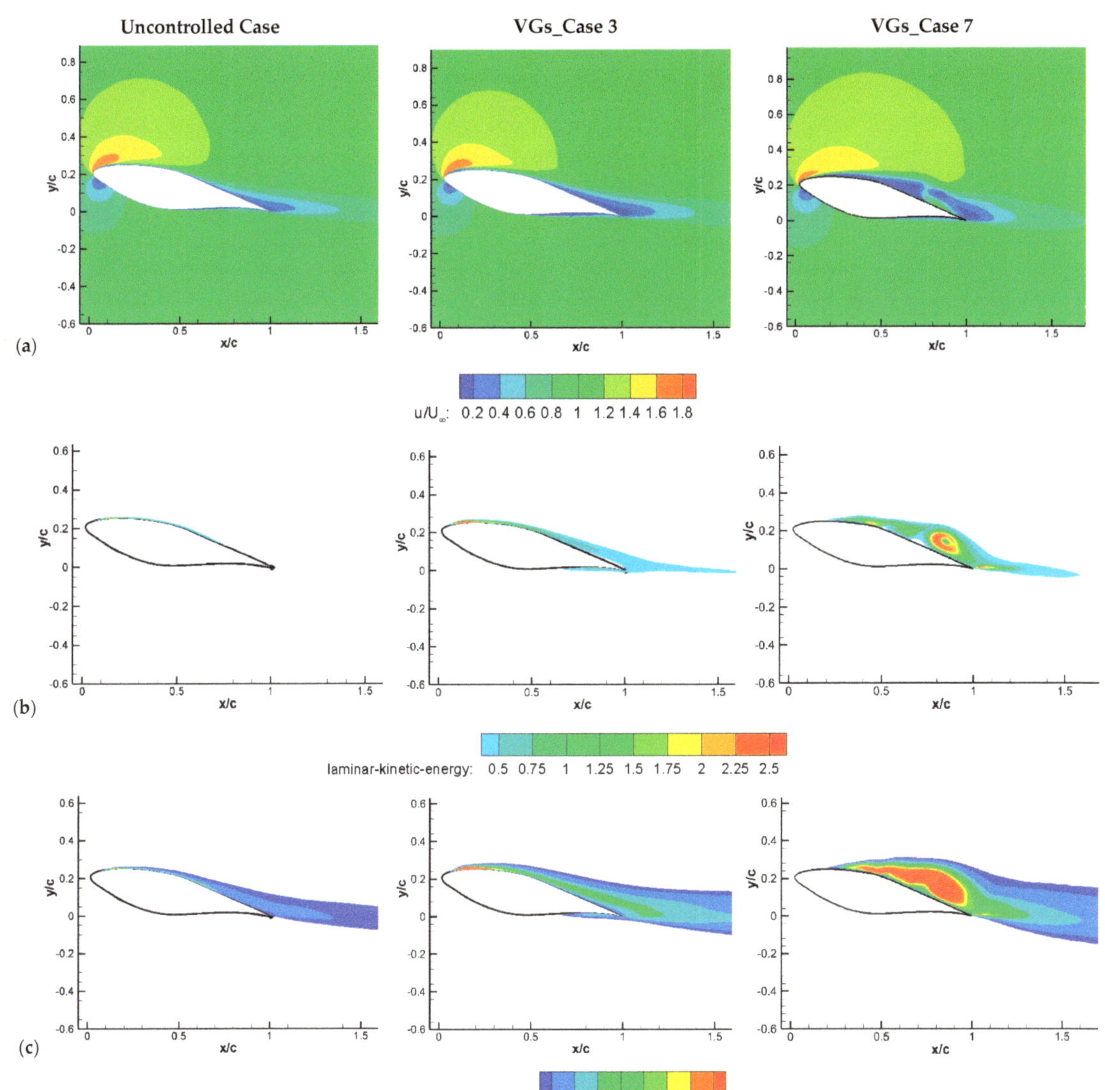

Figure 10. Graphs showing flow statistics at $\alpha = 12°$, Re $= 0.4 \times 10^5$.

In order to understand how VG application impacts the flow characteristics, the graphs belonging to the flow and turbulence statistics, including u/U_∞, laminar kinetic energy and total fluctuation energy at $\alpha = 12°$ and Re $= 0.4 \times 10^5$ are shown in Figure 10. As seen in the graphs of u/U_∞ (Figure 10a), the boundary layer separation occurred at the trailing edge of the airfoil. Those separated flows at the trailing edge were much bigger because of small eddies that were produced by single VGs (VGs_Case 3). In VGs_Case 7, the vortex generator utilized at the pressure surface also generated small eddies, which caused the alteration at the wake region. This alteration caused the flow to be reattached near the trailing edge of the suction surface, resulting in an enhancement in the aerodynamic performance. In the graphs depicting the laminar kinetic energy, which were obtained from the solution of the transport equation by means of the k-k_L-ω transition model (Figure 10b), it can be observed that the VGs induced small eddies that triggered the transition to turbulence, and they forced the flow to bypass the transition, as seen from the increases in the turbulence kinetic energy at x/c = 0.3 in Figure 9b,c. In VGs_Case 7, the kinetic energy at the trailing edge was increased via small eddies produced by the vortex generator at the pressure surface. This led the flow to gain more momentum as larger eddies formed at the trailing edge of the suction surface, resulting in the presence of flow reattachment. These situations were clearly supported by the graphs depicting the fluctuation energy distribution (in Figure 10c); the flow was energized by those vortex generators. Furthermore, it was clearly revealed that there were different pressure distributions at both the suction and pressure sides of the airfoil, as denoted in Figure 10a. In particular, the activity value of the velocity region for the uncontrolled case was less than those that occurred at VGs_Case 7. This shows that utilizing VGs with Case 7 led to the velocity increasing close to the leading edge of the airfoil. Based on Bernoulli's principle, the pressure distribution was lower at the region where the velocity distribution was higher, resulting in more lift force.

4. Conclusions

The electricity demand has expanded even more over the years. Despite many challenges, the renewable energy sector is improving in many countries. This study presents a new method using double vortex generators to increase aerodynamic efficiency. The passive control method using VG application on the surfaces of the S809 airfoil is proposed in this study. The novelty of this paper is the utilization of both single and double VG application on more cambered airfoil, which is a rare study (even absent) in the literature. The effects of the positions of 2D VGs mounted on the pressure and suction surfaces were experimentally and numerically investigated at different Reynolds numbers and angles of attack after the design optimization of VG locations was performed. The novel findings from this study are the following:

-Concerning the appropriate position of VG applications, the numerical results point out that the best optimized position was ensured for both the single and double VG applications when it was employed at a 30% chord length of the S809 airfoil. The CL/CD was 13.27 at VGs_Case 3, while the CL/CD was 11.01 at VGs_Case 7.

-It was found that VGs_Case 3 exhibited better performance at Re = 0.6×105 and lower angles of attack. However, the aerodynamic performance obtained from VGs_Case 7 was better than that obtained by VGs_Case 3 when the angle of attack was $12°$. VGs_Case 7 led the aerodynamic performance to be enhanced by ~17%. This also shows that double VG application was more reliable at higher angles of attack.

-In VGs_Case 7, the kinetic energy at the trailing edge was increased via small eddies produced by the vortex generator at the pressure surface. This led the flow to gain more momentum with larger eddies formed at the trailing edge of the suction surface, resulting in the presence of flow reattachment.

-In the graphs depicting the laminar kinetic energy, it was observed that VG-induced small eddies triggered the transition to turbulence, and they forced the flow to bypass the transition. These situations were clearly supported by the graphs depicting the fluctuation energy distribution.

Author Contributions: Conceptualization, M.Ö. and M.S.G.; methodology, M.Ö. and M.S.G.; software, M.Ö. and M.S.G.; validation, M.Ö. and M.S.G.; formal analysis, M.Ö., M.S.G. and K.K.; investigation, M.Ö., M.S.G. and K.K.; resources, M.Ö., M.S.G. and K.K.; data curation, M.Ö., M.S.G. and K.K.; writing—original draft preparation, M.Ö., M.S.G. and K.K.; writing—review and editing, M.Ö., M.S.G. and K.K.; visualization, M.Ö. and K.K.; supervision, M.S.G.; project administration, M.S.G.; funding acquisition, M.Ö. All authors have read and agreed to the published version of the manuscript.

Funding: This research was supported by the Scientific and Technological Research Council of Turkey (2211/C).

Data Availability Statement: Not applicable.

Acknowledgments: Mustafa Özden thanks the Scientific and Technological Research Council of Turkey (TÜBİTAK) for the Doctoral Scholarship for Priority Areas 2211/C.

Conflicts of Interest: The authors declare no conflict of interest.

References

1. Ung, S.K.; Chong, W.T.; Mat, S.; Ng, J.H.; Kok, Y.H.; Wong, K.H. Investigation into the aerodynamic performance of a vertical axis wind turbine with endplate design. *Energies* **2022**, *15*, 6925. [CrossRef]
2. Çiftci, C.; Erdoğan, A.; Genç, M.S. Investigation of the Mechanical Behavior of a New Generation Wind Turbine Blade Technology. *Energies* **2023**, *16*, 1961. [CrossRef]
3. Özden, M.; Genç, M.S.; Koca, K. Investigation of the effect of hidden vortex generator-flap integrated mechanism revealed in low velocities on wind turbine blade flow. *Energy Convers. Manag.* **2023**, *287*, 117107. [CrossRef]
4. Das, T.K.; Samad, A. Influence of stall fences on the performance of Wells turbine. *Energy* **2020**, *194*, 116864. [CrossRef]
5. Gerrie, C.; Islam, S.Z.; Gerrie, S.; Turner, N.; Asim, T. 3D CFD Modelling of Performance of a Vertical Axis Turbine. *Energies* **2023**, *16*, 1144. [CrossRef]
6. Anup, K.C.; Whale, J.; Urmee, T. Urban wind conditions and small wind turbines in the built environment: A review. *Renew. Energy* **2019**, *131*, 268–283.
7. Açıkel, H.H.; Genç, M.S. Flow control with perpendicular acoustic forcing on NACA 2415 aerofoil at low Reynolds numbers. *Proc. Inst. Mech. Eng. Part G J. Aerosp. Eng.* **2016**, *230*, 2447–2462. [CrossRef]
8. Zhong, J.; Li, J.; Liu, H. Dynamic mode decomposition analysis of flow separation control on wind turbine airfoil using leading-edge rod. *Energy* **2023**, *268*, 126656. [CrossRef]
9. Karasu, İ.; Genç, M.S.; Açıkel, H.; Akpolat, M.T. An experimental study on laminar separation bubble and transition over an aerofoil at low Reynolds number. In Proceedings of the 30th AIAA Applied Aerodynamics Conference, AIAA-3030, New Orleans, LA, USA, 25–28 June 2012.
10. Genç, M.S.; Lock, G.; Kaynak, Ü. An experimental and computational study of low Re number transitional flows over an aerofoil with leading edge slat. In Proceedings of the 26th Congress of ICAS and 8th AIAA ATIO, AIAA-8877, Anchorage, AK, USA, 14–19 September 2008.
11. Koca, K.; Genç, M.S.; Açıkel, H.H.; Çağdaş, M.; Bodur, T.M. Identification of flow phenomena over NACA 4412 wind turbine airfoil at low Reynolds numbers and role of laminar separation bubble on flow evolution. *Energy* **2018**, *144*, 750–764. [CrossRef]
12. Koca, K.; Genç, M.S.; Özkan, R. Mapping of laminar separation bubble and bubble-induced vibrations over a turbine blade at low Reynolds numbers. *Ocean. Eng.* **2021**, *239*, 109867. [CrossRef]
13. Koca, K.; Genç, M.S.; Veerasamy, D.; Özden, M. Experimental flow control investigation over suction surface of turbine blade with local surface passive oscillation. *Ocean. Eng.* **2022**, *266*, 113024. [CrossRef]
14. Das, T.K.; Islam, N.; Samad, A.; Pasha, A.A. Passive flow control via tip grooving and stall fencing mechanisms of a marine energy harvesting turbine. *Sci. Rep.* **2023**, *13*, 2677. [CrossRef] [PubMed]
15. Karasu, İ. Flow control over a diamond-shaped cylinder using slits. *Exp. Therm. Fluid Sci.* **2020**, *112*, 109992. [CrossRef]
16. Qian, Y.; Zhang, Y.; Sun, Y.; Wang, T. Numerical investigations of the flow control effect on a thick wind turbine airfoil using deformable trailing edge flaps. *Energy* **2023**, *265*, 126327. [CrossRef]
17. Genç, M.S.; Kaynak, Ü. Control of flow separation and transition point over an aerofoil at low Re number using simultaneous blowing and suction. In Proceedings of the 19th AIAA Computational Fluid Dynamics, AIAA-3672, San Antonio, TX, USA, 22–25 June 2009.
18. Koca, K.; Genç, M.S.; Bayır, E.; Soğuksu, F.K. Experimental study of the wind turbine airfoil with the local flexibility at different locations for more energy output. *Energy* **2022**, *239*, 121887. [CrossRef]
19. Dong, H.; Xia, T.; Chen, L.; Liu, S.; Cui, Y.D.; Khoo, B.C.; Zhao, A. Study on flow separation and transition of the airfoil in low Reynolds number. *Phys. Fluids* **2019**, *31*, 103601.
20. Tumse, S.; Karasu, I.; Sahin, B. Experimental investigation of ground effect on the vortical flow structure of a 40° swept Delta Wing. *J. Aerosp. Eng.* **2022**, *35*, 04022055. [CrossRef]
21. Özden, K.S.; Karasu, İ.; Genç, M.S. Experimental investigation of the ground effect on a wing without/with trailing edge flap. *Fluid Dyn. Res.* **2020**, *52*, 045504. [CrossRef]
22. Demir, H.; Genç, M.S. An experimental investigation of laminar separation bubble formation on flexible membrane wing. *Eur. J. Mech. B/Fluids* **2017**, *65*, 326–338. [CrossRef]
23. Genç, M.S.; Demir, H.; Özden, M.; Bodur, T.M. Experimental analysis of fluid-structure interaction in flexible wings at low Reynolds number flows. *Aircr. Eng. Aerosp. Technol.* **2021**, *93*, 1060–1075. [CrossRef]
24. Genç, M.S.; Koca, K.; Açıkel, H.H. Investigation of pre-stall flow control on wind turbine blade airfoil using roughness element. *Energy* **2019**, *176*, 320–334. [CrossRef]
25. Genç, M.S.; Koca, K.; Demir, H.; Açıkel, H.H. Traditional and New Types of Passive Flow Control Techniques to Pave the Way for High Maneuverability and Low Structural Weight for UAVs and MAVs. In *Autonomous Vehicles*; IntechOpen: London, UK, 2020.

26. Arcondoulis, E.J.G.; Doolan, C.J.; Zander, A.C.; Brooks, L.A. A Review of Trailing Edge Noise Generated by Airfoils at Low to Moderate Reynolds Number. *Acoust. Aust.* **2010**, *38*, 129–133.
27. Sheikhnejad, Y.; Nassab, S.A.G. Enhancement of Solar Chimney Performance by passive vortex generator. *Renew. Energy* **2021**, *169*, 437–450. [CrossRef]
28. Zhu, C.; Wang, T.; Wu, J. Numerical investigation of passive vortex generators on a wind turbine airfoil undergoing pitch oscillations. *Energies* **2019**, *12*, 654. [CrossRef]
29. Meana-Fernández, A.; Fernández Oro, J.M.; Argüelles Díaz, K.M.; Velarde-Suárez, S. Turbulence-model comparison for aerodynamic-performance prediction of a typical vertical-axis wind-turbine airfoil. *Energies* **2019**, *12*, 488. [CrossRef]
30. Ye, X.; Hu, J.; Zheng, N.; Li, C. Numerical study on aerodynamic performance and noise of wind turbine airfoils with serrated gurney flap. *Energy* **2023**, *262*, 125574. [CrossRef]
31. Taylor, H.D. *The Elimination of Diffuser Separation by Vortex Generators*; Technical Report No. R-4012-3; United Aircraft Corporation: Moscow, Russia, 1947.
32. Gao, L.; Liu, Y.; Han, S.; Yan, J. Aerodynamic performance of a blunt trailing-edge airfoil affected by vortex generators and a trailing-edge wedge. In Proceedings of the 3rd Renewable Power Generation Conference (RPG 2014), Naples, Italy, 24–25 September 2014; pp. 1–6.
33. Gao, L.; Zhang, H.; Liu, Y.; Han, S. Effects of vortex generators on a blunt trailing-edge airfoil for wind turbines. *Renew. Energy* **2015**, *76*, 303–311. [CrossRef]
34. Prince, S.A.; Badalamenti, C.; Regas, C. The application of passive air jet vortex-generators to stall suppression on wind turbine blades. *Wind. Energy* **2017**, *20*, 109–123. [CrossRef]
35. Seshagiri, A.; Cooper, E.; Traub, L.W. Effects of vortex generators on an airfoil at low Reynolds numbers. *J. Aircr.* **2009**, *46*, 116–122. [CrossRef]
36. Baldacchino, D.; Ferreira, C.; Tavernier, D.D.; Timmer, W.A.; Van Bussel, G.J.W. Experimental parameter study for passive vortex generators on a 30% thick airfoil. *Wind. Energy* **2018**, *21*, 745–765. [CrossRef]
37. Sutardi, S.; Nurcahya, A.E. Experimental study on the effect of vortex generator on the aerodynamic characteristics of Nasa ls-0417 airfoil. *Appl. Mech. Mater.* **2015**, *758*, 63–69. [CrossRef]
38. Godard, G.; Stanislas, M. Control of a decelerating boundary layer. Part 1: Optimization of passive vortex generators. *Aerosp. Sci. Technol.* **2006**, *10*, 181–191. [CrossRef]
39. Wang, H.; Zhang, B.; Qiu, Q.; Xu, X. Flow control on the NREL S809 wind turbine airfoil using vortex generators. *Energy* **2017**, *118*, 1210–1221. [CrossRef]
40. Zhao, Z.; Zeng, G.; Wang, T.; Xu, B.; Zheng, Y. Numerical research on effect of transition on aerodynamic performance of wind turbine blade with vortex generators. *J. Renew. Sustain. Energy* **2016**, *8*, 063308. [CrossRef]
41. Tian, Q.Q.; Corson, D.; Baker, J.P. Application of vortex generators to wind turbine blades. In Proceedings of the 34th Wind Energy Symposium, San Diego, CA, USA, 4–8 January 2016; p. 518.
42. Manolesos, M.; Voutsinas, S.G. Experimental investigation of the flow past passive vortex generators on an airfoil experiencing three-dimensional separation. *J. Wind. Eng. Ind. Aerodyn.* **2015**, *142*, 130–148. [CrossRef]
43. Forster, K.J.; White, T.R. Numerical investigation into vortex generators on heavily cambered wings. *AIAA J.* **2014**, *52*, 1059–1071. [CrossRef]
44. Fouatih, O.M.; Medale, M.; Imine, O.; Imine, B. Design optimization of the aerodynamic passive flow control on NACA 4415 airfoil using vortex generators. *Eur. J. Mech. -B/Fluids* **2016**, *56*, 82–96. [CrossRef]
45. Zhu, C.; Feng, Y.; Shen, X.; Dang, Z.; Chen, J.; Qiu, Y.; Feng, Y.; Wang, T. Effects of the height and chordwise installation of the vane-type vortex generators on the unsteady aerodynamics of a wind turbine airfoil undergoing dynamic stall. *Energy.* **2023**, *266*, 126418. [CrossRef]
46. Zhu, C.; Qiu, Y.; Feng, Y.; Wang, T.; Li, H. Combined effect of passive vortex generators and leading-edge roughness on dynamic stall of the wind turbine airfoil. *Energy Conv Manag.* **2022**, *251*, 115015. [CrossRef]
47. Genç, M.S.; Koca, K.; Açıkel, H.H.; Özkan, G.; Kırış, M.S.; Yıldız, R. Flow characteristics over NACA4412 airfoil at low Reynolds number. In *EPJ Web Of Conferences*; EDP Sciences: Les Ulis, France, 2016; Volume 114.
48. Koca, K.; Genç, M.S.; Ertürk, S. Impact of local flexible membrane on power efficiency stability at wind turbine blade. *Renew. Energy* **2022**, *197*, 1163–1173. [CrossRef]
49. Wang, P.; Liu, Q.; Li, C.; Miao, W.; Yue, M.; Xu, Z. Investigation of the aerodynamic characteristics of horizontal axis wind turbine using an active flow control method via boundary layer suction. *Renew. Energy* **2022**, *198*, 1032–1048. [CrossRef]
50. Zhu, C.; Chen, J.; Wu, J.; Wang, T. Dynamic stall control of the wind turbine airfoil via single-row and double-row passive vortex generators. *Energy.* **2019**, *189*, 116272. [CrossRef]
51. Chen, T.Y.; Liou, L.R. Blockage corrections in wind tunnel tests of small horizontal-axis wind turbines. *Exp. Therm. Fluid Sci.* **2011**, *35*, 565–569. [CrossRef]
52. Karasu, İ.; Özden, M.; Genç, M.S. Performance assessment of transition models for three-dimensional flow over NACA4412 wings at low Reynolds numbers. *J. Fluids Eng.* **2018**, *140*, 121102. [CrossRef]

53. Qu, H.; Hu, J.; Gao, X. The impact of Reynolds number on two-dimensional aerodynamic airfoil flow. In Proceedings of the 2009 World Non-Grid-Connected Wind Power and Energy Conference, Nanjing, China, 24–26 September 2009; pp. 1–4.
54. Ducoin, A.; Loiseau, J.C.; Robinet, J.C. Numerical investigation of the interaction between laminar to turbulent transition and the wake of an airfoil. *Eur. J. Mech. -B/Fluids* **2016**, *57*, 231–248. [CrossRef]

 *energies*

Review

Forecasting of Solar and Wind Resources for Power Generation

M. K. Islam [1], N. M. S. Hassan [1,*], M. G. Rasul [2], Kianoush Emami [1] and Ashfaque Ahmed Chowdhury [3]

[1] School of Engineering and Technology, Central Queensland University, Abbott Street, Cairns, QLD 4870, Australia; m.islam2@cqu.edu.au (M.K.I.); k.emami@cqu.edu.au (K.E.)
[2] School of Engineering and Technology, Central Queensland University, Yaamba Rd., Rockhampton, QLD 4701, Australia; m.rasul@cqu.edu.au
[3] School of Engineering and Technology, Central Queensland University, Bryan Jordan Dr., Gladstone, QLD 4680, Australia; a.chowdhury@cqu.edu.au
* Correspondence: n.hassan@cqu.edu.au

Abstract: Solar and wind are now the fastest-growing power generation resources, being ecologically benign and economical. Solar and wind forecasts are significantly noteworthy for their accurate evaluation of renewable power generation and, eventually, their ability to provide profit to the power generation industry, power grid system and local customers. The present study has proposed a Prophet-model-based method to predict solar and wind resources in the Doomadgee area of Far North Queensland (FNQ), Australia. A SARIMA modelling approach is also implemented and compared with Prophet. The Prophet model produces comparatively less errors than SARIMA such as a root mean squared error (RMSE) of 0.284 and a mean absolute error (MAE) of 0.394 for solar, as well as a MAE of 0.427 and a RMSE of 0.527 for wind. So, it can be concluded that the Prophet model is efficient in terms of its better prediction and better fitting in comparison to SARIMA. In addition, the present study depicts how the selected region can meet energy demands using their local renewable resources, something that can potentially replace the present dirty and costly diesel power generation of the region.

Keywords: solar; wind; forecasting; prophet; SARIMA

Citation: Islam, M.K.; Hassan, N.M.S.; Rasul, M.G.; Emami, K.; Chowdhury, A.A. Forecasting of Solar and Wind Resources for Power Generation. *Energies* **2023**, *16*, 6247. https://doi.org/10.3390/en16176247

Academic Editor: Davide Astolfi

Received: 27 June 2023
Revised: 14 August 2023
Accepted: 17 August 2023
Published: 28 August 2023

1. Introduction

Australia, including Far North Queensland (FNQ), is seeking the better utilisation of its abundant sustainable power generation sources, especially solar and wind. Clean energy sources are crucial to combat weather change by diminishing greenhouse gas (GHG) emissions. However, FNQ has many remote regions. Doomadgee is one of them, and it depends on mainly diesel generation for its basic energy needs. So, the solar and wind energy harvesting in Doomadgee is promising, having a mean annual daily solar irradiation of 6.12 kWh/m^2 with average annual daily sunshine hours of 7–8 h [1,2] and average daily wind speed of around 6.0 m/s [3]. Any area that receives solar irradiation greater than 4 kWh/m^2/day can have geographical potential for harnessing solar energy [4] and some previous studies considered the minimum mean daily wind velocity to be 4 m/s and 5 m/s for installing wind farms [5,6].

The forecasting of renewable energy has gained substantial focus since the utilisation of renewable energy from solar and wind is continuously increasing. Other forms of renewable energy may involve some forecasting tasks, but most emphasis has been given to solar and wind over the last decade because of their limited predictability. Numerous researchers are devoted to developing models for predicting long-term daily or monthly average solar radiation and wind velocity. Such solar radiation and wind velocity models for any region are very beneficial in determining the power generation of a particular solar and wind power plant. A wide range of topics of interest in renewable energy forecasting regarding solar and wind can be referred to (see references [7–10]). The forecasting of renewable energy can be divided into two groups: energy market participants and power system operators. Market participants are concerned with the buying and selling of energy, while the system operators are concerned with reliable energy supplies. Both need timely and accurate renewable energy forecasting. Renewable energy forecasting helps in preparing the generation mix that the market delivers [8].

Enormous forecasting models have been proposed to detect an effective method that can be implemented in practical circumstances [11]. These methods can diminish the risk of failure within the energy system and predict its reliability by modelling future scenarios. An analog approach was used for the first time to predict the wind speed distribution [12]. A time series model was proposed to forecast wind power up to several hours ahead of time [13]. A Kalman filter was implemented to forecast short-term wind speed [14]. In earlier times, numerous authors implemented statistical approaches to simulate the historic data features of time series. Financial stock markets can be simulated using ARIMA (Auto-Regression Integrated Moving Average) and AR-ARCH (Auto-Regression—Auto-Regression conditional heteroskedasticity) techniques to project rates of returns [15,16]. The fractional-ARIMA can be implemented to predict a few days of wind speed features [17], which showed accuracy over the persistence model in a case study of a targeted 750 kW wind turbine. An ARIMA model was utilised to forecast global solar irradiance. A novel statistical model, a combination of ARIMA and repeated wavelet transform, developed by Singh and Mohapatra [18], out-performed individual ARIMA forecasting. ARIMA, with an extreme learning model, was implemented by Wang et. al. [19], and was proven to be accurate for wind projection through some case studies. Cadenas and Rivera [20] developed a hybrid model incorporating an ANN (Artificial Neural Network) and ARIMA, which proved to have higher accuracy over individual ones. A more sophisticated model, such as a machine learning technique incorporating ARIMA, can provide more precision and consistency in forecasting wind speed over the individual techniques [21]. Shukur and Lee [22] implemented a novel ARIMA-based model to increase precision and manage uncertainty in wind speed prediction.

All the above-mentioned ARIMA-based models, however, can only work well on stationary time series. Time series, such as solar radiation and wind speed, have the characteristics of seasonality and trend. These features can be handled by Seasonal ARIMA (SARIMA). A SARIMA model implemented by Fang and Ladhelma [23] showed high accuracy in predicting the thermal energy requirement for district heating systems. Another model, named SARIMA-RVFL (Random Vector Functional Link), showed relatively high accuracy in predicting short-term solar photovoltaic (PV) generation [24]. Guo et al. [25] implemented the SARIMA model for monthly wind velocity forecasting, which showed improved accuracy over the ARIMA-based model.

An artificial neural network (ANN) was utilised in various studies [11,26–28]. ANN has the capability to solve a great range of nonlinear equation sets for predicting any future scenario [29]. The time series statistical method and ANN were widely applied in solar and wind forecasting [30]. Benghanem et al. [31] applied ANN techniques to predict solar irradiation, which provided better results than the conventional empirical regression models. In a study [32], three types of ANN, namely feed-forward propagation (FFBP), adaptive linear element (ADALINE) and radial basis function (RBFNN), were implemented. The results showed that different modelling structures and parameters can produce different forecast accuracies. The most broadly used feed-forward neural network (FFNN) can perform wind power prediction with reasonably good accuracy [33]. A unique genetic NN (GNN) was utilised in predicting wind velocity [34]. Instead of back propagation, genetic algorithm was used to determine the weights and biases of the NN layers. The results revealed that GNN can perform better compared to back propagation. Utilising particle swarm optimisation (PSO) during the training process of ANN can produce better results compared to the method involving a training process without PSO [35,36]. A study to predict the solar irradiance one day ahead for a grid-connected solar photovoltaic plant using an ANN model produced a MAE and MBE of

3.21% and 8.54%, respectively [37]. Support vector machines (SVMs) can also model non-linear characteristics in data like ANN techniques. SVMs can demonstrate better prediction performance in terms of multi-layer perception (MLP) NNs [38]. Wavelet networks, a combination of wavelet theory and NN, were also applied in solar irradiance prediction. In a study, a wavelet network modelling technique was developed to predict solar radiation and the model showed reasonably good performance compared to other NN techniques [39]. ANNs and SVMs have the ability to model complex non-linear trends.

However, a study [40] of solar forecasting showed that a statistical method, such as the ARIMA model, compared to the ANN, could perform better because of weather conditions, such as clouds. A study [41] for predicting wind velocity in Mexico utilised both SARIMA and ADALINE NN. The outcomes revealed that the SARIMA model can predict with lower statistical errors than that of ADALINE. Solar irradiation concentration and wind resources partially depend on different weather, location and time factors. So, they present a type of serial correlation, which indicates that time series forecasting is suitable for predicting solar irradiation or wind speed [11]. On the other hand, neural networks have some limitations, such as unknown working principles, which makes them not suitable for the forecasting of solar or wind.

Physical prediction approaches, such as numerical weather projection (NWP) and satellite-based, can be implemented in the renewable sector. NWP can be designed by modelling the relationships among physical variables such as temperature, wind velocity, dew point, humidity, pressure, surface roughness, etc. [42,43]. NWP has some sub models, each of which can solve complex equations [44]. In a study [45], the numerical simulation prediction approach, coupled with the 3D Price-Weller-Pinkel—Ocean model and wind farm parameterisation model, was implemented to analyse and project wake and wind power generation interactions in an offshore wind farm in the Jiangsu Province of China under operational and future development scenarios. Wind turbines interact with the boundary layers. An increment in the number of wind turbines or turbine capacity withing a certain distance may induce a larger wake, resulting in increased turbulence intensity and reduced wind speed, ultimately lowering power generation as well as the revenues of the farm [45]. A study by Wang et al. [46] on the performance of wind farms over complex terrains using a numerical simulation prediction approach showed that the inter-farm wake may persist up to 20 km downstream in a stable layer and wind speed can decrease up to 8% at night, which causes a reduction in power generation by 17%. However, an unstable layer with increased turbulence may result in less deficit and wake may recover within 6km during the daytime when power generation loss is comparatively lower than that of nighttime. In addition, a powerful wake with a streaky structure may form at the downstream adjacent wind farm and result in a greater range of power generation loss. Another study by Wang at al [47] using a numerical prediction approach has proved that the wake induced by an upstream wind farm at the downstream farm is doubled and power generation is experienced with great losses at night, at dawn and in the evening. A study using a meso-scale weather simulation and numerical projection approach showed that there might be a deficit of maximal wind, at the mountainous and isolated areas, of 0.5 m/s and 2.5 m/s, respectively. The farm wake could recover within around 20–35 km at the isolated site and around 6 km at the mountainous site. The performance of wind turbines utilising the terrain acceleration could surpass the turbines sited in the lowland places of the mountain [48]. Although meteorological organisations utilise NWP broadly, it is, however, not suitable for direct application in solar or wind prediction [49] due to its extreme reliance on fluctuating and sensitive air quality and hydrological characteristics [50]. Physical techniques and inherent learning algorithms have large constraints in that they are not flexible enough to capture time series wind data. Although LSTM provides good forecasting accuracy, the processing sequence cannot be longer and the training dataset is very complex [51].

Prophet, an additive modelling approach, has recently drawn attention as a forecasting method. Initially, this model was developed to project 'at scale', bringing together configurable models. This approach is already being used in different fields and the outcome proves that it outperforms other statistical and machine learning approaches [52]. The Prophet model is very convenient for capturing seasonality in data [53]. A study on air pollution using the Prophet technique showed good outcomes with minimum error metrics [54]. Another study on solar irradiation showed that ANN, CNN and LSTM perform better for short term solar prediction, while SARIMAX and Prophet could provide better accuracy for long term solar prediction [55]. The Prophet model can model future hydrogen production through predicting solar energy and can produce better results than stochastic gradient descent and SARIMAX [56]. The Prophet model, with non-linear tree-based ensembles, can provide better prediction for short-term solar and wind energy production compared to baseline persistence models and tree-based regression ensembles [57]. A study on air temperature showed that long short-term memory (LSTM) performs better at minimum temperatures and that the Prophet model performs better at maximum temperatures [58].

Accurate forecasting for solar radiation and wind speeds is a challenging task because of stochastic and uncertain features. However, solar radiation and wind velocity time series may maintain a seasonally periodic distribution. In this study, the SARIMA and Prophet models are implemented to project solar radiation and wind velocity in the Doomadgee region of Far North Queensland, Australia. Both the SARIMA and Prophet models have advantages, the prior figures of which will persist in the future. As solar and wind are the prime of a hybrid generation system, unfailing projections of these resources are crucial to predict the system's performance. Proper projection can help in understanding that local renewable resources are enough or not enough to meet the energy demand of the selected area

From the literature, it is notable that there are ample studies which utilise different modelling approaches to project solar and wind resources. The Prophet model is implemented in a range of topics, including short- and long-term solar radiation, as well as short-term wind velocity, air pollution and energy consumption. However, the Prophet model has never been utilised for predicting long-term wind velocity, including wind power generation, in any study. The SARIMA modelling technique is already broadly used in predicting solar and wind resources. However, SARIMA has never been used for weather prediction in the Doomadgee region. This article will bridge the above knowledge gaps by exploring solar radiation and wind speed projection using the Prophet model and SARIMA technique. The main motivation of the present study is to employ the Prophet model and SARIMA technique for projecting long-term solar radiation and wind speeds, as well as to conduct a performance comparison of Prophet with SARIMA.

The key contributions of present research work can be summerised as:

- Proposing a Prophet-model-based forecasting approach for long-term wind velocity and wind power generation, which is a first in the literature.
- Predicting future solar and wind power generation using the Prophet modelling approach, something that can potentially replace the present diesel power generation of the selected region for meeting energy demand.
- Introducing a seasonal term to project the monthly average daily wind velocity and solar irradiation of the Doomadgee region of FNQ.
- Evaluating the forecasting possibilities of solar radiation and wind speeds, the sample comprises two types of data—one set is for training (from January 1990 to December 2018) and another set is for testing or validating (from January 2019 to December 2022).
- Developing the Prophet and SARIMA models in order to project future solar radiation and wind velocity for any year and compare between them.
- Predicting future solar and wind power generation for any year.

The present document continues as follows. Section 2 illustrates the basics and methods concerned with the suggested models, including in terms of data collection. The Prophet and SARIMA projected model building and validation, as well as a comparison of both models, is presented in Section 3. In this section, an analysis and discussion of the predicted findings are also presented. Section 4 describes the implementation of the predicted results in power generation, followed by a discussion on the results in Section 5. The conclusion is presented in Section 6.

2. Materials and Methods

A time series is a succession of any information collected across a time range. If, in a time series, a variable is random, then it is termed a stochastic time series. In this study, solar radiation and wind velocity data sets have been forecasted using SARIMA and Prophet modelling, which is realised in a number of stages characterised by the adopted procedure. The methodology involves visualising solar radiation and wind speed time series, stationary testing, decomposition of time series, developing a correlogram, estimating and fine-tuning hyperparameters, and residual analysis. The employed method of this study is summarised in Figure 1:

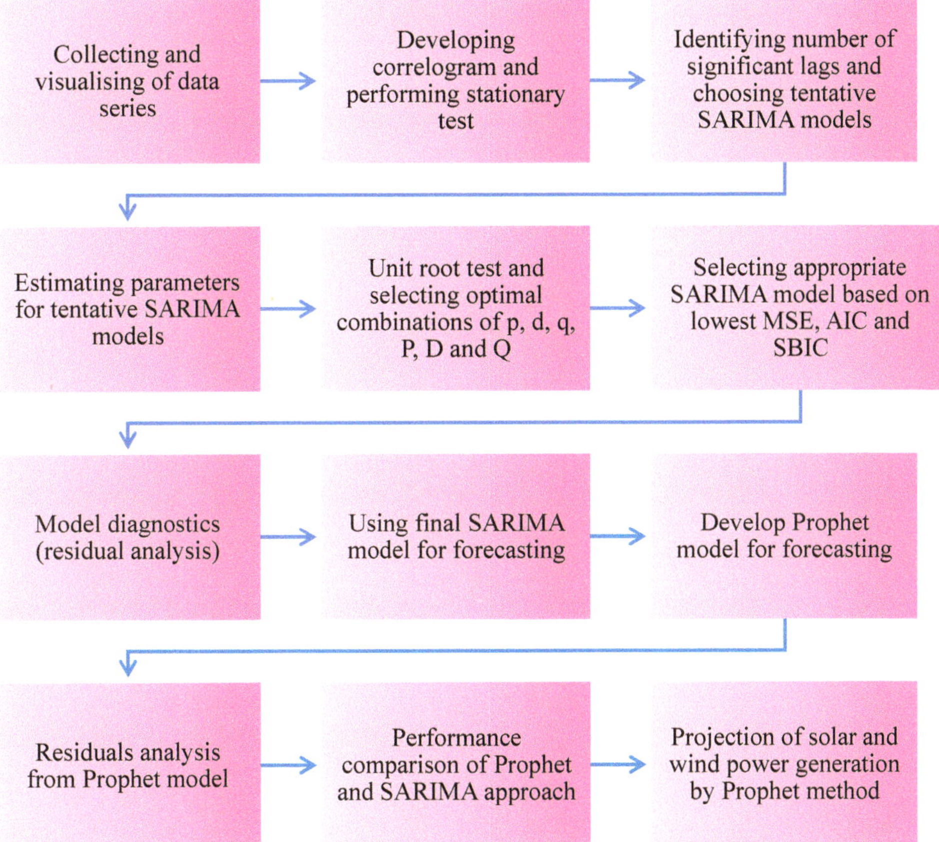

Figure 1. Proposed methodology for implementing Prophet and SARIMA modelling technique.

2.1. Data Collection

The monthly solar radiation data in Doomadgee, Far North Queensland for over 29 years (from January 1990 to December 2018) are utilised in the modelling. The data are captured and recorded by the observation station of the Bureau of Meteorology, situated in Doomadgee, and can easily be accessed online. Data on mean monthly wind speeds for 29 years (from 1990 to 2018) are collected from the NASA database. The records are first executed for any missing data and outliers. The missing data have been substituted by the resembling mean of the same months from the rest of the years. The total quantity of missing data on solar radiation is just 0.29% of the total data points and there are no missing data for wind velocity. Data have been analysed using SPSS Statistics 29.0.0.0 software.

2.2. Series' Stationarity or Non-Stationarity

A time series can possess stationarity when its elements are not affected by the changes in time, that is, if the joint probability distribution n samples of data Ys1, Ys2, ..., Ysn, collected at times s1, s2, ..., sn, is identical to that associated with n samples Ys1 + z, Ys2 + z, ..., Ysn + z, collected at times s1, s2, ..., sn. If the data are affected by shifting the time, then the series is non-stationary.

Testing Series for Stationarity

Unit root test is an important option for checking stationarity of a series. A time series may possess a non-stationary variable, including a unit root. It is important to define the null and alternative hypotheses appropriately in order to determine the stationarity during testing for unit roots. Several techniques, such as the Phillips–Perron (PP) test, Augmented Dickey–Fuller (ADF) test and Kwiatkowski–Phillips–Schmidt–Shin (KPSS) test, are usually used for unit root testing. According to the ADF and PP tests, the hypotheses are:

Hypothesis 1: $\phi1$ = *The series possesses unit root.*

Hypothesis 2: $\phi1$ = *The series does not possess unit root.*

According to KPSS test, the hypotheses are as below:

Hypothesis 1: $\phi1$ = *The series possesses stationarity around a deterministic trend.*

Hypothesis 2: $\phi1$ = *The series does not possess stationarity around a trend.*

According to the ADF and PP tests, the series possesses stationarity in the case of having a p-value less than the significance level (0.05) and, according to the KPSS test, the series has stationarity in the case of the p-value being higher than 0.05.

2.3. SARIMA Modelling

SARIMA, which is an ARIMA (Auto-Regression Integrated Moving Average) variant, can support data with trends and seasonality. To predict the upcoming series points, ARIMA or SARIMA models can be fitted to time series data. In the case of the ARIMA approach, initial differentiation can be implemented to eliminate non-stationarity from a time series, if the series is not stationary. Seasonal differencing can be employed to remove seasonality, if seasonality presents. SARIMA is expressed as ARIMA (p, d, q) (P, D, Q)s; where p, d, q, P, D, Q, refer to the size of autoregressive window, size of differentiation, dimension of moving average window, size of seasonal autoregressive window, size of seasonal differentiation and size of seasonal moving average window, respectively, with 's' referring to the size of seasonal cycles.

In brief, the key parts in the SARIMA modelling are:

— Auto-Regression (AR) deals with the reliance of a viewed data on its prior lagged viewed data. AR model can be stated as follows [25]:

$$Z_s = \alpha_1 Z_{s-1} + \alpha_2 Z_{s-2} + \cdots + \alpha_p Z_{s-p} + \varepsilon_s = \left(\sum_{i=1}^{p} \alpha_i L^i \right) Z_s + \varepsilon_s \tag{1}$$

where $s = p + 1; \ldots; n$; Zs refers to the records at time s; α refers to the coefficient of lags, ε_s is the error part; L refers to the backward shift operator; and i stands for the number of terms considered from the historical data.

— Moving Average (MA) deals with the past errors; a direct grouping of residual errors happened concurrently and at earlier different times. In the case of a separate MA model, Zs relies on the lagged predicted errors, which have been presented as follows [25]:

$$Z_s = \varepsilon_s - \theta_1 \varepsilon_{s-1} - \theta_2 \varepsilon_{s-2} - \cdots - \theta_q \varepsilon_{s-q} = \left(1 - \sum_{i=1}^{q} \theta_i L^i\right) \varepsilon_s \tag{2}$$

where $s = q + 1; \ldots; n; q$ denotes the constant factors estimating by the model; L refers to the backward shift operator; and ε_s refers to the error terms at time s.

— I stands for integrated or trend, which refers to the fact that the raw observations are to be replaced by differentiating their values from previous values.

A combination of AR and MA models including I can be presented as follows [25]:

$$\left(1 - \sum_{i=1}^{p} \alpha_i L^i\right)(1 - L)^d X_s = \left(1 - \sum_{i=1}^{q} \theta_i L^i\right) \varepsilon_s + a \tag{3}$$

If $V = 1 - L$ refers to the trend term, then Equation (3) is simplified as:

$$\alpha_p(L)\nabla^d Z_s = \theta_q(L)\varepsilon_s + a \tag{4}$$

where a denotes the intercept term. Equation (4) can be extended by including the seasonal terms [25]:

$$\alpha_p(L)B_p(L^s)\nabla^d\nabla^D_m X_s = \theta_q(L)\varnothing_Q(L^s)\varepsilon_s + a \tag{5}$$

where α denotes the non-seasonal AR part; B denotes the seasonal AR part; θ denotes the non-seasonal MA part; $\varnothing$ refers to the seasonal MA part; and L refers to the backward shift operator.

An SPSS software tool is utilised for preparing the data and computing the autocorrelation function (ACF) and partial autocorrelation function (PACF). ACF and PACF are necessary in order to find the optimal orders p, q, P and Q for the autoregressive and moving average operators.

2.3.1. ACF

The first-order autocorrelation coefficient is the simple coefficient of the first $N - 1$ observations, $s = 1, 2, \ldots, N - 1$ and the next $N - 1$ observations, $Xs, s = 2, 3, \ldots, N$. The correlation between Xs and $Xs + 1$ is given by [25]:

$$r_1 = \frac{\sum_{t=1}^{N-1}\left(X_s - \overline{X_1}\right)\left(X_{s+1} - \overline{X_2}\right)}{\left[\sum_{s=1}^{N-1}\left(X_s - \overline{X_1}\right)^2\right]\left[\sum_{s=1}^{N-1}\left(X_s - \overline{X_1}\right)^2\right]} \tag{6}$$

where X_1 denotes the mean of the first $N - 1$ observation and X_2 denotes the mean of the last $N - 1$ observation. The correlation coefficient presented above, for measuring the correlation between successive observations, is called the autocorrelation coefficient or serial correlation coefficient. When N is reasonably large, the difference between the sub-period means X_1 and X_2 can be ignored and r_1 can be approximated as follows [25]:

$$r_1 = \frac{\sum_{s=1}^{N-1}\left(X_s - \overline{X}\right)\left(X_{s+1} - \overline{X}\right)}{\left[\sum_{s=1}^{N}\left(X_s - \overline{X}\right)^2\right]} \tag{7}$$

Equation (7) can be generalised to give the correlation between observations separated by k years:

$$r_k = \frac{\sum_{s=1}^{N-k}\left(X_s - \overline{X}\right)\left(X_{s+k} - \overline{X}\right)}{\left[\sum_{s=1}^{N}\left(X_s - \overline{X}\right)^2\right]} \tag{8}$$

The quantity is called the autocorrelation coefficient at lag k.

2.3.2. PACF

PACF computes the degree of association between Ys and $Ys + k$ when the effect of other time lags on Y are held constant. The PACF denoted by the set of partial autocorrelations at various lags k are defined by (r_{kk}, k = 1, 2, 3 ...). The set of partial autocorrelations at various lags k are defined by [25]:

$$r_{kk} = \frac{r_k - \sum_{j=1}^{k-1} r_{k-1,j} r_{k-1}}{1 - \sum_{j=1}^{k-1} r_{k-1,j} r_j} \tag{9}$$

where $rk,j = rk - 1, j - rkkrk - 1, k - 1$ and $j = 1, 2, \ldots \ldots \ldots k - 1$.

In the present work, the standard methodology of Box–Jenkins theorem is employed in conducting the SARIMA modelling, which applies a repetitive diagnosis to derive optimised hyperparameters and involves three core phases:

- Model identification: identifying seasonal cycles, trends (weekly, monthly or yearly; linear or quadratic) and autocorrelation patterns in the wind and solar time series (see Sections 3.1 and 3.2);
- Evaluation of model parameters: evaluation of model parameters refers to the estimation of parameters using an auto-regressive model or moving average model (see Section 3.3);
- Diagnostics of model: diagnosing a model is how well the predicted model fits the data, which can be performed through residual analysis (see Section 3.4).

2.4. Prophet Modelling Approach

The Prophet modelling approach is an additive forecasting approach based on the Bayesian curve fitting procedure. This technique can incorporate any types of non-linear trend and multiple seasonality such as hourly, daily, weekly, monthly and yearly with holiday effects [59]. The model has four main components such as trend, seasonality, holiday and error term, as seen in equation [59]:

$$y(t) = g(t) + s(t) + h(t) + e(t) \tag{10}$$

where $y(t)$, $g(t)$, $s(t)$, $h(t)$ and $e(t)$ represent outcome, trend, seasonality, holiday function and error term, respectively. Since solar irradiation and wind velocity are not affected by any holidays, the $h(t)$ function is not used.

A piecewise linear growth model can be expressed as follows [59]:

$$g(t) = \left(k + a(t)^T \partial\right) t + \left(m + a(t)^T \gamma\right) \tag{11}$$

where k, t, ∂, m, γ represent growth rate, time steps, adjustment rate, offset parameters and trend changes points, respectively. Trend changes points can be set to $-S_j \partial_j$ with $a(t)$ expressed in Equation (12) [59]:

$$a_j(t) = \begin{cases} 1, & t \geq S_j, \\ 0, & otherwise \end{cases} \tag{12}$$

The growth model can alter trends after a certain number of time steps using trend change points in order to improve model flexibility. Trend change points are either ascertained by the model automatically or as described explicitly by the modeller.

In this study, the solar radiation and wind velocity data for January 1990 to December 2018 are trained using the Prophet model. The modelling approach is developed in the Python environment using numpy, pandas and statsmodels library. After being trained using the Prophet method, prediction is conducted up to 2030 and compared with the outcome of the SARIMA technique.

3. Results

3.1. Shape of the Collected Average Solar Irradiation and Wind Velocity

The monthly mean solar irradiation and mean wind velocity from January 1990 to December 2018 are presented in Figure 2. From Figure 2, it is noted that both the series shows sharp rises and falls in solar radiation and wind velocity, and that there is no particular trend showing the average solar radiation and mean wind velocity fluctuating around a horizontal line. So, both series can be counted as stationary. However, the highest mean radiation and mean wind speed were recorded in the year 1990 (7.9 kWh/m^2) and 2014 (7.53 m/s), respectively, and the least recorded, respectively, in 2000 (4.5 kWh/m^2) and 2008 (3.98 m/s). Both series further split into trends—seasonal and irregular elements—which are presented in Figures 3 and 4. It is noticeable that the trends, seasonal variants and irregular parts, are steady over time and, hence, both the series seem stationary.

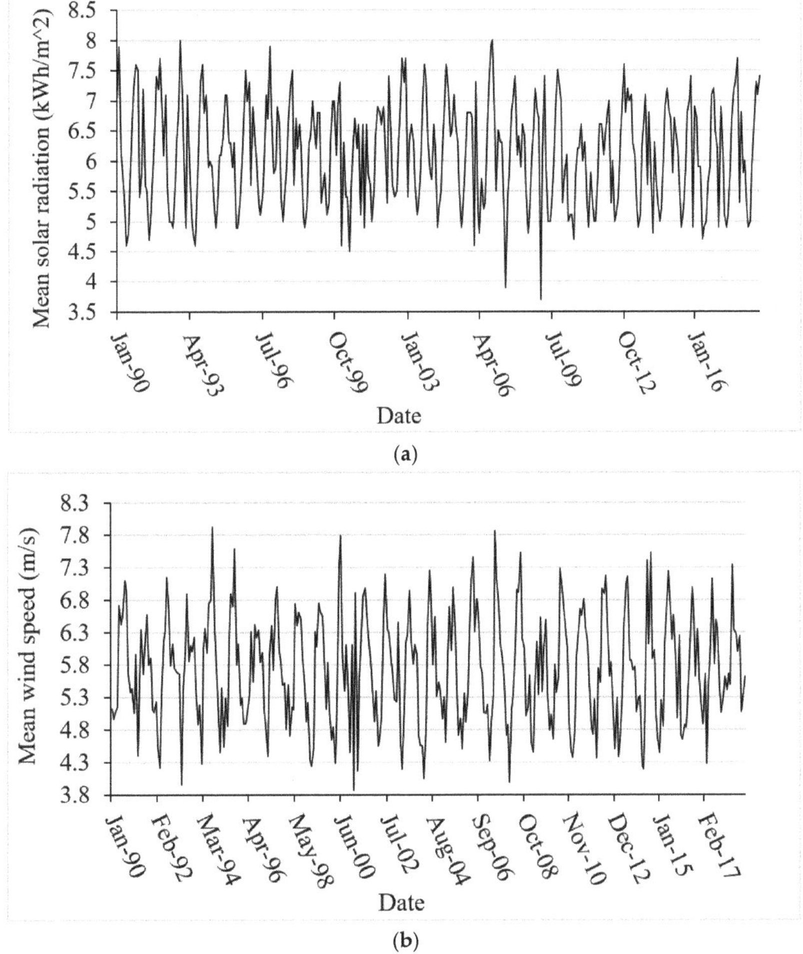

Figure 2. Time series plot of mean solar radiation (**a**) and mean wind velocity (**b**) recorded from January 1990 to December 2018.

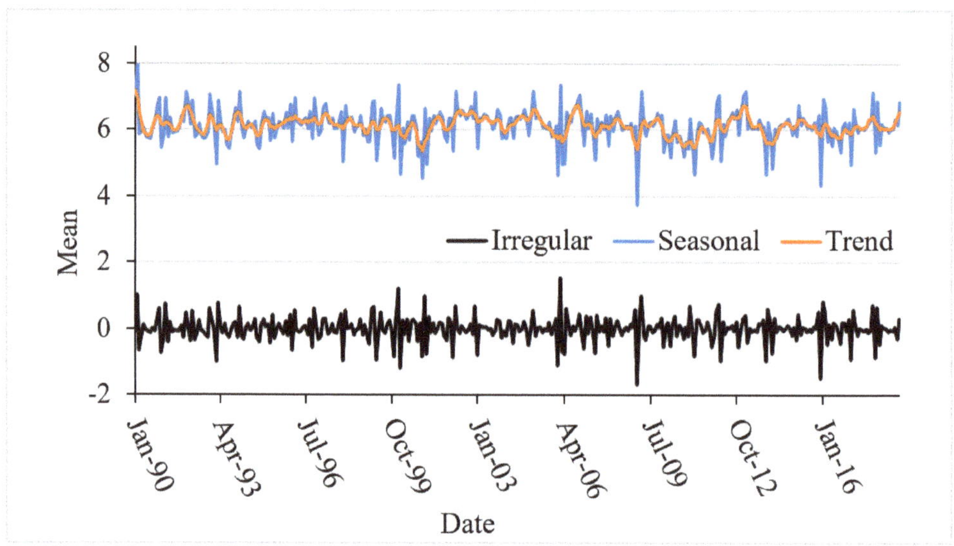

Figure 3. Plot of seasonal, trend and irregular parts of mean solar radiation.

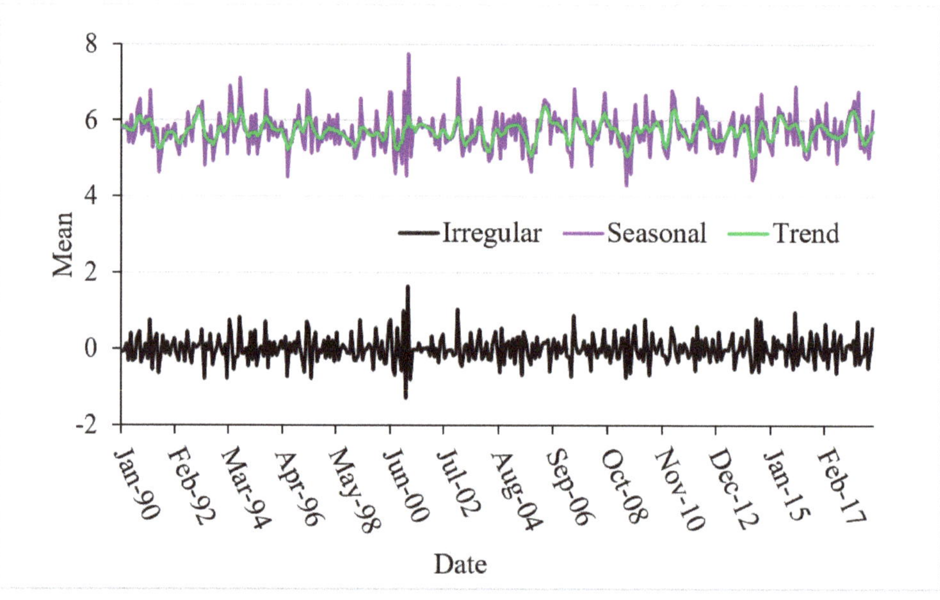

Figure 4. Plot of seasonal, trend and irregular parts of mean wind velocity.

3.2. Correlations

The model development procedure starts with the plotting of the recorded information with time, ACF and PACF, as well as the testing of raw data, in order to identify the series as being stationary or non-stationary. ACF and PACF plots are also helpful for investigating stationarity. Figures 5 and 6 depict the sample ACF and PACF plots for respective mean solar radiation and mean wind speed series with 95% confidence interval. Most of the spikes of the correlogram, in both the ACF and PACF plots of either the mean solar radiation or wind speed series, have been found to be outside of the confidence level. The ACF for both series presents a seasonal repetition of the correlations, thereby exhibiting sinusoidal ripple and changes. Furthermore, the slow decay of the spikes in the ACF and PACF plots for both series reveals that both the series have no trend and are therefore stationary.

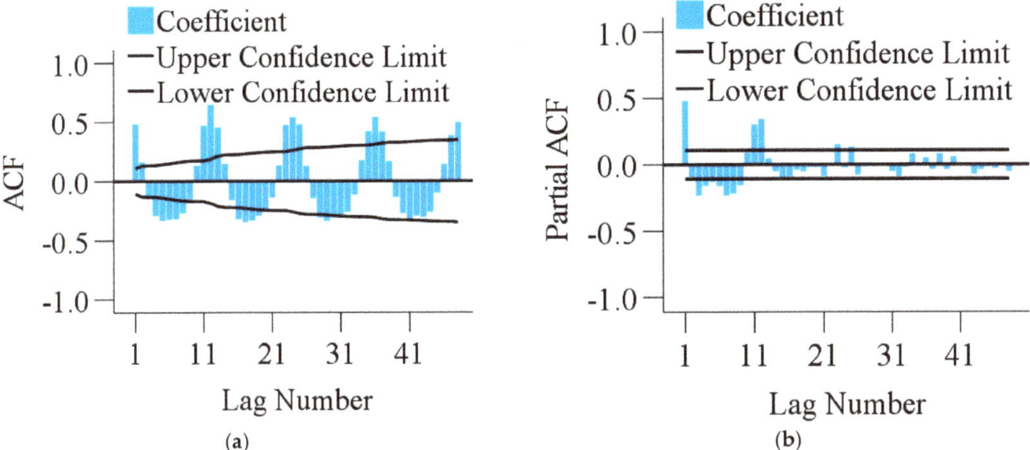

Figure 5. ACF (**a**) and PACF (**b**) plot of the series for mean solar irradiation.

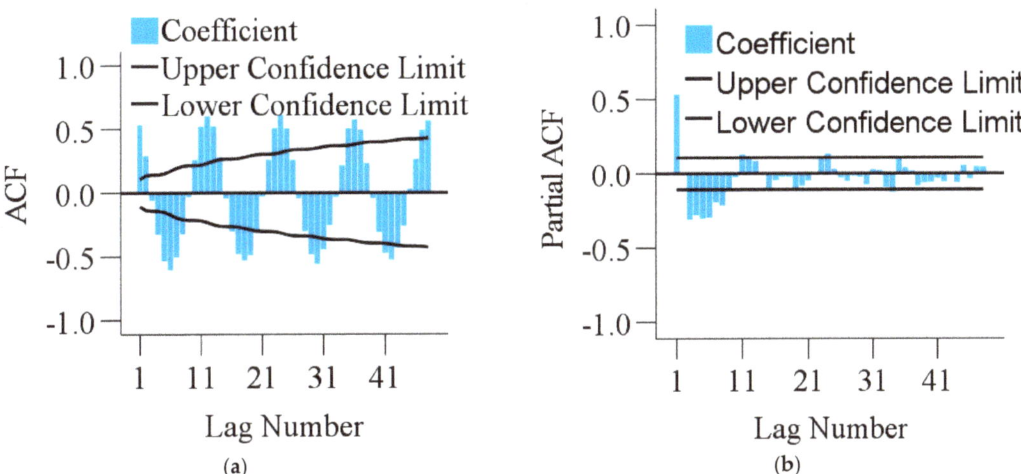

Figure 6. ACF (**a**) and PACF (**b**) plot of the series for mean wind speed.

Furthermore, ADF, PP and KPSS tests have been carried out for both series. *p*-values of 0.01 were found for the ADF and PP tests in both series, as well as and a *p*-value of 0.1 was found for the KPSS test, which is proof of the series' stationarity.

3.3. Parameter Estimation

This step is to differentiate the series by removing seasonal components including the estimation of non-seasonal and seasonal orders. Figure 7 represents the ACF and PACF plot of one season's differencing for a solar series. From the plot of ACF, as seen in Figure 7, the spikes touch the confidence limit at lags 2 and 3, which suggests a non-seasonal MA order of 3. Seasonal MA terms result from lags which are multiples of 12. A seasonal MA order of two can be considered due to two significant spikes at lags 12 and 24. Likewise, according to the PACF plot, the non-seasonal AR order is three because of the significant spikes at lags 2 and 3, as well as the seasonal AR order being two due to the spikes at lags 12 and 24. Therefore, the suggested model for mean solar radiation series can be (3,0,3) (2,1,2) (12).

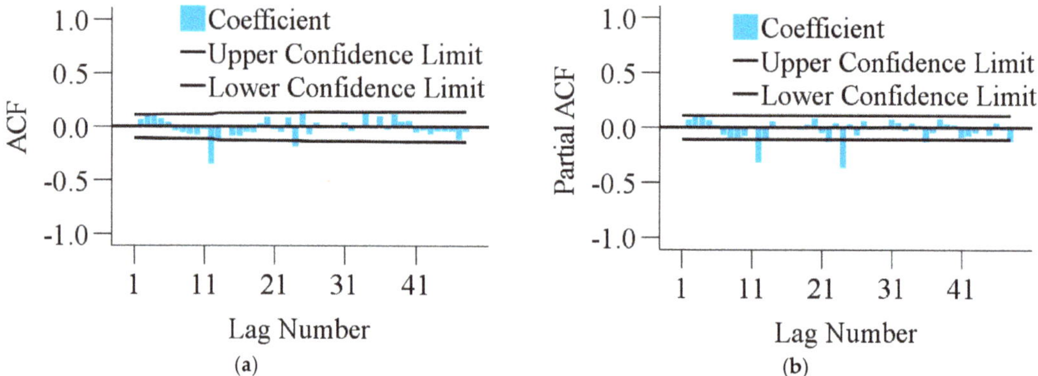

Figure 7. ACF (**a**) and PACF (**b**) plots of one season's differencing (mean solar irradiation).

Figure 8 depicts the ACF and PACF plot for one season's differencing of wind series. No spikes touch or cross the confidence level at low lags, which suggests non-seasonal AR and MA orders of 0. A seasonal MA order of one can be considered due to one significant spike at lag 12 in the ACF plot. Similarly, significant spikes at lags 12, 24 and 36 in the PACF plot suggest a seasonal AR order of three. Therefore, the suggested model for mean wind speed series can be (0,0,0) (3,1,1) (12).

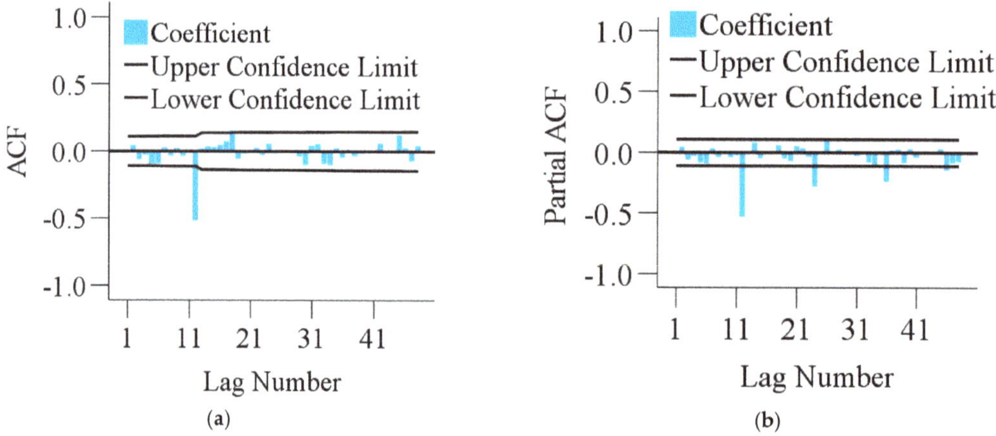

Figure 8. ACF (**a**) and PACF (**b**) of one season's differencing (mean wind speed).

Now, for both solar and wind series, different models including different combinations of non-seasonal or seasonal AR and MA orders have been executed in order to check the unit root in the AR and MA terms (the sum of non-seasonal or seasonal AR coefficients and the sum of MA coefficients cannot be almost equal to one). In addition, the p-values for AR and MA terminology should be significant (<0.05). So, those models which meet the criteria that all the AR and MA terminology are significantly different from zero, for which there is no unit root in the non-seasonal and seasonal AR or MA parts and for which the p-values for all AR or MA terms are significant (<0.05), have been included. The procedure to choose a best model relies on the minimum AIC (Akaike information criterion), SBC

(Bayesian information criterion) and MSE (Mean squared error). Tables 1 and 2 depict the proposed SARIMA models for the respective solar and wind series, including their corresponding AIC, SBC and MSE. SARIMA (2,0,2) (1,1,1) (12) with no constant for solar and (1,0,1) (0,1,1) (12) excluding a constant for wind have been found with the smallest AIC, SBC and MSE, indicating that they are the finest models for projecting the mean solar radiation and wind speed.

Table 1. Suggested SARIMA models for mean solar radiation.

Model	RMSE	AIC	SBC
(2,0,2) (1,1,0) (12)	0.636	652.026	674.911
(2,0,2) (1,1,1) (12)	0.582	597.752	624.451
(2,0,2) (1,1,1) (12) no constant	0.543	557.605	580.508
(2,0,2) (0,1,1) (12)	0.589	602.316	625.201
(1,0,2) (1,1,0) (12)	0.653	666.093	685.163
(2,0,1) (1,1,0) (12)	0.648	666.721	685.792

Table 2. Suggested SARIMA models for mean wind speed.

Model	RMSE	AIC	SBC
(1,0,1) (0,1,1) (12)	0.577	549.721	564.989
(1,0,1) (0,1,1) (12) no constant	0.543	547.550	559.001
(0,0,0) (0,1,2) (12)	0.592	553.789	565.241
(3,0,1) (1,2,2) (12)	0.861	726.145	756.391

3.4. Model Diagnostics

Residual analysis is the major tool of model diagnostics, which involves checking how well a model fits or how well and adequately a model can capture information in the data. A model is counted as an adequate model if the residuals have zero mean and are not correlated. If correlations are present among the residuals, then the information left in the residuals needs to be utilised in the computing forecast. If the residuals' mean is other than zero, then the forecasting will be biased. Furthermore, having constant variance and normally distributed residuals is very useful in forecasting.

Figures 9 and 10 depict the plots of ACF and PACF for respective models (2,0,2) (1,1,1) (12) no constant and (1,0,1) (0,1,1) (12) no constant, which shows that all residuals are within the 95% confidence limits with no significant spikes. So, the residuals are uncorrelated, and they have the property of white noise. The time plot of residuals in Figures 11 and 12 shows that the change in residuals remains almost identical throughout the historic data and, hence, the variance of residuals is considered constant. Furthermore, the p-value of 0.903 and 0.376 (greater than 0.05) for respective solar and wind series models from the Lung-Box Q statistics indicates that the residuals have the property of white noise and both models are well specified and sufficient.

Hyperparameters significantly influence the projections, especially in the time series models. Table 3 presents the optimised parameters used for Prophet modelling. These optimal parameters were found using trial-and-error to find the best fitted model. Figures 13 and 14 show a comparison of the fitted values using the SARIMA and Prophet models for solar radiation and wind velocity, respectively. It can be seen that both the models project the randomness of data reasonably and provide almost the same trend as the real data. Both models follow the pattern of the real data.

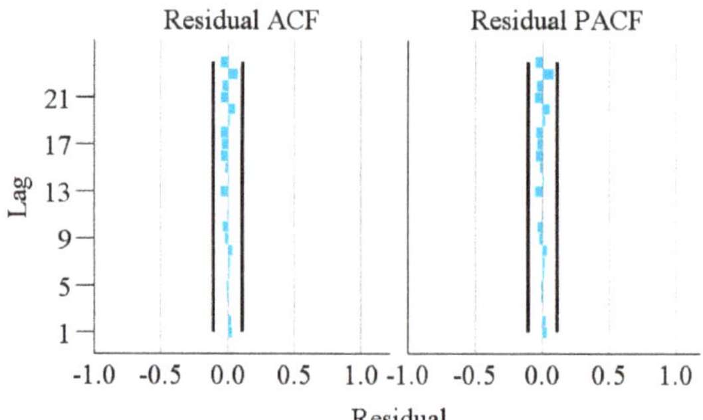

Figure 9. Residual autocorrelation and partial autocorrelations for mean solar radiation: SARIMA (2,0,2) (1,1,1) (12) no constant.

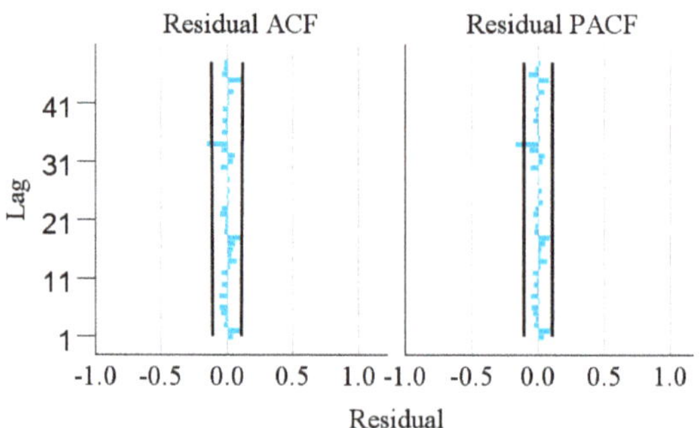

Figure 10. Residual autocorrelation and partial autocorrelations for mean wind speed: SARIMA (1,0,1) (0,1,1) (12) no constant.

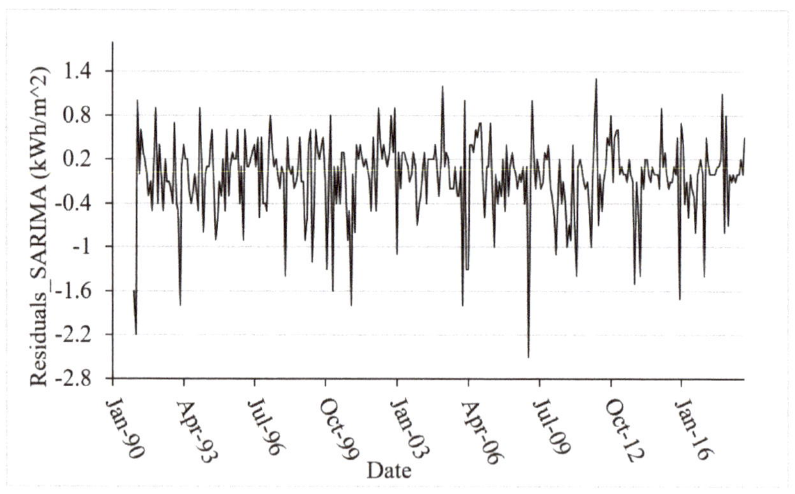

Figure 11. Time plot of residuals of model SARIMA (2,0,2) (1,1,1) (12) no constant for solar.

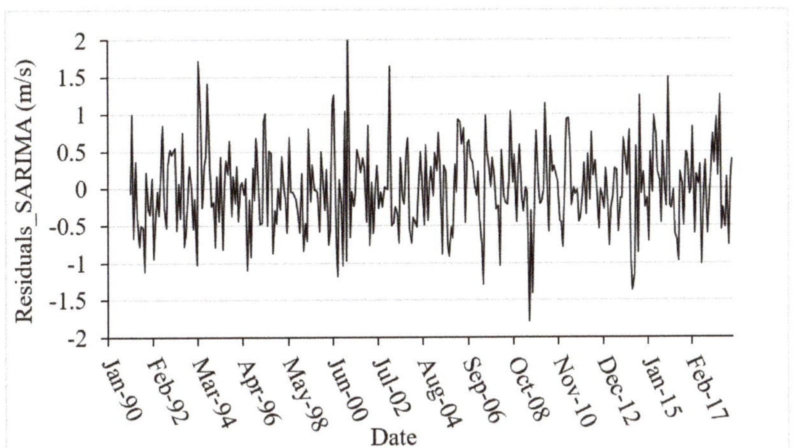

Figure 12. Time plot of residuals of model SARIMA (1,0,1) (0,1,1) (12) no constant for wind.

Table 3. Chosen parameters for the Prophet model.

Parameters	Solar Radiation	Wind Velocity
Changepoint_prior scale	1.3	1.6
Changepoint_range	0.8	0.90
Seasonality_prior scale	0.9	1.3
Seasonality_mode	Additive	Additive
Yearly_seasonality	True	True

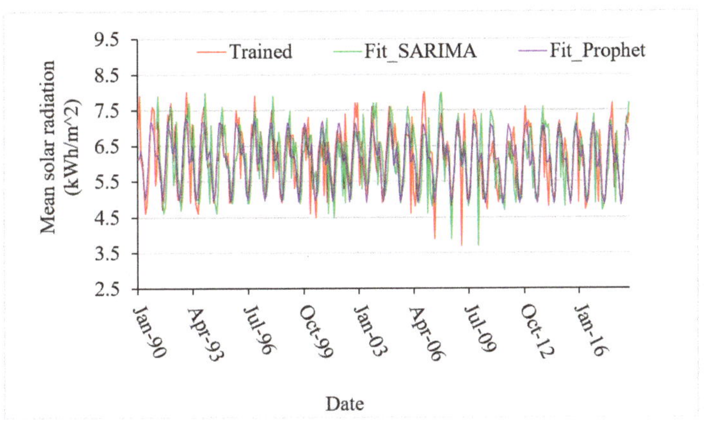

Figure 13. Fitted values using SARIMA and Prophet models for mean solar radiation.

Similar to SARIMA, the residuals of solar radiation and wind velocity for the Prophet model are depicted in Figures 15 and 16, respectively. From both figures, it can be seen that the variation of the residuals is almost identical and constant throughout the historic data and that it fluctuates around the horizontal line. So, the Prophet model can also be considered good for prediction. To find the better fit model from two models, the error metrics (MAE and RMSE) can be compared, as can be seen in Table 4. Table 4 also presents the findings of some reviewed sources. According to the findings of two error metrics such as MAE and RMSE, Prophet provides better results than SARIMA. Accurate projecting of uncertain and intermittent nature of solar and wind resources is challenging. Truthfully, the Prophet model succeeded with in achieving more precise prediction than the SARIMA. The Prophet model provides a MAE of 0.284 and RMSE of 0.395 for solar radiation, which is lower than that of the SARIMA. Similarly, the MAE of 0.427 and RMSE of 0.527 for wind velocity from the Prophet model is also lower than that of the SARIMA. In addition, the present study's error metrics for both the SARIMA and Prophet models are also below the errors of previous studies, as can be seen in Table 4. So, the proposed SARIMA and Prophet models are adequate for prediction. Since the Prophet model provides better results, this model is used for solar and wind power generation predictions. Figures 17 and 18 present the predicted results of the SARIMA and Prophet models for solar and wind up to 2030. The forecast patterns follow the recorded true values.

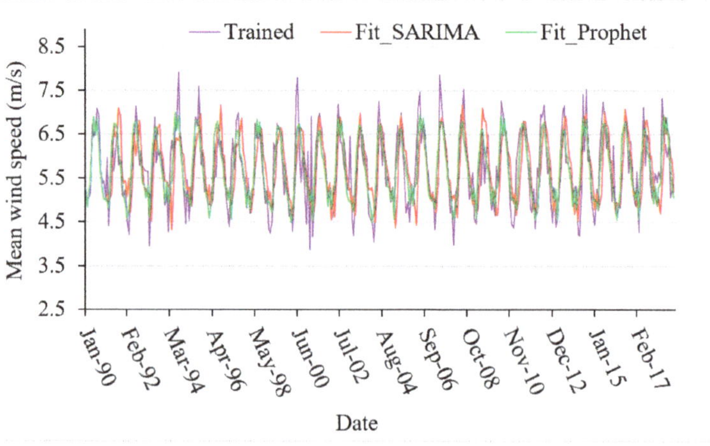

Figure 14. Fitted values using SARIMA and Prophet models for mean wind speed.

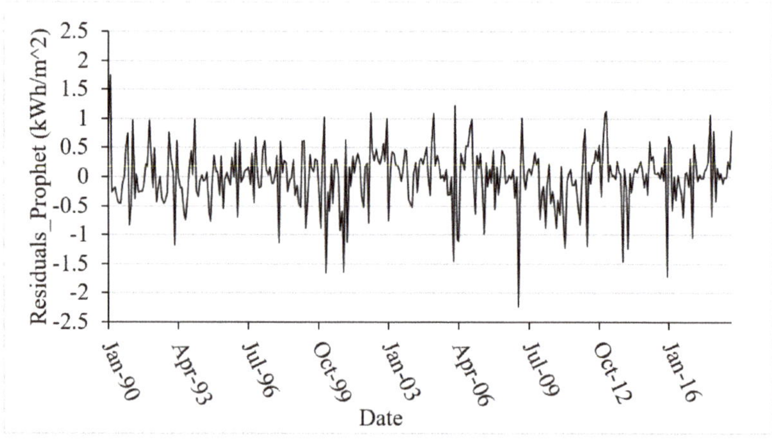

Figure 15. Time plot of residuals of Prophet model for mean solar radiation.

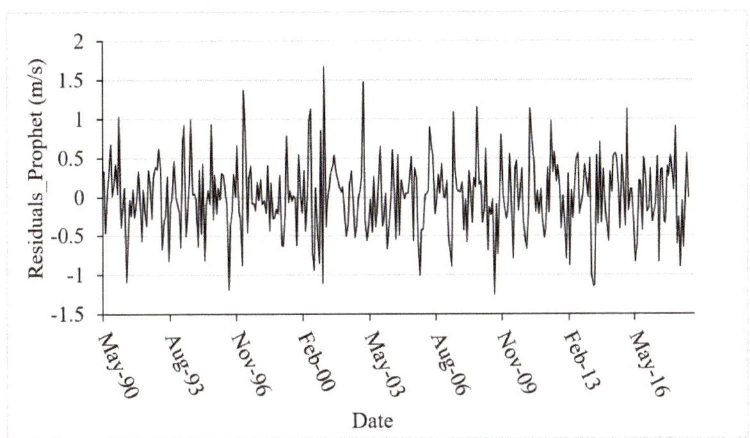

Figure 16. Time plot of residuals of Prophet model for mean wind velocity.

Table 4. Error metrics of forecasting modelling techniques including some reviewed sources.

Method	MAE	RMSE	Data Type
SARIMA (Current study)	0.406	0.582	Solar radiation
Prophet (Current study)	0.284	0.394	Solar radiation
SARIMA (Current study)	0.431	0.543	Wind velocity
Prophet (Current study)	0.427	0.527	Wind velocity
ARIMA	0.935	1.115	Wind velocity [60]
ARIMA	-	1.125	Wind velocity [61]
SARIMA	1.777–5.386	2.560–6.243	Wind velocity [62]
Prophet	1.908–8.411	2.331–8.603	Wind velocity [62]
Prophet	-	4.28	Energy demand [63]
SARIMA	-	5.39	Energy demand [63]
LSTM RNN	-	7.92	Energy demand [63]

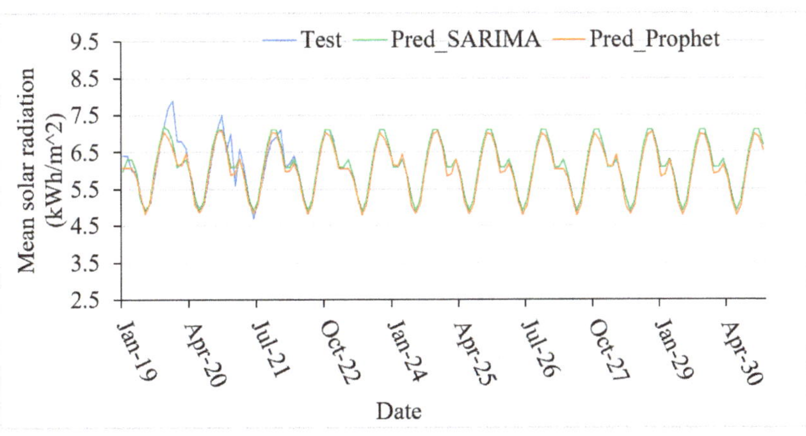

Figure 17. Time plot of predicted values from SARIMA and Prophet models for mean solar radiation.

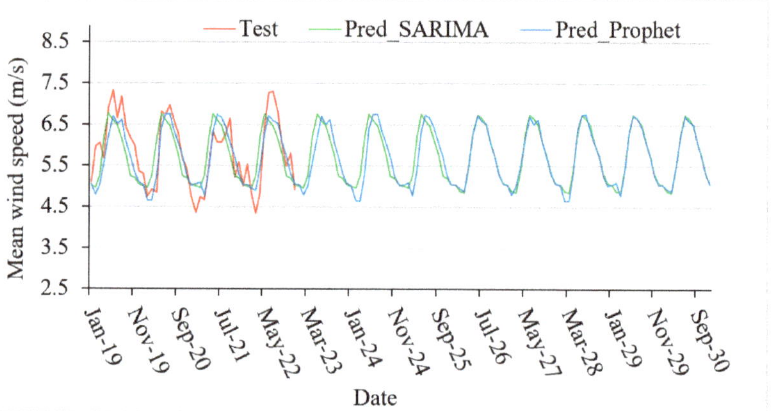

Figure 18. Time plot of predicted values from SARIMA and Prophet models for mean wind velocity.

4. Power Generation

The solar irradiation and wind velocity predicted using the Prophet model are implemented in a calculation of power generation in order to examine whether or not the solar or wind generation are enough for meeting the power demand. The Doomadgee region's load profile for 2021 is managed by the Ergon Energy company and the load here is increased 3–5% yearly. The load profile for 2030 is calculated by applying a yearly increment of maximum 5%. To compare with the load, the predicted values of solar radiation and wind speed of 2030 are employed for calculation. Solar energy generation is determined using the equation [64]:

$$P_{solar} = IR \times A \times AF \times \eta_s \tag{13}$$

where P_{solar} refers to solar power generation (kW), IR refers to the yearly average horizontal solar irradiation in kWh/m^2/day, A refers to the land area (m^2), AF refers to the area factor (%), and η_s is the solar panel efficacy (%). In this study, only 0.055% of the total Doomadgee area, with an area factor of 0.70, is used in calculation. The study used the first solar series 4^{TM} PV module (advanced thin film solar module) with maximum efficiency of 17% for calculating solar power generation.

Wind power generation is determined using the below equation in terms of air density, swept area and performance coefficient [65]:

$$P_{wind} = \frac{1}{2} \times \rho \times A \times V_{hub}{}^3 \times C_p \times \eta_w \tag{14}$$

and wind velocity at hub height can be determined using the following equation:

$$V_{hub} = V_{forecast} \times \frac{\ln(Z_{hub}/Z_0)}{\ln(Z_{anem}/Z_0)} \tag{15}$$

where Z_{hub}, Z_{anem} and Z_0 refer to the turbine hub height, anemometer height and surface roughness length, respectively. The wind turbine 'Enercon E-138 EP3', of 3500 kW rated capacity with hub height of 160m, is considered here. HOMER Pro 3.15.3 software is used to estimate the power produced by the wind turbine. In total, 10 Pcs wind turbines are sufficient to fulfil the power demand. From Figure 19, it can be seen that, individually, solar photovoltaics or wind turbines have huge potential to meet the energy demand. In addition, extra power can be stored to be used later, because solar and wind are erratic in nature and, at nighttime, solar power generation is not possible.

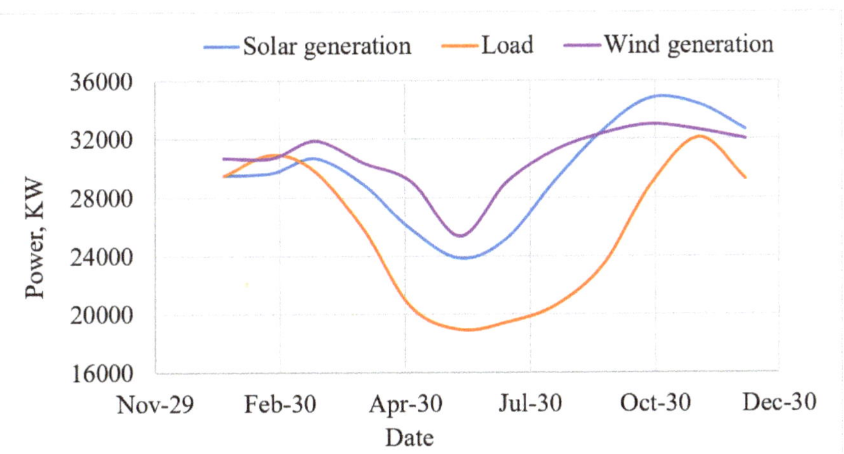

Figure 19. Predicted solar and wind power generation with load.

Solar and wind generation are calculated to meet the maximum load or power demand. During November to February, the load is high and in April to August, the load is very low. So, to meet the maximum load, either 122.5 W capacity first solar series 4^{TM} PV module in 0.055% of the total Doomadgee area, or 10 pcs wind turbine (Enercon E-138 EP3 of 3500 kW rated capacity from ENERCON GmbH company of Germany), need to be installed. From Figure 19, it can also be noticed that there is a huge possibility for excess power generation from solar or wind during the maximum time of the year, especially March to October, and that excess power can be stored to meet the load in the case of unavailability of solar or wind resources.

5. Discussion

The SARIMA model provides an acceptable performance for projecting solar and wind resources with slight errors (MAE of 0.406 and RMSE of 0.582 for solar, and MAE of 0.431 and RMSE of 0.543 for wind). The SARIMA model captures trend and seasonality well. However, the Prophet model performs comparatively better in capturing the trends and seasonality that present in the real time series data. The performance of the Prophet model can be realised by verifying the error metrics. The Prophet model produces MAE of 0.284 and RMSE of 0.394 for solar, as well as MAE of 0.427 and RMSE of 0.527 for wind, which are lower than that of SARIMA. The error metrics of the present study are also compared with previous studies (as seen in Table 4). The errors of the present study using the SARIMA and Prophet methods are well below the errors considered acceptable in previous studies. The Prophet method is recommended in this study in order to capture the time series temporal features precisely. The proposed Prophet model can be utilised to project future average solar irradiation and wind velocity as well as future solar or wind power generation, which shows the great potential of solar and wind resources for meeting energy demand. The selected region utilises diesel generation for their basic energy demand, but diesel is associated with high costs and climate pollution. Solar and wind power generation have already become the cheapest options, which can lower power generation costs and the need for diesel generation. Furthermore, the exploitation of local resources can enhance economic growth and improve human well-being in order to build a better future. The present study may provide a better understanding and scientific basis for decision makers in any region to identify local, clean and cost-effective renewable resources.

The Prophet model produces the lowest errors, yielding the closest match with the real data. Hence, the Prophet method should be adopted to project solar and wind resources. The present study shows that the Prophet model has the ability to train huge datasets fast and efficiently in order to generate precise projections. The additive submodular part of the Prophet technique helps in effectively handling the periodicity, anomalies and abrupt jumps present in the time series data. In addition, the Prophet method can effectively handle any seasonal variations.

6. Conclusions

Precise long term solar irradiation and wind velocity are significant in the renewable energy sector. This study proposes long term solar and wind resources estimation methods to predict future power generation for any year based on Prophet. In addition, the Prophet method has not yet been implemented and studied for long-term wind velocity and wind power generation prediction in the literature. The patterns of both solar and wind have been found to be stationary, including following no trend. A seasonality effect is observed in the raw data. The proposed Prophet model is compared with SARIMA and found to be superior in projecting solar and wind resources. The significant details of the modelling outcome can be summarised as follows:

- SARIMA, a commonly used powerful modelling approach, produces good results to project solar and wind. However, the careful hyperparameter tuning and comprehensive understanding of data are challenging tasks.
- The Prophet model is constructed to predict solar and wind resources, including power generation, in the present study. Additionally, the Prophet technique needs less hyperparameter tuning.
- The error metrics, such as the MAE and RMSE results in Table 4, prove that the Prophet model better fits the true data.
- The predicted solar and wind power generation shows huge potential for fulfilling the region's power demand, using local resources that can replace the present diesel power generation in order to combat climate change.

Future study is encouraged in order to use different hybrid methods, such as machine learning approaches and optimisation techniques, and to compare outcomes with the Prophet model's results for improving modelling accuracy. It is recommended that the Prophet modelling approach should be used in renewable energy prediction and other disciplines, since this approach is found to be simple, flexible and reliable. The significant scientific contribution of present study is that many different approaches, such as physical, statistical and hybrid methods, are utilised in predicting long-term wind velocity and wind power generation in the literature; the Prophet modelling approach has never before been studied or implemented for long-term wind energy prediction. Finally, the Prophet modelling approach can potentially be considered for the planning and developing of a sustainable system with a better environmental policy.

Funding: This research received no external funding.

Data Availability Statement: Data are not available for this research.

Conflicts of Interest: The authors declare that they have no known competing financial interests or personal relationships that could have appeared to influence the work reported in this paper.

References

1. Bureau of Meteorology. Monthly Mean Daily Global Solar Irradiation. 2021. Available online: http://www.bom.gov.au/climate/maps/averages/solar-exposure/ (accessed on 20 December 2021).
2. Bureau of Meteorology. Average Daily Sunshine Hours. 2005. Available online: http://www.bom.gov.au/watl/sunshine/ (accessed on 8 June 2021).
3. Bureau of Meteorology. Renewable Energy Atlas of Australia, Mean Wind Speed at 80 m above Ground Level. 2008. Available online: www.environment.gov.au/renewable/atlas (accessed on 15 February 2021).
4. Mishra, T.; Rabha, A.; Kumar, U.; Arunachalam, K.; Sridhar, V. Assessment of solar power potential in a hill state of india using remote sensing and geographic information system. *Remote Sens. Appl. Soc. Environ.* **2020**, *19*, 100370. [CrossRef]
5. Ali, S.; Taweekun, J.; Techato, K.; Waewsak, J.; Gyawali, S. Gis based site suitability assessment for wind and solar farms in Songkhla, Thailand. *Renew. Energy* **2019**, *132*, 1360–1372. [CrossRef]

6. Noorollahi, E.; Fadai, D.; Akbarpour Shirazi, M.; Ghodsipour, S.H. Land suitability analysis for solar farms exploitation using gis and fuzzy analytic hierarchy process (fahp)—A case study of Iran. *Energies* **2016**, *9*, 643. [CrossRef]

7. Inman, R.H.; Pedro, H.T.; Coimbra, C.F. Solar forecasting methods for renewable energy integration. *Prog. Energy Combust. Sci.* **2013**, *39*, 535–576. [CrossRef]

8. Sweeney, C.; Bessa, R.J.; Browell, J.; Pinson, P. The Future of Forecasting for Renewable Energy. *Wiley Interdisc. Rev. Energy Environ.* **2020**, *9*, e365. [CrossRef]

9. Notton, G.; Nivet, M.-L.; Voyant, C.; Paoli, C.; Darras, C.; Motte, F.; Fouilloy, A. Intermittent and stochastic character of renewable energy sources: Consequences, cost of intermittence and benefit of forecasting. *Renew. Sustain. Energy Rev.* **2018**, *87*, 96–105. [CrossRef]

10. Corizzo, R.; Ceci, M.; Fanaee, T.H.; Gama, J. Multi-aspect renewable energy forecasting. *Inform. Sci.* **2021**, *546*, 701–722. [CrossRef]

11. Alsharif, M.H.; Younes, M.K.; Kim, J. Time series arima model for prediction of daily and monthly average global solar radiation: The case study of Seoul, South Korea. *Symmetry* **2019**, *11*, 240. [CrossRef]

12. Wendell, H.; Wegley, L.A.; Verholek, M. *Report from a Working Group Meeting on Wind Forecasts for Wecs Operation*; Technical Report; United States Department of Commerce: Springfield, VA, USA; Pacific Northwest Laboratory: Washington, DC, USA, 1978.

13. Brown, R.W.; Katz, B.G.A.; Murphy, A.H. Time series models to simulate and forecast wind speed and wind power. *J. Clim. Appl. Meteorol.* **1984**, *23*, 1184–1195. [CrossRef]

14. Bossanyi, E. Short-term wind prediction using kalman filters. *Wind Eng.* **1985**, *9*, 1–8.

15. Corba, B.S.; Egrioglu, E.; Dalar, A.Z. Ar–arch type artificial neural network for forecasting. *Neural Process. Lett.* **2020**, *51*, 819–836. [CrossRef]

16. Li, Z.; Han, J.; Song, Y. On the forecasting of high-frequency financial time series based on arima model improved by deep learning. *J. Forecast.* **2020**, *39*, 1081–1097. [CrossRef]

17. Kavasseri, R.G.; Seetharaman, K. Day-ahead wind speed forecasting using f-arima models. *Renew. Energy* **2009**, *34*, 1388–1393. [CrossRef]

18. Singh, S.; Mohapatra, A. Repeated wavelet transform based arima model for very short-term wind speed forecasting. *Renew. Energy* **2019**, *136*, 758–768.

19. Wang, L.; Li, X.; Bai, Y. Short-term wind speed prediction using an extreme learning machine model with error correction. *Energy Convers. Manag.* **2018**, *162*, 239–250. [CrossRef]

20. Cadenas, E.; Rivera, W. Wind speed forecasting in three different regions of Mexico, using a hybrid arima–ann model. *Renew. Energy* **2010**, *35*, 2732–2738. [CrossRef]

21. Wang, J.; Hu, J. A robust combination approach for short-term wind speed forecasting and analysis—Combination of the arima (autoregressive integrated moving average), elm (extreme learning machine), svm (support vector machine) and lssvm (least square svm) forecasts using a gpr (gaussian process regression) model. *Energy* **2015**, *93*, 41–56.

22. Shukur, O.B.; Lee, M.H. Daily wind speed forecasting through hybrid kf-ann model based on arima. *Renew. Energy* **2015**, *76*, 637–647. [CrossRef]

23. Fang, T.; Lahdelma, R. Evaluation of a multiple linear regression model and sarima model in forecasting heat demand for district heating system. *Appl. Energy* **2016**, *179*, 544–552. [CrossRef]

24. Kushwaha, V.; Pindoriya, N.M. A sarima-rvfl hybrid model assisted by wavelet decomposition for very short-term solar pv power generation forecast. *Renew. Energy* **2019**, *140*, 124–139. [CrossRef]

25. Guo, Z.; Zhao, J.; Zhang, W.; Wang, J. A corrected hybrid approach for wind speed prediction in hexi corridor of China. *Energy* **2011**, *36*, 1668–1679. [CrossRef]

26. Koca, A.; Oztop, H.F.; Varol, Y.; Koca, G.O. Estimation of solar radiation using artificial neural networks with different input parameters for mediterranean region of Anatolia in Turkey. *Expert Syst. Appl.* **2011**, *38*, 8756–8762. [CrossRef]

27. Voyant, C.; Muselli, M.; Paoli, C.; Nivet, M.-L. Optimization of an artificial neural network dedicated to the multivariate forecasting of daily global radiation. *Energy* **2011**, *36*, 348–359. [CrossRef]

28. Ramedani, Z.; Omid, M.; Keyhani, A.; Shamshirband, S.; Khoshnevisan, B. Potential of radial basis function based support vector regression for global solar radiation prediction. *Renew. Sustain. Energy Rev.* **2014**, *39*, 1005–1011. [CrossRef]

29. Baghaee, H.R.; Mirsalim, M.; Gharehpetian, G.B.; Talebi, H.A. Three-phase ac/dc power-flow for balanced/unbalanced microgrids including wind/solar, droop-controlled and electronically-coupled distributed energy resources using radial basis function neural networks. *IET Power Electron.* **2017**, *10*, 313–328. [CrossRef]

30. Chowdhury, S.; Rahman, B.H. Forecasting sub-hourly solar irradiance for prediction of photovoltaic output. In Proceedings of the 19th IEEE Photovoltaic Specialists Conference, New Orleans, LA, USA, 4–8 May 1987.

31. Benghanem, M.; Mellit, A.; Alamri, S. Ann-based modelling and estimation of daily global solar radiation data: A case study. *Energy Convers. Manag.* **2009**, *50*, 1644–1655. [CrossRef]

32. Li, G.; Shi, J. On comparing three artificial neural networks for wind speed forecasting. *Appl. Energy* **2010**, *87*, 2313–2320. [CrossRef]

33. Xingpei, L.; Yibing, L.; Weidong, X. Wind speed prediction based on genetic neural network. In Proceedings of the 4th IEEE Conference on Industrial Electronics and Applications, ICIEA 2009, Xi'an, China, 25–27 May 2009; pp. 2448–2451.

34. Amjady, N.; Keynia, F.; Zareipour, H. Wind power prediction by a new forecast engine composed of modified hybrid neural network and enhanced particle swarm optimization. *IEEE Trans. Sustain. Energy* **2011**, *2*, 265–276. [CrossRef]

35. Jursa, R. Variable selection for wind power prediction using particle swarm optimization. In Proceedings of the Nineth Annual Genetic and Evolutionary Computation Conference (GECCO-2007), London, UK, 7–11 July 2007; pp. 2059–2065.
36. Zhang, N.; Behera, P.K. Solar radiation prediction based on recurrent neural networks trained by Levenberg-Marquardt back-propagation learning algorithm. In Proceedings of the Innovative Smart Grid Technologies (ISGT), 2012 IEEE PES, Washington, DC, USA, 16–20 January 2012; pp. 1–7.
37. Mohandes, M.A.; Halawani, T.O.; Rehman, S.; Hussain, A.A. Support vector machines for wind speed prediction. *Renew. Energy* **2004**, *29*, 939–947. [CrossRef]
38. Mellit, A.; Benghanem, M.; Kalogirou, S. An adaptive wavelet-network model for forecasting daily total solar-radiation. *Appl. Energy* **2006**, *83*, 705–722. [CrossRef]
39. Reikard, G. Predicting solar radiation at high resolutions: A comparison of time series forecasts. *Sol. Energy* **2009**, *83*, 342–349. [CrossRef]
40. Baghaee, H.R.; Mirsalim, M.; Gharehpetian, G.B.; Talebi, H.A. Generalized three phase robust load-flow for radial and meshed power systems with and without uncertainty in energy resources using dynamic radial basis functions neural networks. *J. Clean. Prod.* **2018**, *174*, 96–113. [CrossRef]
41. Cadenas, E.; Rivera, W. Wind speed forecasting in the south coast of Oaxaca, Mexico. *Renew. Energy* **2007**, *32*, 2116–2128. [CrossRef]
42. Lange, M.; Focken, U. *Physical Approach to Short-Term Wind Power Prediction*; Springer: Berlin/Heidelberg, Germany, 2006.
43. Watson, S.; Landberg, L.; Halliday, J. Application of wind speed forecasting to the integration of wind energy into a large scale power system. *IEEE Proc. Gener. Transm. Distrib.* **1994**, *141*, 357–362. [CrossRef]
44. Candy, B.; English, S.J.; Keogh, S.J. A comparison of the impact of quikscat and windsat wind vector products on met office analyses and forecasts. *IEEE Trans. Geosci. Remote Sens.* **2009**, *47*, 1632–1640. [CrossRef]
45. Wang, Q.; Luo, K.; Wu, C.; Tan, J.; He, R.; Ye, S.; Fan, J. Inter-farm cluster interaction of the operational and planned offshore wind power base. *J. Clean. Prod.* **2023**, *396*, 136529. [CrossRef]
46. Wang, Q.; Luo, K.; Wu, C.; Mu, Y.; Tan, J.; Fan, J. Diurnal impact of atmospheric stability on inter-farm wake and power generation efficiency at neighboring onshore wind farms in complex terrain. *Energy Convers. Manag.* **2022**, *267*, 115897. [CrossRef]
47. Wang, Q.; Luo, K.; Yuan, R.; Zhang, S.; Fan, J. Wake and performance interference between adjacent wind farms: Case study of Xinjiang in China by means of mesoscale simulations. *Energy* **2019**, *166*, 1168–1180. [CrossRef]
48. Shamim, M.; Remesan, R.; Bray, M.; Han, D. An improved technique for global solar radiation estimation using numerical weather prediction. *J. Atmos. Sol.-Terr. Phys.* **2015**, *129*, 13–22. [CrossRef]
49. Bauer, P.; Thorpe, A.; Brunet, G. The quiet revolution of numerical weather prediction. *Nature* **2015**, *525*, 47–55. [CrossRef] [PubMed]
50. Wang, Q.; Luo, K.; Wu, C.; Zhu, Z.; Fan, J. Mesoscale simulations of a real onshore wind power base in complex terrain: Wind farm wake behavior and power production. *Energy* **2022**, *241*, 122873. [CrossRef]
51. Baghaee, H.R.; Mirsalim, M.; Gharehpetian, G.B. Power calculation using rbf neural networks to improve power sharing of hierarchical control scheme in multi-der microgrids. *IEEE J. Emerg. Sel. Top. Power Electron.* **2016**, *4*, 1217–1225. [CrossRef]
52. Baghaee, H.R.; Mirsalim, M.; Gharehpetan, G.B.; Talebi, H.A. Nonlinear load sharing and voltage compensation of microgrids based on harmonic power-flow calculations using radial basis function neural networks. *IEEE Syst. J.* **2017**, *12*, 2749–2759. [CrossRef]
53. Maxwell, J.C. *A Treatise on Electricity and Magnetism*, 3rd ed.; Clarendon: Oxford, UK, 1892; Volume 2, pp. 68–73.
54. Zheng, Y.; Liu, Y.; Jiang, Z.; Tang, Q.; Xiang, Y. Wind Power Forecasting Based on Prophet Model. In Proceedings of the 2022 IEEE/IAS Industrial and Commercial Power System Asia (I&CPS Asia), Shanghai, China, 8–11 July 2022; pp. 1544–1548.
55. Arslan, S. A hybrid forecasting model using LSTM and Prophet for energy consumption with decomposition of time series data. *PeerJ Comput. Sci.* **2022**, *8*, e1001. [CrossRef]
56. Samal, K.K.R.; Babu, K.S.; Das, S.K.; Acharaya, A. Time series based air pollution forecasting using SARIMA and prophet model. In Proceedings of the International Conference on Information Technology and Computer Communications, Virtual, 27–28 September 2019; pp. 80–85.
57. Haider, S.A.; Sajid, M.; Sajid, H.; Uddin, E.; Ayaz, Y. Deep learning and statistical methods for short-and long-term solar irradiance forecasting for Islamabad. *Renew. Energy* **2022**, *198*, 51–60. [CrossRef]
58. Haider, S.A.; Sajid, M.; Iqbal, S. Forecasting hydrogen production potential in Islamabad from solar energy using water electrolysis. *Int. J. Hydrogen Energy* **2021**, *46*, 1671–1681. [CrossRef]
59. Vartholomaios, A.; Karlos, S.; Kouloumpris, E.; Tsoumakas, G. Short-term renewable energy forecasting in greece using prophet decomposition and tree-based ensembles. In Proceedings of the International Conference on Database and Expert Systems Applications, Virtual, 27–30 September 2021; pp. 227–238.
60. Shivani, K.S.; Nair, A.R. A comparative study of ARIMA and RNN for short term wind speed forecasting. In Proceedings of the 10th International Conference on Computing, Communication and Networking Technologies (ICCCNT), Kanpur, India, 6–8 July 2019; pp. 1–7.
61. Yuan, D.; Qian, Z.; Jing, B.; Pei, Y. Short-Term wind speed forecasting using STLSSVM hybrid model. In Proceedings of the 2018 International Conference on Power System Technology (POWERCON), Guangzhou, China, 6–9 November 2018; pp. 1661–1667.

62. Atasever, S.; Öztürk, B.; Bilgiç, G. A new approach to short-term wind speed prediction: The prophet model. *Energy Sources Part A Recovery Util. Environ. Eff.* **2022**, *44*, 8826–8841. [CrossRef]
63. Chaturvedi, S.; Rajasekar, E.; Natarajan, S.; McCullen, N. A comparative assessment of SARIMA, LSTM RNN and Fb Prophet models to forecast total and peak monthly energy demand for India. *Energy Policy* **2022**, *168*, 113097. [CrossRef]
64. Toharudin, T.; Pontoh, R.S.; Caraka, R.E.; Zahroh, S.; Lee, Y.; Chen, R.C. Employing long short-term memory and Facebook prophet model in air temperature forecasting. *Commun. Stat.-Simulat. Comput.* **2021**, *52*, 1–24. [CrossRef]
65. Ma, T.; Yang, H.; Lu, L. A feasibility study of a stand-alone hybrid solar–wind–battery system for a remote island. *Appl. Energy* **2014**, *121*, 149–158. [CrossRef]

Article

The Effect of Leading-Edge Wavy Shape on the Performance of Small-Scale HAWT Rotors

Riad Morina [1] and **Yahya Erkan Akansu [2],***

[1] Faculty of Mechanical Engineering, University of Prishtina "Hasan Prishtina", 10000 Prishtina, Kosovo; riad.morina@uni-pr.edu
[2] Faculty of Mechanical Engineering, Niğde Ömer Halisdemir University, Niğde 51240, Turkey
* Correspondence: akansu@ohu.edu.tr

Abstract: The purpose of this experimental work was to investigate the role of the leading-edge wavy shape technique on the performance of small-scale HAWT fixed-pitch rotor blades operating under off-design conditions. Geometric parameters such as amplitude and wavelength were considered design variables to generate five different wavy shape blade models in order to increase the aerodynamic performance of the rotor with a diameter of 280 mm. A dedicated airfoil type S822 for small wind turbine application from the NREL Airfoil Family was chosen to fulfil both the aerodynamic and structural aspects of the blades. Rotor models were tested in a wind tunnel for different wind speeds while maintaining constant rotational speed to provide the blade-tip chord Reynolds number of 4.7×10^4. The corrected tunnel data, in terms of power coefficients and tip-speed ratios, were compared first with the literature to validate the experimental approach, and then among themselves. It was observed that for minimal sizes of tubercles, the performance of the rotor increases by about 40% compared to the RB1 baseline rotor model for a low tip-speed ratio. Conversely, for the maximum size of the tubercles, there is a marked decrease of about 51% of the rotor performance for a moderate tip-speed ratio compared to the RB1 rotor model. Among these models, specifically, the RB2 rotor model with the smallest values of amplitude and wavelength provides a 2.8% higher peak power coefficient compared to the RB1 rotor model, and at the same time preserves higher performance values for a broad range of tip-speed ratios.

Keywords: small HAWT rotor; blade design; power coefficient; passive flow control; bio-inspired technique; leading-edge tubercles; low Reynolds number; fixed-pitch rotor; wind tunnel

Citation: Morina, R.; Akansu, Y.E. The Effect of Leading-Edge Wavy Shape on the Performance of Small-Scale HAWT Rotors. *Energies* **2023**, *16*, 6405. https://doi.org/10.3390/en16176405

Academic Editor: Davide Astolfi

Received: 4 August 2023
Revised: 23 August 2023
Accepted: 28 August 2023
Published: 4 September 2023

1. Introduction

1.1. Turbine Efficiency for Sustainable Energy: An Overview

In accordance with Emeis [1], increasing global attention is being focused on the application of renewable resources due to its effects in reducing greenhouse effects and global warming. One of these sources is wind energy, as it is estimated that it can sufficiently provide the energy required and can be harnessed on the Earth's surface. At present, the challenge is to determine how to utilize this energy at the largest possible scale, with the highest efficiency and minimum cost.

There are two types of wind-converting machines: the horizontal axis wind turbine (HAWT) and the vertical axis wind turbine (VAWT) [2]. However, in terms of performance, horizontal axis wind turbines surpass vertical shaft turbines, which is why it is more common to observe the former with extraordinary dimensions and capacities. In this aspect, small wind turbines, which had a global installed capacity of over 1.73 gigawatts by 2018 [3], are foreseen as a very significant part of the future sustainable energy mix and represent an efficient method for reducing greenhouse gas emissions [4]. According to the IEC 61400-2: 2013 standard, wind turbines having a rotor swept area less than 200 m^2, corresponding to a maximum power output of 70 kW, are considered to be small wind turbines [5]. Due to their construction characteristics, reliability, and low cost, they can be utilized in both urban and rural areas, and even for on-grid application. However, these machines are most commonly fixed-pitch regulated, because conditions in which they operate are characterized by poor performance that may be several times lower than the Betz limit (~0.593) [5,6]. Such characteristic results are mainly due to the mismatch of the blade profile and the operational speed of the rotor with the extremely severe and sensitive working conditions as a consequence of increased viscosity forces. These forces stimulate the formation of laminar separation bubbles, particularly when the Reynolds number is lower than 1×10^5, which has a significant impact on the performance of airfoils [7–14]. Therefore, a unique shape of the rotor blade balanced with the above requirements is necessary to effectively master these highly viscous forces in the fluid.

1.2. Optimizing Rotor Blade Performance Using the Leading-Edge Tubercle Technique

The key step to designing an efficient rotor blade shape is choosing the right airfoil profile for the given Reynolds number flow since it significantly affects the rotor's performance [6,8]. The Reynolds number is a crucial non-dimensional parameter that usually varies throughout the rotor radius by reaching minimum values in the vicinity of the hub, whereas the maximum values are reached close to the tip of the blade. As the tip region is considered to be the part where most of the torque of the power is produced [5,15,16], it is often the case that the representative Reynolds number for a small wind turbine is determined for the blade-tip chord. This is one of the main reasons why other parts of the blade closer to the hub suffer from a significant drop in performance. In addition to the large number of traditional existing airfoils, several other profiles have been designed specifically for small wind turbines. Studies have shown that chambered thin airfoils are the most common profiles for low Reynolds number flows [7,8,17,18]. However, for Reynolds numbers lower than 1×10^5, the data for airfoils are limited. Therefore, using an undedicated airfoil profile for these operating conditions was reported in many studies to be associated with a significant decrease in performance [5,8,16,19]. This issue was also addressed by Burdett et al. [20], which suggests the necessity of using methods to control the flow, and that passive flow devices may present an alternative, significant contribution to both the aerodynamic and economic aspects. A solution inspired by aquatic animals, which share the same operational conditions as small wind turbines, is known as the leading-edge tubercle technique. This technique, proven to be effective by several studies, improves the aerodynamic characteristics of airfoils without consuming additional energy. It particularly excels under specific operating conditions, such as low Reynolds number regimes, which are characterized by highly severe and sensitive working conditions due to increased viscosity forces and high angles of attack, and ensures smoothed stall characteristics. This concept derives from the idea that animals have evolved over time to adapt to the terms and challenges of nature [21]. For this reason, this natural aerodynamic mechanism added to the leading edges of the rotor blades aims to imitate a certain mass, such as the humpback whale's wing shape. This generates the so-called counter-rotating streamwise vortices, with the purpose of effectively controlling the laminar separation bubbles over the wings and preventing the stall occurrence or progression along the spanwise direction, akin to the effective swimming pattern observed in aquatic animals [12,13,22].

1.3. Enhancing Airfoil Performance with Tubercles

Many studies of the effects of leading-edge tubercles on the aerodynamic performance of airfoil wings and wind turbine blades under different conditions have been conducted numerically and experimentally.

Usually, an optimal airfoil is selected for the maximun value of the glide ratio (C_L/C_D), which corresponds to an optimal angle of attack. The maximum lift coefficient, whose value is achieved for larger angles of attack, is another very important aerodynamic quantity as it determines the stall angle of an airfoil. Beyond this value, the lift is reduced and the performance is reduced. Studies indicate that the tubercle geometry significantly affects aerodynamic characteristics of airfoils. According to Fish and Battle [13] and Miklosovic et al. [12], the humped shape of the whale's wings, along with their leading edges, are responsible for the efficiency of the maneuverability employed when catching its prey. They pointed out that this natural mechanism improves the maximum lift coefficient of the whale's wings at high angles of attack, especially when these animals take narrow turns, and thus improves their stall characteristics by smoothing and delaying the occurrence. In addition, the authors suggest that this morphology of the humpback flippers may serve as an alternative to improving engineering designs. This technique is observed to improve the glide ratio of a swept wing at pre-stall angles of attack by varying the main parameters such as the amplitude and the wave length [23]. Miklosovic et al. [12] conducted an experimental study of the effects of the tubercles along the leading edge on two scaled models similar to humpback whale flippers using an NACA 0020 airfoil at angles of attack ranging from $-2°$ to $20°$. The study found that models with tubercles were able to delay the stall occurrence to a significant degree compared to clean models at Reynolds numbers between 5.05×10^5 and 5.2×10^5. According to the authors, this was attributed to the presence of bumps which delayed the stall angle by nearly 40% and improved the post-stall characteristics. However, for an angle of attack lower than $8.5°$, no effect was observed. Haque et al. [24] also carried out an experimental study of a wing consisting of an NACA 4412 airfoil with curved and straight leading edges at an air speed of around 23.71 m/s corresponding to $Re = 1.82 \times 10^5$ and at several angles of attack ranging from $-4°$ to $24°$. It was found that the wavy-shaped model performed better at an angle of attack below $12°$. Hansen et al. [25] conducted an experimental study to determine the influence of leading-edge tubercles on the performance of two different airfoil types at angles of attack ranging from $-4°$ to $25°$ and $Re = 1.2 \times 10^5$. NACA 65-021 and NACA 0021 airfoils were used to generate 10 wing models by taking several values of the amplitude, and multiplying the mean chord length by 3%, 6%, and 11%; and the wavelength, and multiplying the mean chord length by 11%, 21%, 43%, and 86%. These values for amplitude and wavelength were taken from Fish and Battle [13]. It was found that, as the amplitude decreased, a higher maximum lift coefficient and larger stall angle and improved post-stall characteristics were obtained. Furthermore, with the reduction in the wavelength of tubercles, benefits were found in all aspects of aerodynamic performance. They also suggested that models with tubercles having the smallest amplitude performed better compared to an unmodified airfoil. Sudhakar et al. [26] conducted an experimental investigation of the impact of leading-edge tubercles on different airfoil types at low Reynolds numbers. The test was conducted at three Reynolds numbers, $Re = 1 \times 10^5$, 1.5×10^5, and 2×10^5, and for several angles of attack ranging from $-6°$ to $24°$. They found that tubercles on the S1223 airfoil positively affected the aerodynamic characteristics in all three flow regimes, resulting in delayed stall, an increase in lift, and elimination of the hysteresis loop compared to the baseline model. On the other hand, no significant improvements were observed regarding the NACA 4415 airfoil. Johari et al. [27] performed an experimental study to investigate the effects of sinusoidal tubercles on the performance of the NACA 63_4-021 airfoil type. Several models were generated by varying the amplitude from 2.5% to 12% of the chord length and wavelengths from 25% to 50% of the chord length for $Re = 1.83 \times 10^5$. They found that improvements were noted in the post-stall regime where the lift coefficient increased nearly 50% compared to the baseline airfoil. They also suggested that the amplitude of tubercles can significantly influence the resultant force and moment coefficient, but the wavelength and the leading-edge radius have only a minor impact. In their wind tunnel experiments, Guerreiro and Sousa [28] investigated the effects of the aspect ratio, leading-edge geometry, and Reynolds number on the micro-air vehicle wing lift forces. The tests were conducted on several wing models having amplitudes of 6% and 12% of the mean chord length, and wavelengths of 25% and 50% of the mean chord length, at angles of attack from $0°$ to $30°$ and at the Reynolds numbers of 7×10^4 and 14×10^4. It was found that a proper combination of the amplitude and the wavelength resulted in an increase of 45% of the maximum lift coefficient for the highest Reynolds number and for an angle of attack larger than the straight baseline stall angle. For the lowest Reynolds number, the benefits were for low angles of attack. According to Wei et al. [29], regarding the impact of protuberances on hydrofoil performance at a Reynolds number of 1.4×10^4, it was observed that, in this flow condition, a greater amplitude and shorter wavelength were favorable in managing flow separation. Furthermore, according to the study by Bolzon et al. [30] regarding the impact of bumps located on the leading edge of an airplane's wings at a low angle of attack of $1°$ to $8°$ and at Re = 2.2×10^5, the observations indicated that this passive mechanism resulted in a reduction in aerodynamic coefficients while concurrently enhancing the lift-to-drag ratio. Natarajan et al. [31] carried out an experimental study on the leading edge of a NACA 4415 airfoil for different angles ranging from $6°$ to $18°$ at Re = 1.2×10^5. They revealed that the presence of the tubercles resulted in the formation of several smaller separation bubbles, instead of the single long bubble along the blade length that was noticed in the unmodified airfoil. They also found that tubercles were very effective beyond the stall condition of the modified airfoil.

1.4. Advancing Wind Turbine Performance with Tubercles

The rotational rotor speed defined by the tip-speed ratio plays a crucial role in its performance because only one of its values gives the maximum efficiency [10,11,17,32]. This non-dimensional parameter is closely related to the solidity of the rotor, which varies depending on the size of the wind turbine rotors and their purpose [32,33]. For the purposes of electricity generation, small wind turbine rotors are characterized by a relatively low solidity as they have to spin faster. Furthermore, due to the spinning work principle of the wind turbine rotor, the flow over their blade surfaces is three-dimensional in nature, where the root and the tip regions are strongly affected by trailing vortices due to the complexity of the flow that develops in these two extreme parts of the blade [5,34–36]. Addressing these issues can considerably influence the increase in the blade productivity and minimization of power loss, where only tip losses can represent a nearly 10% decrease in the Annual Energy Production (AEP) [35]. On the other hand, the root region is characterized by larger solidity and pitch angle than the tip region because this part of the blade is considered the most significant contributor to ensuring the strength and rigidity of the blade structure, as well as the factor initiating system at low wind speeds [5,33]. However, it is claimed that even the root part of the rotor blades affects the increase in the performance of the wind turbine [36]. Due to these specifics, the root region is more prone to the onset of stall occurrence and spread towards the tip of the blade, thus causing a decrease in its performance. Therefore, a challenge remains the early prevention of the expansion in this negative phenomenon, both in the streamwise and spanwise directions of the blade. By contrast with the airfol wings, there are a limited number of research studies dealing with the application of the leading-edge tubercle technique for improving wind turbine performance [37]. Pedro and Kobayashi [38] found that the presence of tubercles on an idealized humpback whale flipper at an angle of attack of $15°$, in addition to changing the vorticity distribution along the model span, leads to the increase in momentum exchange and consequently stalls prevention. Moreover, a reduction in the tip vortex strength was noticed as a consequence of the compartmentalization of the flow due to the differences in chord lengths of the crests and troughs of tubercles. Van Nierop et al. [39] suggested that the higher the pressure gradient at the troughs, the later the flow separation will be initiated behind the tubercles crests. In the approach proposed by the authors, the conditions are created for the separation to be initiated in this zone so as to ensure gradual global stall characteristics. Abate et al. [37] carried out a numerical study of tubercles' effect on the performance of the two-bladed NREL Phase VI wind turbine with a radius of 5 m and a rated power of 20 kW. Several blade configurations were generated using an S809 airfoil section by varying the non-dimensional amplitude from 1% to 5% and the non-dimensional wavelength from 1.6% to 7.5%. These fixed-pitch wind turbines having a tip pitch angle of $3°$ were tested for constant RPM using CFD code in a fully turbulent flow condition at four different flow speeds ranging from 5 to 20 m/s, corresponding to Reynolds number between 10^4 and 10^6 along their blade lengths. It was noticed that the maximum effect was achieved at the highest wind speed or in off-design conditions for high amplitude and low wavelength; for the lowest wind speed, the effect was significantly reduced. They also highlighted the important role of the last tubercle situated at the tip of the blade and its geometric shape, since it can affect the tip vortex's intensity. A computational study conducted by Abate and Mavris [40] analyzed the effects of different leading-edge tubercle positions on wind turbine blade performance. An NREL Phase VI wind turbine blade was used as the baseline model, which was compared with the six blade configurations generated by varying the tubercles' spanwise location from 40% to 95% of the blade span. They found that the presence of tubercles within the range between 62% and 95% of the blade span showed better performance at a high wind speed of 20 m/s compared to other models, resulting in improved Annual Energy Production and shaft torque. Zhang and Wu [41] investigated the aerodynamics of wind turbine blades with sinusoidal leading edges. They generated the baseline model as a reference for comparison, which was the original NREL Phase VI blade model with a radius of 5.029 m and a root chord length of 0.737 m, with a straight leading edge, along with five blade model configurations with different amplitudes ranging from 1.25% to 3.75% of the aforementioned root chord length, and wavelengths ranging from 17% to 42% of the same root chord length. All the models were built using an S809 NREL airfoil section having a tip pitch angle of 3°. The blades were tested via CFD code under five dissimilar flow speeds ranging from 7 to 25 m/s. The results showed that, compared to the reference blade model, the wavy-shaped blade models improved the wind turbine blades' aerodynamic performance under high course speeds or above the rated wind speed. Based on the findings, the key region along the blade where tubercles can be used to improve the aerodynamic performance extends from 60% of the rotor radius to the tip of the blade, resulting in high values of both amplitude and wavelength. They also suggested that this technique may be a useful mechanism even for pitch control wind turbines (modern wind turbines) with variable rotating speed. In their experimental study, Huang et al. [42] compared the performance of a smoothed leading-edge blade with different sinusoidally shaped blade models having a variable rotational speed at a wind speed of 6 to 10 m/s and a Reynolds number between 1.0×10^5 and 3.0×10^5. Four models were generated by combining the amplitude of 1.5% and 8.5% of the mean chord length of 9 cm with the wavelength of 6.5% and 15% of the same mean chord length. They reported that the presence of tubercles considerably improved the performance of the rotor models by effectively delaying the stall, but only for low wind speeds and a low tip-speed ratio. It was concluded that by increasing the wind speed, the rotational speed of the rotor also increased, which resulted in the deterioration of the small-scale wind turbine performance. Kim et al. [43] performed a numerical study on the effects of the leading-edge tubercles on the flow structure on a three-dimensional wing. They considered wing models previously studied by Miklosovic et al. [12] at a Reynolds number of 1.8×10^5, based on free-stream velocity and mean chord length. They reported positive effects of tubercles since the stall angle was delayed by 7° (with an angle of attack from 8° to 15°), while the maximum lift coefficient was increased by nearly 22%. Ng et al. [44] studied the impact of leading-edge tubercles on fatigue loadings on wind turbine blades. They noticed that tubercles with an amplitude of 20% of the chord and a wavelength of 50% of the chord, distributed from the root (20% R) to the blade tip (95% R), decreased the flapwise root-bending moment by 6% compared to the unmodified configuration. The authors suggest that positioning the tubercles near the tip of the blade improved the performance during stall occurrences from large tip deflections. An improvement was also observed when tubercles were closer to the root of the blade, as this area tends to stall earlier due to the lower rotational speed of the blade. Herráz et al. [34] studied the impact of the blade root flow on the performance of the wind turbine rotor. According to the authors, the flow at the root and tip regions is considered to be three-dimensional in nature and strongly influenced by the trailing vortices. This was attributed to the large difference in the angle of attack at the root region and at the tip, thus representing the most two critical zones where the flow separation may take place.

1.5. Problem Statement and Aims of the Research

Preliminary studies highlight a highly unstable operating environment in which small wind turbines usually operate, and they suggest the necessity of using flow control mechanisms, especially for profiles that operate below their design point. A very promising technique that does not require additional energy has been proven by many researchers as an acceptable solution for the treatment of this unfavorable condition, which affects the aerodynamic characteristics of profiles when subjected to such severe working conditions.

Therefore, the focus of this experimental study was to investigate the role of the leading-edge wavy shape on the performance of small-scale wind turbine rotor blades operating under off-design conditions at a Reynolds number of 4.7×10^4, which is somewhat lower than the Reynolds number for which the selected airfoil type is designed. The wavy shape will characterize the entire length of the blade, from the root to the tip, with the purpose of enabling all the parts of the flow to be easily controlled as it is compartmented between the crests of the tubercles. Other objectives were to understand the effects of expanding the application range of the airfoil, beside the tip of the blade for which it is designed, and changing the wind speed while maintaining a constant rotational speed of the rotor blades. This was based on the suggestion from Abate et al. [37] that rotors may perform even better when they operate with variable rotational speed and with a pitch control mechanism. Geometric parameters of the leading-edge tubercles, such as amplitude and wavelength, are considered design variables to generate five different wavy-shaped rotor blades. The results measured in a wind tunnel, in terms of power coefficients and tip-speed ratios, were first corrected using the Van Treuren [9] approach for comparison purposes. A dedicated type S822 airfoil for small wind turbine application from the NREL Airfoil Family was chosen to fulfil both the aerodynamic and the structural aspects of the blades. It was observed that a specific tubercle geometry with smaller amplitude and wavelength applied to rotor blades with a fixed-pitch angle for the given operating conditions resulted in a better performance for the entire range of tip-speed ratios compared to the unmodified rotor blade model.

2. Material and Methodology

2.1. Rotor Blade Geometry Generation

2.1.1. Airfoil Selection

In general, the selection of an airfoil for the wind turbine rotor blades was made by taking into account the factor that defines its operating environment, which is the Reynolds number. This non-dimensional number is defined as follows:

$$Re = \frac{\rho \cdot V_{rel} \cdot c_{avg}}{\mu} \tag{1}$$

where ρ—air density, V_{rel}—relative velocity, c_{avg}—average chord length, μ—dynamic viscosity of the air.

The relative velocity at the tip of the blade was calculated as follows [16]:

$$V_{rel} = \sqrt{V_{\infty}^2 + V_{tip}^2} \tag{2}$$

where V_{∞}—wind speed, V_{tip}—blade-tip speed.

The efficient functioning of the system is enabled by the aerodynamic factor, strength, and rigidity of the airfoil. However, for small wind turbines, the choice of a very thin airfoil profile with a thickness of 5% of the chord, which is considered more favorable for very low Reynolds number flows by Sunada et al. [45], would significantly affect the structural aspect of the blade, specifically the root part. For this reason, it is necessary to find a solid balance between the sizes of this profile in relation to the large centrifugal forces that characterize small wind turbines [5]. In order to satisfy the above requirements, the S822 airfoil with a thickness of 16% at a 39.2% chord and a maximum camber of 1.8% at a 59.5% chord from the NREL Airfoil Family was chosen as a suitable solution, since choosing a thicker profile than 25% would greatly affect the aerodynamic aspect [5,46] (Figure 1). This dedicated profile for a small horizontal axis wind turbine application with a rotor diameter of 1 to 5 m is designed for a Reynolds number of 6×10^5, specifically for the blade-tip region [8,47].

Figure 1. Reference airfoil geometry (NREL S822).

In this study, the aerodynamic quantities for the S822 airfoil, such as the maximum lift and minimum drag coefficients, and the corresponding angle of attack, were obtained using XFoil 6.9 for a Reynolds number of 1×10^5 and $N_{crit} = 9$, which are presented in the following sub-section.

2.1.2. Blade Design Procedure

The geometry of the actual optimum baseline rotor blade model with a fixed-pitch angle and a radius of 140 mm was derived via a small-scale geometry of the original rotor blade model with a radius of 300 mm [46,48], which was designed according to Blade Element Momentum (BEM) equations. The Reynolds number for the original rotor design was selected to coincide with the limit value of 1×10^5 mentioned previously in the literature. Below this value, the operating conditions begin to become significantly harsher, which significantly affects the aerodynamic performance of airfoils and the rotors. The wind speed was taken to be very close to the rated wind speed for small wind turbines (above 10 m/s) [5,6], while the optimum tip-speed ratio was taken to be in the range between 2 and 4 based on the suggestion for improved reliability of rotor performance and lower noise level [6]. The rotor solidity was chosen to be in the range between 15 and 30% to maintain the tip-speed ratio between its optimum values in order to achieve better power performance [6,49]. The body of the blade was made from a single S822 airfoil profile from root to tip. In this study, the blade-tip design has a simple rounded tip cap geometry. In Table 1, the input data for the original design of the rotor are introduced.

Table 1. Original rotor parameters [46].

Parameter	Value
Rotor radius (R), mm	300
Wind speed (V_∞), m/s	9
Air density (ϱ), kg/m^3	1.225
Reynolds number (Re)	1×10^5
Number of blades (B)	3
Tip-speed ratio (TSR, λ)	3.658
Rotor solidity (σ), %	22.1
Lift coefficient (C_L)	0.9256
Drag coefficient (C_D)	0.02168
Angle of attack (α), °	8.5
Lift/drag ratio (C_L/C_D)	42.7
No. blade elements (N)	10

The Schmitz Method (Equations (3)–(6)) was used to obtain the optimum blade geometry such as chord and twist angle distributions [32,50]:

The local tip-speed ratio is calculated as

$$\lambda_i = \lambda \left(\frac{r_i}{R} \right) \tag{3}$$

The local optimum relative inflow angle is calculated as

$$\varphi_i = \left(\frac{2}{3} \right) \cdot \text{atan} \left(\frac{1}{\lambda_i} \right) \tag{4}$$

The local optimum chord length is calculated as

$$c_i = \frac{16 \cdot \pi \cdot r_i}{B \cdot C_L} \cdot \sin \left(\frac{1}{3} \cdot \arctan \left(\frac{1}{\lambda_i} \right) \right)^2 \tag{5}$$

The local optimum twist angle is calculated as

$$\beta_i = \varphi_i - \alpha \tag{6}$$

In Figure 2, chord and twist angle values and other quantities for the original blade geometries (R = 300 mm) for each station along its length are presented.

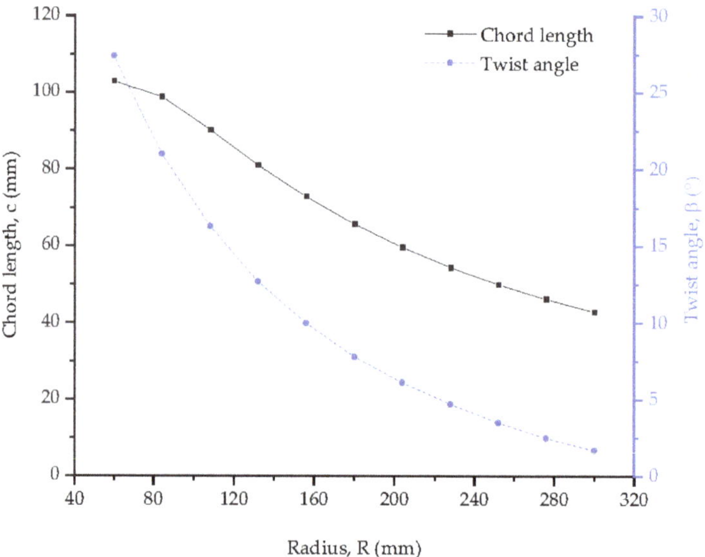

Figure 2. Arrangement of chord lengths and angles of twist of the original blade model.

Figure 3 represents the chord and twist angle values for the small-scale blade geometry. Figure 4 shows the three-dimensional geometry of the RB1 baseline blade model.

The rotor blade models were drawn in CAD software (Solidworks 2019) and printed on a Wanhao Duplicator 9 3D printer using ABS material, and were sanded with 1000-grit sandpaper after the printing process, but not dyed. The surface roughness of the baseline blade model and the wavy-shape models was approximately 0.001 and 0.011 mm, respectively.

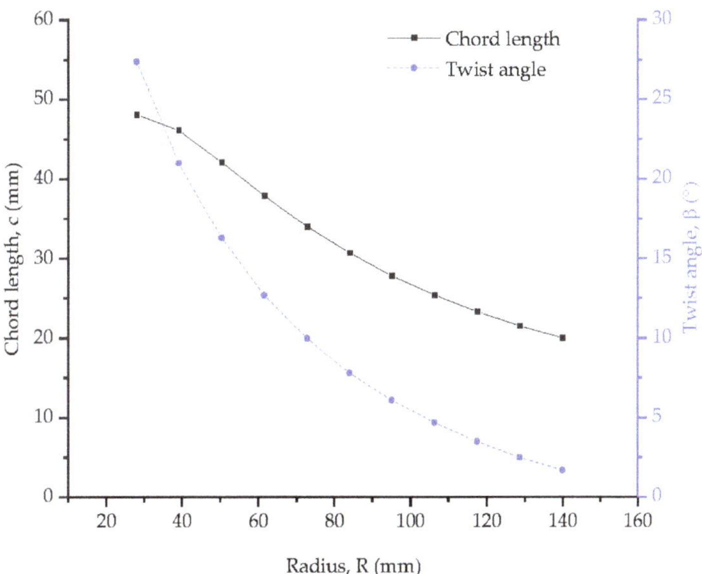

Figure 3. Arrangement of chord lengths and angles of twist of the small-scale blade model.

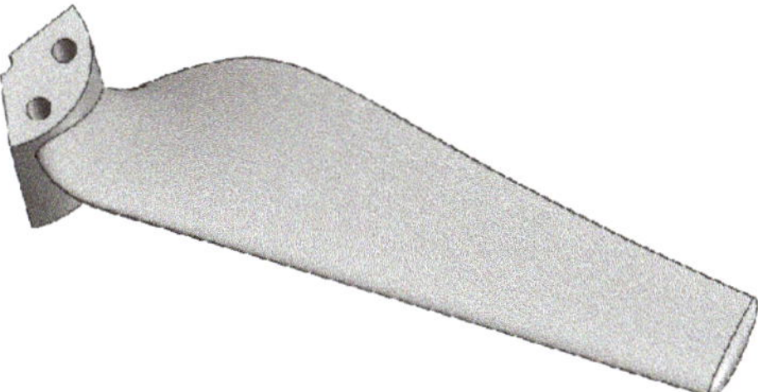

Figure 4. Three-dimensional RB1 baseline blade model.

2.2. Generation of Wavy-Shape Blade Models

Using the leading-edge tubercle technique, following the methodologies outlined by Abate et al. [37] and Hansen et al. [25], five different configurations were produced by modifying the amplitude and wavelength parameters (Table 2) in accordance with Equation (7) (Figure 5). The wave's amplitude (A) was determined by multiplying the average chord length of the blade by the values of 0.03, 0.06, 0.09, 0.11, and 0.14 [46]. Conversely, the wavelength (λ) was derived by multiplying the average blade chord length with the factors of 0.11, 0.14, 0.22, 0.29, and 0.43 [46]. Tubercles span from 20% of the rotor radius to the tips of the blades. The present rotor blades have an average chord length (c_{avg}) of 32.4 mm.

The tubercles' shape along the leading edge of the blade was calculated using the following equation [37]:

$$y = y_0 + \frac{A}{2} \cdot \sin\left(\frac{2 \cdot \pi \cdot x}{\lambda}\right) \tag{7}$$

where y_0 is the coordinate of the starting point of the sinusoidal path, which is equal to 18.66 mm, A is the amplitude, λ is the wavelength, and x is the extension of the bumps from the root (0.2 R = 28 mm) to the tip of the blade.

In Table 2, all rotor blade models are introduced, distinguished by the letter RB, which stands for the rotor blade model, and an ordinal number from 1 to 6.

Table 2. Rotor blade models [46].

RB1—Base Model	RB2—A1λ3.5

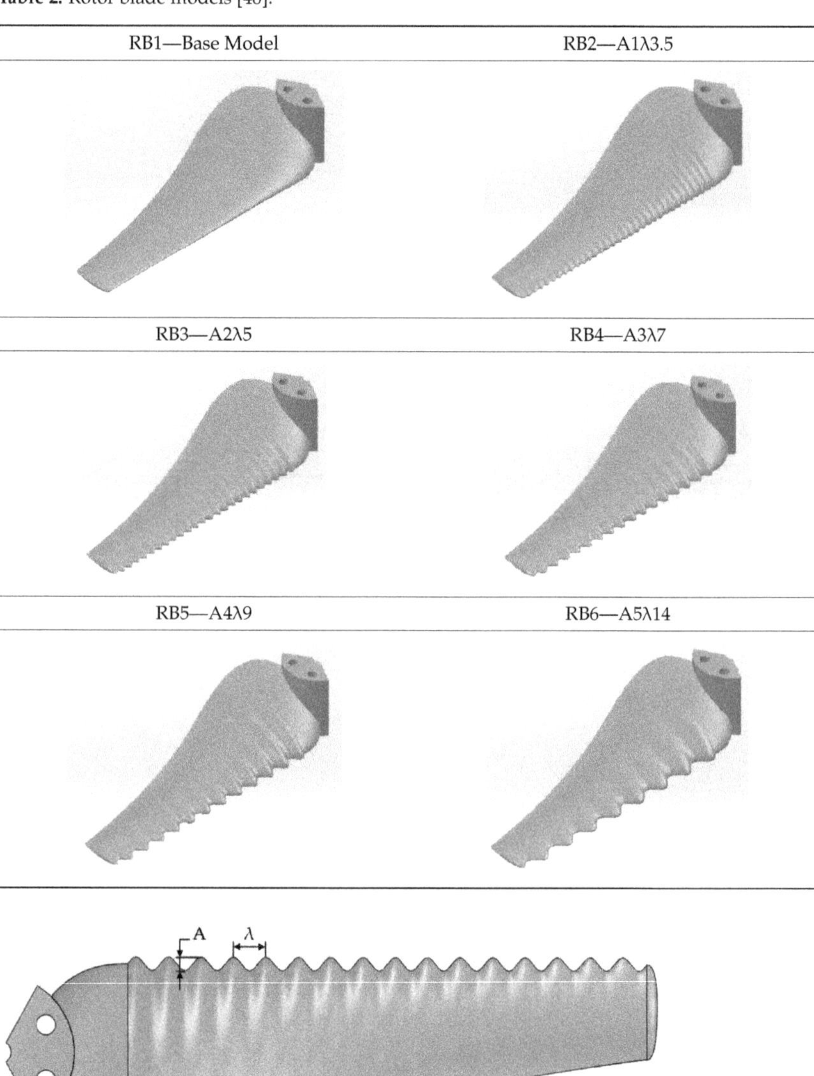

RB3—A2λ5	RB4—A3λ7

RB5—A4λ9	RB6—A5λ14

Figure 5. Leading-edge tubercle parameters.

2.3. Experimental Setup

For this experimental investigation, which originates from a PhD thesis [46], a wind tunnel with a square cross-section test section was utilized, capable of achieving velocities up to 30 m/s (depicted in Figure 6). The test section possessed a square cross-sectional shape measuring 0.58 m × 0.58 m at the entry. Constructed from Plexiglas material, the side walls of the test section are slightly angled (0.3°) towards the exit, ensuring a uniform static pressure throughout the entirety of the test area. The turbulence intensity within the wind tunnel remains below 1% for flow velocities between 3 and 20 m/s.

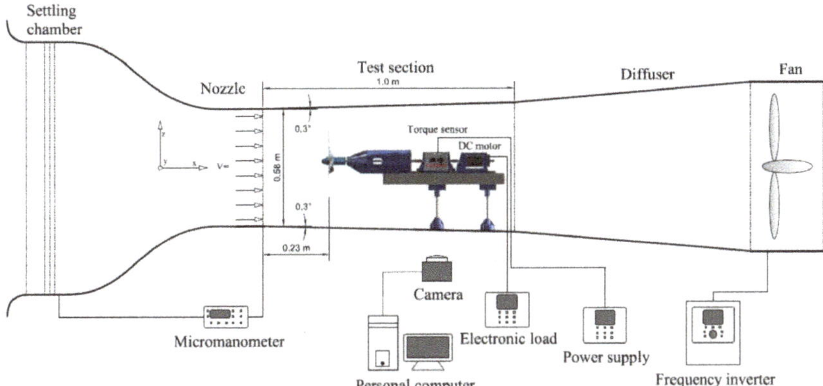

Figure 6. A schematic view of the wind tunnel setup [46].

Velocity, temperature, and pressure measurements were conducted using a Pitot-static tube micro-manometer (ManoAir 500, Schiltknecht Messtechnik AG, Schaffhausen, Switzerland). The speed of the rotor shaft was monitored using an optical laser sensor (Monarch Instrument ROS, USA), which was connected to an external data acquisition and analysis device known as OROS OR35 (USA). To regulate the wind tunnel's adjustable flow velocity, a frequency inverter (Schneider electric, Altivar 71, 4 kW, France) was employed. To gauge the mechanical torque and rotational speed of the rotor shaft, a rotary torque sensor (DYN-200, China) was utilized.

As shown in Figures 6 and 7, the wind rotor is horizontally mounted in the test area, connected via two rod supports to the lower wall of the test section. The torque sensor is positioned between the wind rotor and the DC motor (Maxon RE50 Ø 50 mm, Switzerland) and is powered by the DC power supply (TT T-ECHNI-C YH-605D, China). Additionally, an electronic load device known as Rigol DL3021 Precision (China) is connected with the DC motor, which functions as a braking mechanism. This setup ensures that the desired rpm of the rotor shaft is upheld, regardless of changing flow conditions.

For each test session, the procedures were replicated three times, and the resulting power coefficient curves illustrate the average values derived from these measurements. Wind tunnel data was gathered using the NI PCle-6323 data acquisition card (DAQ) (National Instruments, Austin, TX, USA) to acquire voltage signals by means of the MiniCTA software (Dantec Dynamics, Denmark), which is integrated into the computer system.

Figure 7. Turbine rotor mounted in the wind tunnel.

2.3.1. Testing Method

It is suggested by several authors that when downscaling small wind turbine models for wind tunnel testing, due to the high sensitivity of the flow regime, matching the Reynolds number in addition to the tip-speed ratio and geometric scaling is a necessity [51,52]. However, scaling in such a way led to impractical test conditions as it required high wind speeds of around 19.3 m/s, which manifested in vibrations of the entire turbine structure in the wind tunnel test section. Thus, taking into account the current laboratory conditions and the suggestion by Post and Boirum [53], who emphasize the vital importance of maintaining the same tip-speed ratio for such conditions and bypassing the achievement of matching the Reynolds number, it was decided to conduct experiments at lower wind speeds.

The appropriate method for testing and measuring the aerodynamic characteristics of the proposed rotors was implemented using the baseline rotor blade model (RB1).

In this study, considering the constraints of the laboratory setup [46], it was advantageous to conduct experimental investigations by maintaining the tip chord-based Reynolds number unchanged while varying flow speeds. This approach involved keeping the relative velocity constant at the blade tip within a range of tip-speed ratios from 2 to 5, as recommended by Kishore et al. [6].

The testing methodology, according to the reference literature [46], was executed while considering a rotor speed of 2866 rpm and its corresponding blade-tip speed of 42 m/s, given a rotor radius of 140 mm. This setup resulted in a calculated relative velocity of 43.29 m/s at the blade tip, achieved for a tip-speed ratio of 4 and wind speed of 10.5 m/s, in accordance with Equation (2). The chosen tip-speed ratio was an intermediate value, falling between 3 and 5, and was coupled with wind speeds ranging from 8.4 m/s to 14 m/s. The resulting blade-tip-chord-based Reynolds number for this specific scenario was 4.7×10^4, based on Equation (1). This computation considered a blade-tip chord length of 20 mm, an approximate air density of 1 kg/m^3, a relative velocity of 43.29 m/s, and an air dynamic viscosity of 18.56×10^{-6}. The atmospheric pressure was taken as 86 kPa, representing the conditions in Niğde town (Turkey), accompanied by an air temperature of 27 °C.

In the initial testing phase, it was observed that the rotor models did not initiate rotation at flow velocities below approximately 10 m/s due to the resistive loads generated by the bearings and the DC electric motor. Consequently, a higher initial flow velocity was employed to initiate rotor motion, which was then subsequently adjusted to the desired operating flow speed.

2.3.2. Tunnel Blockage Correction

In the present study, wind tunnel blockage was calculated using Equation (8) and was about 18.9% [46]; this value is about twice as large as the limit of 10% suggested by the literature [9,54], below which correction of the measured results is not a necessity.

$$B_T = \frac{B \cdot A_{rotor}}{A_{tunnel}} \tag{8}$$

where B—number of blades, A_{rotor}—rotor swept area, A_{tunnel}—cross-sectional area of test section.

The power coefficients for each case computed from the mechanical torque and tip-speed ratios were corrected for tunnel blockage effects according to the Van Treuren approach [9]. The blockage factor was determined separately for each rotor model being tested in the wind tunnel for each wind speed, with and without the rotor in the test section.

2.3.3. Uncertainty in the Measurement

The uncertainty associated with design parameters was determined using the Akbıyık approach [55] under the maximum performance condition of the baseline rotor model (RB1). This condition corresponds to a tip-chord-based Reynolds number of 4.7×10^4, which in turn corresponds to a wind speed of 11.29 m/s, a blade-tip speed of 42 m/s, a rotor shaft speed of 2866 rpm, an air density of approximately 1 kg/m^3, a local atmospheric pressure of 86 kPa, and a temperature of 27 °C [36,37].

Given these values, the uncertainty in flow speed was calculated to be approximately 1.3%, while the uncertainty in the tip-speed ratio was around 1.4%. The uncertainty in the power coefficient was estimated at 4.1%, and that of the Reynolds number was about 1.8%.

2.4. Validation of the Proposed Rotor Model

Due to the specific size of the scaled rotor, the authors were unable to find a model with the same characteristics, that is, designed with the same parameters and tested under the same conditions. Therefore, to validate the proposed baseline rotor model, the experimental results obtained in the wind tunnel were compared to the study of Lanzafame et al. [56], following the suggestion of Bakırcı and Yılmaz [50], who found that the power coefficient and the optimal tip-speed ratio does not depend on the radius of the wind turbine but on the number of blades. For this reason, the results of the wind turbine rotor from the literature with the approximate diameter and tip-speed ratio, and an equal number of blades, were obtained.

The characteristics of the selected rotor model for comparison purposes were as follows: rotor radius 112.5 mm, three blades, operational Reynolds number lower than 8×10^4, optimal tip-speed ratio ~3.3, NACA 4415 airfoil, rotor solidity ~19%, rotational speed 2450 rpm, wind speed range from 5 to 30 m/s, and blockage ratio 0.159.

In the following figure (Figure 8), the reference rotor blade geometry, such as chord and twist angle distributions, is depicted.

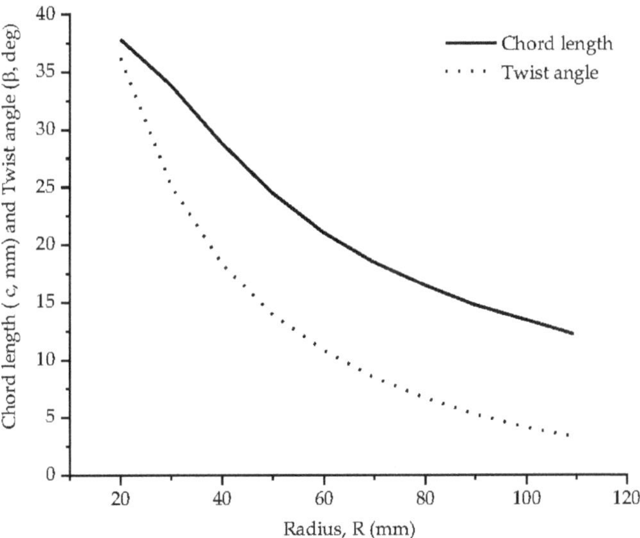

Figure 8. Chord and twist angle distributions from the literature [56].

3. Results and Discussion

The laboratory tests were performed to understand the effects of leading-edge tubercles on rotor blade models for several wind speeds between 8.5 and 15.5 m/s at the entry point of the test section. Measurements were taken at various tip-speed ratios, spanning the range from 2 to 5.

Firstly, a reference curve was determined from the experimental data obtained by using the RB1 rotor model as a base, which will serve later for the comparison of the obtained curves of the modified rotor models. Furthermore, the same rotor model was used to determine the proper performance testing method for the given conditions. In Figure 9, the RB1 baseline rotor model mounted on the support structure is presented.

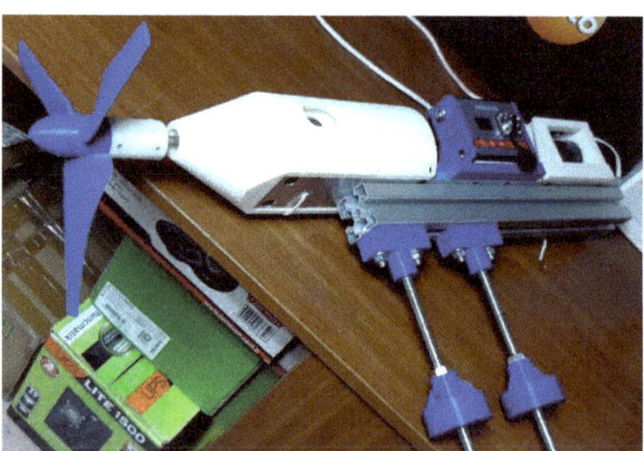

Figure 9. RB1 baseline rotor model mounted on the support structure.

3.1. Representative Power Coefficient Curve

Three experimental measurement cycles were derived to determine the representative average power coefficient curve of the corrected RB1 rotor model to be used later as a reference, in comparison with modified rotor blade models under the same flow conditions. During each measurement, the mechanical torque and rpm values were captured three times via a digital camera. Subsequently, the average power coefficient was calculated by correcting the recorded data. Figure 10 shows the representative power coefficient curve, which was obtained using a non-linear polynomial fit. The total error in the experimental data represents a scatter of $\pm 4.1\%$. As shown in the figure, the stability of the rotor was affected by the tip-speed ratio, especially for higher values.

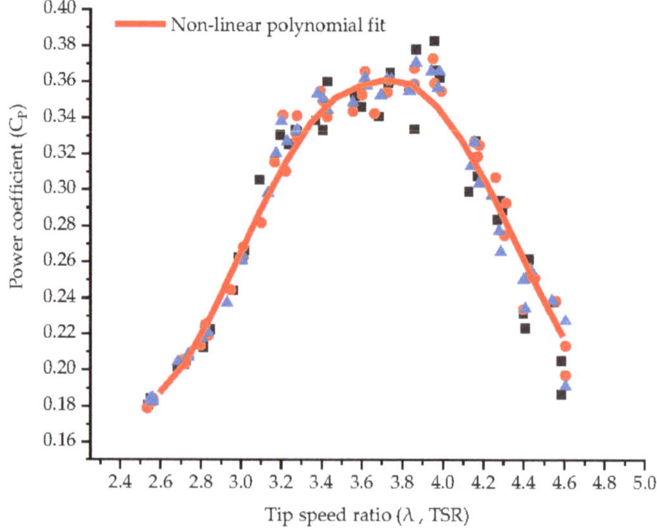

Figure 10. Representative power coefficient (C_p) against tip speed ratio (λ, TSR) curve with all measurement points [46].

As is evident from the figure, the curve's shape is distinctly characterized by a steep incline, a maximum value, and then a steep decline, indicating that the rotor ran fast at small tip-speed ratios, while decreasing quickly after reaching its maximum power coefficient. The highest power coefficient, of 0.361, was attained at a tip-speed ratio of 3.717, which closely aligns with the design tip-speed ratio of 3.658.

This graph was used as a reference model for comparison with the other rotor model configurations.

3.2. RB1 Baseline Rotor Model and Validation

Figure 11 depicts the experimental result of the rotor model from the literature [56], in terms of the power coefficient and tip-speed ratio, obtained for approximately similar test conditions. By comparing the results from Figures 10 and 11, it can be seen that the rotor model discussed in the literature demonstrates improved performance at lower tip-speed ratios, specifically up to 3.2, which is near its optimal tip-speed ratio. At this range, it achieves a peak power coefficient of approximately 0.294. After this point, the performance of the rotor drops significantly with the increase in the tip-speed ratio up to around 4.5. This could be expected due to the different rotor design parameters, one of which is the airfoil profile. It is therefore acceptable to validate the chosen approach when there are no data for rotor models designed with the same parameters and tested for the same conditions.

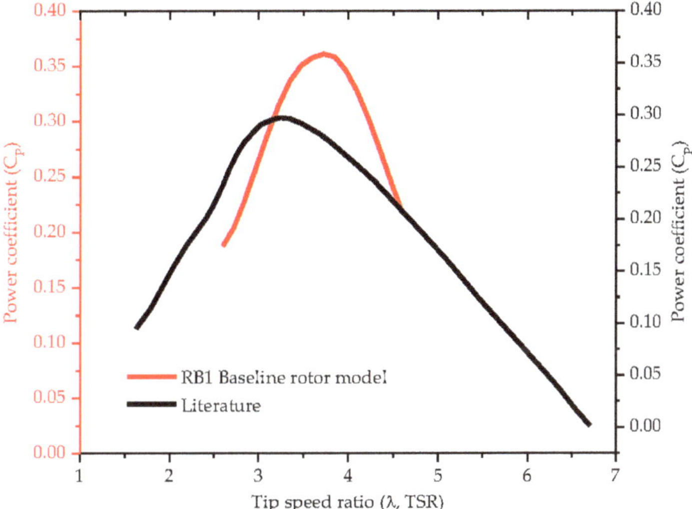

Figure 11. Power coefficient (C_p) against tip-speed ratio (λ, TSR) of the RB1 baseline rotor model and the rotor from the reference literature [46,56].

3.3. Wind Tunnel Blockage Correction

Due to the considerable presence of the rotor within the test section, it became essential to implement blockage corrections to account for the influence of interfering objects. A blockage factor was separately determined for all rotor models for each free stream velocity as the ratio of the measured wind speed with and without the presence of the rotor inside the test section [9]. Figure 12 shows the situation before and after application of the tunnel blockage factor to the power coefficient curve for the RB1 rotor model. It is evident that at lower tip-speed ratios, the impact of the blockage ratio is minimal.

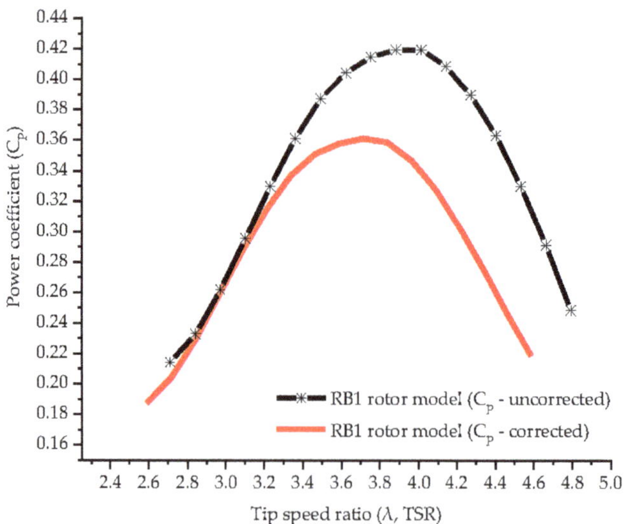

Figure 12. Power coefficient curves (C_p) against tip-speed ratio (λ, TSR) for tunnel blockage [46].

3.4. RB2 Rotor Model

Figure 13 illustrates the power coefficient curves for the RB1 and RB2 rotor models. This particular model configuration is characterized by the smallest wave parameters, having an amplitude of 1 mm and a wavelength of 3.5 mm. The graph shows that by changing the shape of the leading edge of the RB1 rotor model for the same working conditions, the efficiency increases and consequently fulfills the purpose of the experiment for improving the effectiveness of the blade shape.

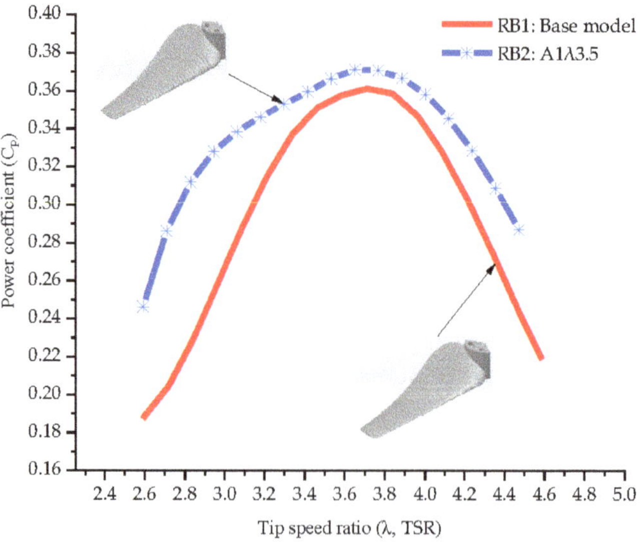

Figure 13. Power coefficient (C_p) against tip-speed ratio (λ, TSR) for the RB2 rotor model [46].

According to the graph, the RB2 rotor model has a wider shape of the power coefficient curve profile, which extends over the RB1 rotor model. This means that this model exhibited better performance characteristics compared to the RB1 rotor model over the entire range of tip-speed ratios, especially from tip-speed ratios from 2.5 to 3.5, where the wind speed is high, and from 4 to 4.5 for lower flow speeds. As depicted in Figure 13, the RB2 rotor model showed an increase in performance of about 40% compared to the RB1 rotor model for low tip-speed ratios. The highest power coefficient of 0.371 was obtained for $\lambda = 3.654$, which is close to the design tip-speed ratio. Meanwhile, this model exhibits a maximum power coefficient that is approximately 2.8% higher than that of the RB1 rotor model. Furthermore, it maintains its advantages across the entire range of tip-speed ratios.

3.5. RB3 Rotor Model

As the amplitude and wavelength values are increased from 1 to 2 mm and from 3.5 to 5 mm, respectively, a notable decrease in the model's performance is observed. However, within the range of tip-speed ratios from 3.46 to 4.1, this model exhibited poorer performance compared to the RB1 rotor model. Figure 14 visually represents the power coefficient curves for the RB1 and RB3 rotor models. Notably, the maximum power coefficient of 0.352 is achieved at a wavelength of 3.46, which is approximately 2.6% lower than that of the RB1 rotor model and 5.4% lower than that of the RB2 rotor model. However, an important characteristic can be noticed, namely, that the maximum power coefficient is shifted to a smaller value of the tip-speed ratio and there is a rapidly growing trend beyond the tip-speed ratio of 4.1.

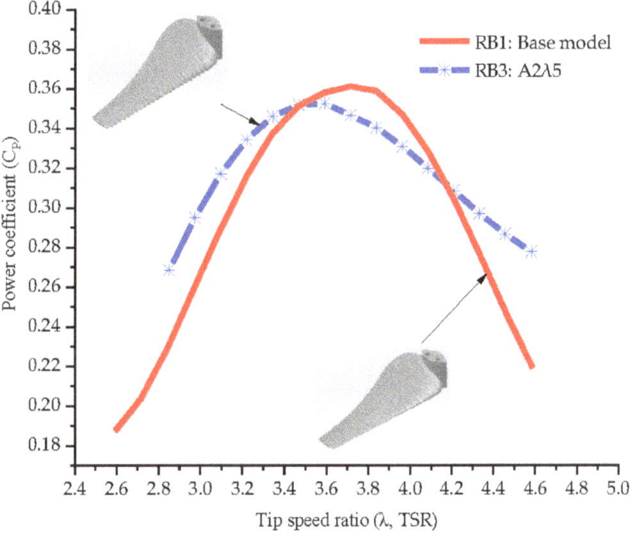

Figure 14. Power coefficient (C_p) against tip-speed ratio (λ, TSR) for the RB3 rotor model [46].

3.6. RB4 Rotor Model

Figure 15 displays the power coefficient curves for the RB1 and RB4 rotor models. The power coefficient curves of RB4 and RB1 rotor models show remarkable similarity for tip-speed ratios up to 3.15, with only a slight divergence beyond 4.21. In the range between these two values, there is a discernible disadvantage of 9.1% of the RB3 rotor model compared to the RB1 rotor model, and approximately 12.1% compared to the RB2 rotor model. The highest power coefficient of 0.331 is obtained at a wavelength of 3.465. It is also apparent that the highest performance of this profile is attained at nearly the same tip-speed ratio as the RB3 rotor model.

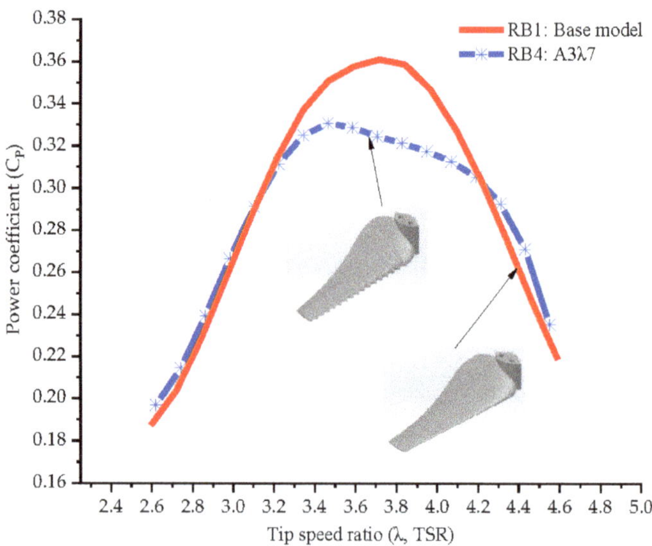

Figure 15. Power coefficient (C_p) against tip-speed ratio (λ, TSR) for the RB4 rotor model [46].

3.7. RB5 Rotor Model

As the amplitude quadruples and the wavelength almost triples, a trend of continuous decline in the model efficiency is observed. Figure 16 illustrates the power coefficient of RB1 and RB5 rotor models. The graph depicted indicates that the RB5 rotor model exhibited lower performance compared to the RB1 rotor model up to a tip-speed ratio of 4.35. However, beyond this point, an upward trend in performance becomes evident. The peak power coefficient of 0.321 is achieved at a wavelength of 3.679.

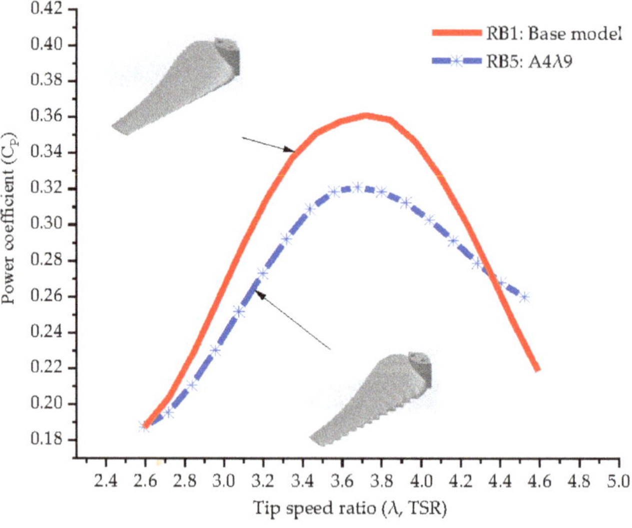

Figure 16. Power coefficient (C_p) against tip-speed ratio (λ, TSR) for the RB5 rotor model [46].

3.8. RB6 Rotor Model

As shown in Figure 17, the RB6 rotor model did not exhibit any improvement throughout the examined range of tip-speed ratios. Referring to the graph, there is a significant decrease of approximately 51% in rotor performance for intermediate tip-speed ratios in comparison to the RB1 rotor model. The peak power coefficient achieved by this model is approximately 0.299, occurring at a wavelength of 3.813.

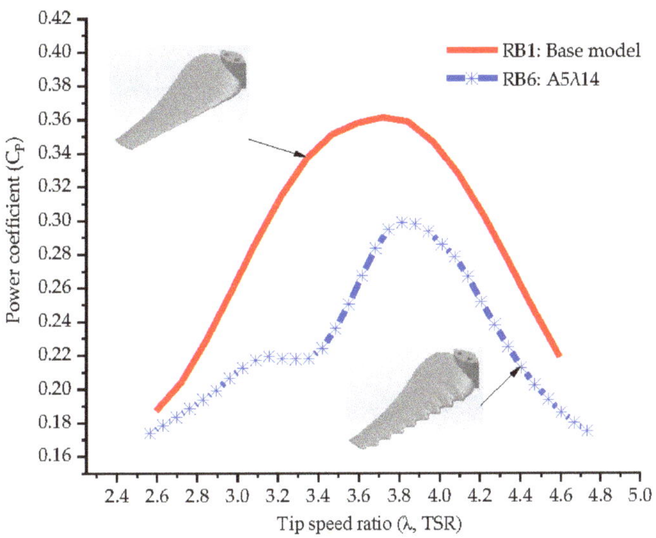

Figure 17. Power coefficient (C_p) against tip-speed ratio (λ, TSR) for the RB6 rotor model [46].

4. Conclusions

This study focused on examining the aerodynamic efficiency of different designs of small-scale horizontal axis wind turbine rotor blades, applying a passive flow control approach. The leading-edge tubercle technique was employed to produce various configurations by modifying the blade's geometry. The rotor blades were constructed using the NREL S822 airfoil, encompassing the entire span from the blade's root to its tip. This design aimed to satisfy both structural integrity and aerodynamic performance considerations. In this paper, five distinct configurations (RB2, RB3, RB4, RB5, and RB6) were generated, incorporating tubercles along the entire length of the leading edge. These configurations were generated by varying the amplitude and the wavelength of the tubercles. The baseline rotor model (RB1) underwent validation and served as a reference point for comparison with the modified rotor blade models. All tests were conducted under stable operating conditions.

Based on the results obtained from the wind tunnel experiments, the specific conclusions are as follows:

- The baseline rotor model (RB1) exhibited a higher peak power coefficient in comparison to the rotor model referenced from the literature. This higher coefficient was achieved for a nearly identical tip-speed ratio. Nonetheless, when considering relatively similar design parameters, such as rotor solidity, number of blades, tunnel blockage, and design tip-speed ratio, but with distinct airfoil profiles, the study's results did not reveal any consistent patterns or trends.

- Among the various leading-edge tubercle configurations investigated, the RB2 rotor model with an amplitude of 1 mm and a wavelength of 3.5 mm outperformed all other rotor models, including the baseline rotor model (RB1), across the entire range of tip-speed ratios examined in this study. For low tip-speed rations, this model demonstrated a significant performance improvement of around 40% compared to the baseline rotor model (RB1). The reduction in this advantage to approximately 2.8% occurred when the RB2 rotor model reached its peak power coefficient. The improved performance of the RB2 rotor model compared to the RB1 rotor model can be attributed to factors such as the increased number of bumps, their subtle dimensions, and enhanced surface roughness. These characteristics collectively altered the airflow structure, influencing both chordwise and spanwise directions. This effect was particularly pronounced during high flow speeds, aligning with findings highlighted in the literature review. Furthermore, it is apparent that the configuration, along with the subtle disparity in the amplitudes of the initial and terminal bumps situated at both blade ends, play a crucial role. According to the existing literature, these specific characteristics can influence the intensity of root and tip vortices of the blade, particularly under conditions of elevated flow speeds.

- The RB3 rotor model demonstrated its effectiveness primarily within the range of smaller values of tip-speed ratios, approximately up to 3.5, when compared to the RB1 rotor model.
- A marginal advantage of the RB4 rotor model over RB1 was observed at tip-speed ratios up to 3.15 and beyond 4.21.
- As the amplitude and wavelength are further increased, the rotor's efficiency experiences a relatively gradual decline, while still maintaining the general shape of the curve.

According to the experimental findings, it was evident that the chosen methodology, which involved utilizing the S822 airfoil to fulfill both aerodynamic and structural considerations, along with maintaining consistent profile usage across the blade length, successfully achieved the objectives set forth in this study.

Given the experimental nature of this study, it would be interesting to further extend the research by incorporating Computational Fluid Dynamics (CFD) simulations. By integrating CFD code, a more comprehensive insight into the flow physics surrounding the blade and the rotor in general could be obtained. This enhanced understanding would shed light on the precise effects of the applied technique, offering a more comprehensive perspective on the subject.

Author Contributions: Conceptualization, R.M.; Methodology, R.M. and Y.E.A.; Software, R.M.; Validation, Y.E.A.; Investigation, R.M.; Resources, R.M.; Data curation, R.M.; Writing—original draft, R.M.; Writing—review & editing, Y.E.A.; Visualization, R.M.; Supervision, Y.E.A. All authors have read and agreed to the published version of the manuscript.

Funding: This research received no external funding.

Data Availability Statement: Not applicable.

Conflicts of Interest: The authors declare no conflict of interest.

Nomenclature

α	angle of attack
A	wave amplitude
A_{rotor}	rotor swept area
A_{tunnel}	cross-sectional area of test section
β_i	local twist angle
B	number of blades
B_T	tunnel blockage
c_{avg}	average chord length of the blade
c_i	local chord
C_L	lift coefficient
C_D	drag coefficient
C_P	power coefficient

φ_i	local inflow angle
λ	tip-speed ratio
λ_i	local tip-speed ratio
λ_w	wavelength
μ	dynamic viscosity of the air
N	number of blade elements
r_i	local radius
R	rotor radius
Re	Reynolds number
ρ	air density
σ	rotor solidity
y	sine wave equation
y_0	location of the starting point of the sine wave path
V_{tip}	blade-tip speed
V_{rel}	relative velocity
V_∞	wind speed
x	extension of the bumps from root to the blade tip

References

1. Emeis, S. *Wind Energy Meteorology*; Springer: Berlin/Heidelberg, Germany, 2013; ISBN 978-3-642-30522-1.
2. Gharaati, M.; Xiao, S.; Wei, N.J.; Martínez-Tossas, L.A.; Dabiri, J.O.; Yang, D. Large-eddy simulation of helical- and straight-bladed vertical-axis wind turbines in boundary layer turbulence. *J. Renew. Sustain. Energy* **2022**, *14*, 053301. [CrossRef]
3. Statista.com. Available online: https://www.statista.com/statistics/269885/capacity-of-small-wind-turbines-worldwide/ (accessed on 15 January 2023).
4. Shah, H.; Mathew, S.; Lim, C.M. A novel low reynolds number airfoil design for small horizontal axis wind turbines. *Wind Eng.* **2014**, *38*, 377–391. [CrossRef]
5. Wood, D. *Small Wind Turbines: Analysis, Design, and Application (Green Energy and Technology)*; Springer: Berlin/Heidelberg, Germany, 2011.
6. Kishore, R.; Stewart, C. Priya *Wind Energy Harvesting: Micro-to-Small Scale Turbines*; Walter de Gruyter GmbH & Co KG: Berlin, Germany, 2018.
7. Winslow, J.; Otsuka, H.; Govindarajan, B.; Chopra, I. Basic Understanding of Airfoil Characteristics at Low Reynolds Numbers (104–105). *J. Aircr.* **2018**, *55*, 1050–1061. [CrossRef]
8. Giguère, P.; Selig, M.S. Low Reynolds number airfoils for small horizontal axis wind turbines. *Wind Eng.* **1997**, *21*, 367–380.
9. Van Treuren, K.W. Small-scale wind turbine testing in wind tunnels under low reynolds number conditions. *J. Energy Resour. Technol. Trans. ASME* **2015**, *137*, 051208. [CrossRef]
10. Hau, E. *Wind Turbines: Fundamentals, Technologies, Application, Economics*; Springer Science & Business Media: Berlin/Heidelberg, Germany, 2013; ISBN 9783642271519.
11. Schmitz, S. *Aerodynamics of Wind Turbines: A Physical Basis for Analysis and Design*; Wiley: Hoboken, NJ, USA, 2020; ISBN 9781119405597.
12. Miklosovic, D.S.; Murray, M.M.; Howle, L.E.; Fish, F.E. Leading-edge tubercles delay stall on humpback whale (*Megaptera novaeangliae*) flippers. *Phys. Fluids* **2004**, *16*, L39–L42. [CrossRef]
13. Fish, F.E.; Battle, J.M. Hydrodynamic design of the humpback whale flipper. *J. Morphol.* **1995**, *225*, 51–60. [CrossRef]
14. Lissaman, P.B.S. Low-Reynolds-number airfoils. *Annu. Rev. Fluid Mech.* **1983**, *15*, 223–239. [CrossRef]
15. Burton, T.; Jenkins, N.; Sharpe, D.; Bossanyi, E. *Wind Energy Handbook*, 2nd ed.; John Wiley & Sons: Hoboken, NJ, USA, 2011; ISBN 9780470699751.
16. Manwell, J.F.; McGowan, J.G.; Rogers, A.L. *Wind Energy Explained: Theory, Design and Application*; John Wiley & Sons: Hoboken, NJ, USA, 2010; ISBN 9780470015001.
17. Hansen, M.O.L. Aerodynamics and Design of Horizontal-Axis Wind Turbines. In *Wind Energy Engineering: A Handbook for Onshore and Offshore Wind Turbines*; Academic Press: Cambridge, MA, USA, 2017; ISBN 9780128094297.
18. McLean, D. *Understanding Aerodynamics: Arguing from the Real Physics*; Wiley: Hoboken, NJ, USA, 2012; ISBN 9781118454220.
19. Mathew, S.; Geeta Susan, P. *Advances in Wind Energy Conversion Technology*; Springer Science & Business Media: Berlin/Heidelberg, Germany, 2011; ISBN 978-3-540-88257-2.
20. Burdett, T.; Gregg, J.; Van Treuren, K. A Numerical Study on Improving Airfoil Performance at Low Reynolds Numbers for Small-Scale Wind Turbines using Intentional Roughness. In Proceedings of the 50th AIAA Aerospace Sciences Meeting Including the New Horizons Forum and Aerospace Exposition, Nashville, TN, USA, 9–12 January 2012; American Institute of Aeronautics and Astronautics: Reston, VA, USA, 2012.
21. Primrose, S.B. *Biomimetics: Nature-Inspired Design and Innovation*; Wiley: Hoboken, NJ, USA, 2020; ISBN 9781119683322.
22. Miklosovic, D.S.; Murray, M.M.; Howle, L.E. Experimental evaluation of sinusoidal leading edges. *J. Aircr.* **2007**, *44*, 1404–1408. [CrossRef]

23. Bolzon, M.D.P.; Kelso, R.M.; Arjomandi, M. Formation of vortices on a tubercled wing, and their effects on drag. *Aerosp. Sci. Technol.* **2016**, *56*, 46–55. [CrossRef]
24. Haque, M.N.; Ali, M.; Ara, I. Experimental Investigation on the Performance of NACA 4412 Aerofoil with Curved Leading Edge Planform. *Procedia Eng.* **2015**, *105*, 232–240. [CrossRef]
25. Hansen, K.L.; Kelso, R.M.; Dally, B.B. Performance variations of leading-edge tubercles for distinct airfoil profiles. *AIAA J.* **2011**, *49*, 185–194. [CrossRef]
26. Sudhakar, S.; Venkatakrishnan, L.; Ramesh, O.N. The influence of leading-edge tubercles on airfoil performance at low Reynolds numbers. In Proceedings of the AIAA SciTech 2020 Forum, Orlando, FL, USA, 6–10 January 2020.
27. Johari, H.; Henoch, C.; Custodio, D.; Levshin, A. Effects of leading-edge protuberances on airfoil performance. *AIAA J.* **2007**, *45*, 2634–2642. [CrossRef]
28. Guerreiro, J.L.E.; Sousa, J.M.M. Low-Reynolds-Number Effects in Passive Stall Control Using Sinusoidal Leading Edges. *AIAA J.* **2012**, *50*, 461–469. [CrossRef]
29. Wei, Z.; New, T.H.; Cui, Y.D. An experimental study on flow separation control of hydrofoils with leading-edge tubercles at low Reynolds number. *Ocean Eng.* **2015**, *108*, 336–349. [CrossRef]
30. Bolzon, M.D.; Kelso, R.M.; Arjomandi, M. The effects of tubercles on swept wing performance at low angles of attack. In Proceedings of the 19th Australasian Fluid Mechanics Conference (AFMC 2014), Melbourne, Australia, 8–11 November 2014.
31. Natarajan, K.; Sudhakar, S.; Paulpandian, S. Experimental Studies On The Effect Of Leading Edge Tubercles On Laminar Separation Bubble. In Proceedings of the 52nd Aerospace Sciences Meeting, National Harbor, MD, UDA, 13–17 January 2014; American Institute of Aeronautics and Astronautics: Reston, VA, USA, 2014.
32. Gasch, R.; Twele, J. *Wind Power Plants: Fundamentals, Design, Construction and Operation*, 2nd ed.; Springer Science & Business Media: Berlin/Heidelberg, Germany, 2012; ISBN 9783642229381.
33. Rosato, M.A. *Small Wind Turbines for Electricity and Irrigation*; CRC Press: Boca Raton, FL, USA, 2018.
34. Herráez, I.; Akay, B.; van Bussel, G.J.W.; Peinke, J.; Stoevesandt, B. Detailed analysis of the blade root flow of a horizontal axis wind turbine. *Wind Energy Sci.* **2016**, *1*, 89–100. [CrossRef]
35. Branlard, E. Wind Turbine Tip-Loss Corrections: Review, Implementation and Investigation of New Models. Master's Thesis, Technical University of Denmark, Lyngby, Denmark, 2011.
36. Døssing, M.; Madsen, H.A.; Bak, C. Aerodynamic optimization of wind turbine rotors using a blade element momentum method with corrections for wake rotation and expansion. *Wind Energy* **2012**, *15*, 563–574. [CrossRef]
37. Abate, G.; Mavris, D.N.; Sankar, L.N. Performance effects of leading edge tubercles on the NREL Phase VI wind turbine blade. *J. Energy Resour. Technol. Trans. ASME* **2019**, *141*, 051206. [CrossRef]
38. Carreira Pedro, H.; Kobayashi, M. Numerical Study of Stall Delay on Humpback Whale Flippers. In Proceedings of the 46th AIAA Aerospace Sciences Meeting and Exhibit, Reno, NV, USA, 7–10 January 2008; American Institute of Aeronautics and Astronautics: Reston, VA, USA, 2008.
39. van Nierop, E.A.; Alben, S.; Brenner, M.P. How Bumps on Whale Flippers Delay Stall: An Aerodynamic Model. *Phys. Rev. Lett.* **2008**, *100*, 054502. [CrossRef]
40. Abate, G.; Mavris, D.N. Performance Analysis of Different Positions of Leading Edge Tubercles on a Wind Turbine Blade. In Proceedings of the 2018 Wind Energy Symposium, Kissimmee, FL, USA, 8–12 January 2018; American Institute of Aeronautics and Astronautics: Reston, VA, USA, 2018.
41. Zhang, R.-K.; Wu, V.D.J.-Z. Aerodynamic characteristics of wind turbine blades with a sinusoidal leading edge. *Wind Energy* **2012**, *15*, 407–424. [CrossRef]
42. Huang, G.-Y.; Shiah, Y.C.; Bai, C.-J.; Chong, W.T. Experimental study of the protuberance effect on the blade performance of a small horizontal axis wind turbine. *J. Wind Eng. Ind. Aerodyn.* **2015**, *147*, 202–211. [CrossRef]
43. Kim, H.; Kim, J.; Choi, H. Flow structure modifications by leading-edge tubercles on a 3D wing. *Bioinspir. Biomim.* **2018**, *13*, 066011. [CrossRef] [PubMed]
44. Ng, B.F.; New, T.H.D.; Palacios, R. Bio-inspired Leading-Edge Tubercles to Improve Fatigue Life in Horizontal Axis Wind Turbine Blades. In Proceedings of the 35th Wind Energy Symposium, Grapevine, TX, USA, 9–13 January 2017; American Institute of Aeronautics and Astronautics: Reston, VA, USA, 2017.
45. Sunada, S.; Sakaguchi, A.; Kawachi, K. Airfoil Section Characteristics at a Low Reynolds Number. *J. Fluids Eng.* **1997**, *119*, 129–135. [CrossRef]
46. Morina, R. Development of a Horizontal Axis Wind Turbine Blade Profile Based on Passive Flow Control Methods. Ph.D. Thesis, Niğde Ömer Halisdemir University, Niğde, Turkey, 2021.
47. Tangler, J.; Somers, D. *NREL Airfoil Families for HAWTs*; National Renewable Energy Lab. (NREL): Golden, CO, USA, 1995.
48. Brusca, S.; Lanzafame, R.; Messina, M. Flow similitude laws applied to wind turbines through blade element momentum theory numerical codes. *Int. J. Energy Environ. Eng.* **2014**, *5*, 313–322. [CrossRef]
49. Duquette, M.M.; Swanson, J.; Visser, K.D. Solidity and blade number effects on a fixed pitch, 50W horizontal axis wind turbine. *Wind Eng.* **2003**, *27*, 299–316. [CrossRef]
50. Bakırcı, M.; Yılmaz, S. Theoretical and computational investigations of the optimal tip-speed ratio of horizontal-axis wind turbines. *Eng. Sci. Technol. Int. J.* **2018**, *21*, 1128–1142. [CrossRef]

51. Burdet, T.A.; Van Treuren, K.W. Scaling small-scale wind turbines for wind tunnel testing. In Proceedings of the ASME Turbo Expo 2012: Turbine Technical Conference and Exposition, Copenhagen, Denmark, 11–15 June 2012.

52. Burdett, T.A.; Van Treuren, K.W. A Theoretical and Experimental Comparison of Optimizing Angle of Twist Using BET and BEMT. In Proceedings of the ASME Turbo Expo 2012: Turbine Technical Conference and Exposition, Copenhagen, Denmark, 11–15 June 2012; Volume 6: Oil and Gas Applications; Concentrating Solar Power Plants; Steam Turbines; Wind Energy. American Society of Mechanical Engineers: New York, NY, USA, 2012; pp. 797–809.

53. Post, S.; Boirum, C. A Model Wind Turbine Design-Build-Test Project. In Proceedings of the ASME 2011 International Mechanical Engineering Congress and Exposition, Denver, CO, USA, 11–17 November 2011; Volume 5: Engineering Education and Professional Development. ASME: New York, NY, USA, 2011; pp. 213–218.

54. Katz, J. *Automotive Aerodynamics*; Kurfess, T., Ed.; Wiley: Hoboken, NJ, USA, 2016; ISBN 9781119185727.

55. Coleman, H.W.; Steele, W.G. *Experimentation, Validation, and Uncertainty Analysis for Engineers: Fourth Edition*; John Wiley & Sons: Hoboken, NJ, USA, 2018; ISBN 9781119417989.

56. Lanzafame, R.; Mauro, S.; Messina, M. Numerical and experimental analysis of micro HAWTs designed for wind tunnel applications. *Int. J. Energy Environ. Eng.* **2016**, *7*, 199–210. [CrossRef]

Review

Photothermal and Photovoltaic Utilization for Improving the Thermal Environment of Chinese Solar Greenhouses: A Review

Gang Wu [1,2], Hui Fang [1,2,*], Yi Zhang [1,2], Kun Li [1,2] and Dan Xu [3,*]

[1] Institute of Environment and Sustainable Development in Agriculture, Chinese Academy of Agriculture Sciences, Beijing 100081, China; wugang@caas.cn (G.W.); zhangyi@caas.cn (Y.Z.); likun@caas.cn (K.L.)
[2] Key Laboratory of Energy Conservation and Waste Management of Agricultural Structures, Ministry of Agriculture, Beijing 100081, China
[3] College of Water Resources and Civil Engineering, China Agricultural University, Beijing 100083, China
* Correspondence: fanghui@caas.cn (H.F.); xudan@cau.edu.cn (D.X.)

Abstract: A Chinese solar greenhouse (CSG) is an agricultural facility type with Chinese characteristics. It can effectively utilize solar energy during low-temperature seasons in alpine regions. The low construction and operation costs make it a main facility for agricultural production in the northern regions of China. It plays an extremely important role in "Chinese vegetable basket projects". Energy is one of the important issues faced by CSGs. The better climate resources in the northern regions of China make it possible to apply solar energy as a green and sustainable energy source in the production of CSGs. Faced with the increasingly serious environmental problems of the new century, the Chinese government has made a decision and put it into practice to improve the rate and efficiency of solar energy utilization in agricultural facilities. In this paper, we summarize the research on the application of photovoltaic power generation and solar thermal technology in CSGs. The application of these advanced solar technologies has made great progress. With the further improvement of economic benefits and the establishment of relevant support policies and incentive mechanisms, the combination of CSG and solar energy technology will have certain application prospects and satisfy China's requirements for long-term sustainable development.

Keywords: Chinese solar greenhouse; sustainable agriculture; photothermal; photovoltaic

Citation: Wu, G.; Fang, H.; Zhang, Y.; Li, K.; Xu, D. Photothermal and Photovoltaic Utilization for Improving the Thermal Environment of Chinese Solar Greenhouses: A Review. *Energies* **2023**, *16*, 6816. https://doi.org/10.3390/en16196816

Academic Editor: Ignacio Mauleón

Received: 15 August 2023
Revised: 11 September 2023
Accepted: 18 September 2023
Published: 26 September 2023

1. Introduction

In recent years, the global greenhouse horticulture field has always focused on "energy and water resources", especially when the continuous increase in energy prices has brought tremendous operational pressure to the growers [1]. Meanwhile, the implementation of the Kyoto Protocol and the Paris Climate Agreement has increased the pressure on the production of greenhouses in terms of environmental and ecological protection. Energy conservation and the broadening of the renewable energy range have become the subjects of greenhouse energy research [2]. Since 2006, renewable energy sources, such as solar energy, wind energy, biomass energy, etc., have been put into commercial operation in part of the greenhouses in Europe and America [3]. Related application research is also being comprehensively developed. In the "New energy greenhouse design contest" of 2007 in the Netherlands, a total of three schemes were awarded, including a Sunwind greenhouse and two other greenhouses designed by the universities of Wageningen and Benk [4]. These three greenhouses used solar energy, geothermal energy, and wind energy with related technologies to explore the technical mechanism and commercialization approach of reducing the future dependence on fossil energy in greenhouses. Related projects have been demonstrated in Bleiswijk with the government's support.

Facility horticulture in China started relatively late, due to strong and rigid demand for agricultural products (particularly vegetables) in the northern regions during the winter.

The facility horticulture area in China has developed rapidly over the past 30 years, and the current facility horticulture area reaches 3.7 million hectares, which makes it the country with the largest facility horticulture area in the world [5]. In facility horticulture, the development of plastic greenhouses and Chinese solar greenhouses (CSGs) is the most rapid, while the multi-span greenhouse is slow (Figure 1). The light-transmitting covering material of the multi-span greenhouse comprises a film, glass, and a plastic plate. Research showed that the coal consumption of the multi-span greenhouse is about 90–150 kg/m^2 in a year, and the coal cost accounts for 30–50% of the whole production cost [6]. Therefore, the multi-span greenhouse is difficult to develop on a large scale, even if the economy of the century develops to a certain level. The plastic-covered tunnel mostly uses bamboo and steel as a framework and is a simple arched greenhouse covered with plastic film. It is built when in use and dismantled when not in use. The little requirement for additional energy makes it mostly used in southern areas of China [7].

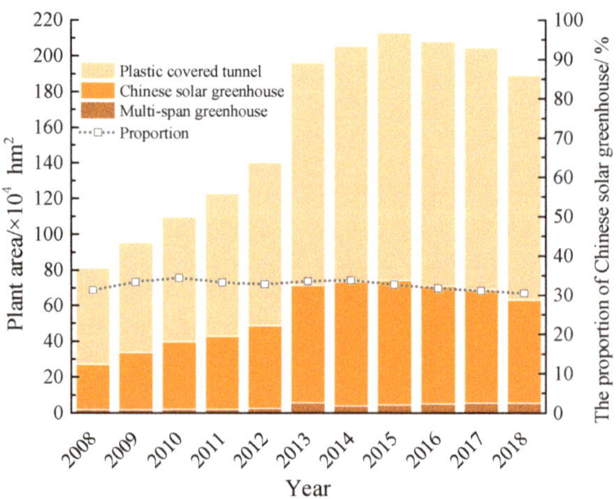

Figure 1. Facility for vegetable plant area and structure type in China [8].

Chinese agriculture is changing from traditional agriculture to modern agriculture. Although plastic-covered tunnels and CSGs are still important production facilities [9], high-yield and high-efficiency modern greenhouses are one of the directions of future Chinese agriculture development. With the continuous promotion of the modernization and industrialization of agriculture in China, the degree of dependence of agricultural production on energy has continuously increased. Compared with traditional agriculture, facility agriculture has a stronger dependence on energy. As shown in Figure 2, the climate and energy use patterns of agricultural facilities in China are very different from those in countries with advanced facility horticulture (http://solargis.com/maps-and-gis-data/download (accessed on 12 January 2023)). It is difficult to reduce energy consumption under complicated climate conditions in China without modifying Venlo-type multi-span greenhouses introduced from abroad.

A CSG is a facility type independently developed in China. It is a passive greenhouse with three wall sides and one light-transmitting side. Generally, solar energy is its only source of heat. At present, the overwintering production of vegetables in the northern regions of China completely depends on greenhouses, but the heat accumulated by the north walls of traditional CSGs is very limited [10]. Therefore, most CSGs need to be heated by coal. However, this heated mode faces problems of low efficiency, high pollution, and high investment and operation costs. Thus, the economic benefit of greenhouse cultivation is restricted. Especially in recent years, many cities in the north of China have come up with relevant policies to cope with haze disasters. As a result, the traditional coal-fired small boiler is banned. In this regard, it is important to select solar energy to replace coal for heating greenhouses [11]. Generally, illumination and temperature are considered the main environmental factors of the CSG, so the future development trend focuses on the research of structural optimization and heat insulation performance of the CSG.

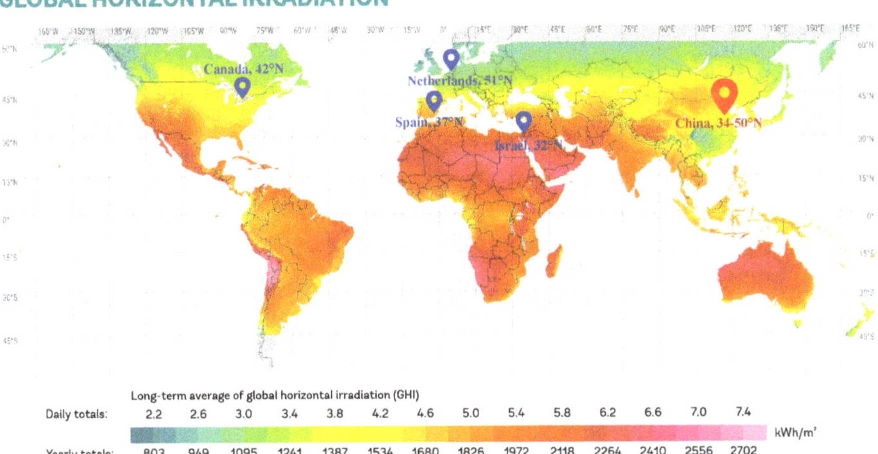

Figure 2. Overview of global horizontal solar radiation and typical regional facility agriculture distribution. This map is published by the World Bank Group, funded by ESMAP, and prepared by Solargis (http://globalsolaratlas.info), the time of browsing this page is March of 2023.

In China today, the capacity of solar water heaters and photovoltaic cells ranks first in the world. In recent years, the solar cell industry has faced the problem of excess capacity under the anti-dumping complaints of Europe and America on Chinese photovoltaic export products. The solar cell industry in China develops diversified consumption markets for destocking by combining related industries. The application demonstration of distributed photovoltaic power generation and the photovoltaic poverty relief policy promoted by the National Energy Bureau make the solar energy and agriculture industries complement each other [12]. After combining photovoltaic cells with CSG, the power generated by the solar energy system can be used for the functions of intelligent temperature control, irrigation, and light supplementation. Moreover, the residual electric quantity can be sold back to a power grid company to create additional benefits [13]. In addition, the photovoltaic CSG can effectively utilize land resources and avoid the problem of farmland occupation. In the critical period of vigorously promoting the development of modern agriculture in China, the development of the photothermal/photovoltaic CSG not only facilitates agricultural production, but also improves the agricultural technology level to a great extent.

This review discusses the limitations of conventional technologies and how hybrid configurations can help to overcome those barriers to boost the implementation of solar systems in protected agriculture. Section 1 introduces the motivation and objectives of this review article. Section 2 presents the different kinds of solar technology systems in CSG. Section 3 presents the most common conventional solar heating systems in CSG. Section 4 presents the most common conventional solar photovoltaic systems in CSG. Section 5 discusses the solar energy supply and energy demand in CSG. Finally, Section 6 presents the conclusions of this article.

The aim of the present work is to analyze the potential of hybrid configurations of PV and solar heating technologies to improve the light and thermal environment in CSG. This work is also a contribution for solar project developers; it helps to encourage the proliferation of solar systems in agricultural engineering.

2. Basic Conditions of CSGs

2.1. Conditions Related to Solar Greenhouses in the Representative Area

At present, in North China, vegetables consumed in winter are mainly supplied by means of facility agriculture, while in all the facilities, more than 90% are produced from CSGs [14]. The continuous popularization and construction of CSGs can increase the labor income of farmers, playing an important role in supplying fresh vegetables in northern areas in the winter. However, the distribution of the CSGs in China is extremely unbalanced; the CSGs in North China, Northeast China, and the Huanghuai area develop faster, while the CSGs in Shandong, Henan, and Liaoning account for the vast majority of the CSGs in China. The proportions of the provinces of Xinjiang, Qinghai, Gansu, Ningxia, and Shaanxi in the northwest are very small compared with the eastern region of China. However, the northwest regions are vast, and the undeveloped, uncultivated land has abundant resources and sufficient photothermal.

The distribution of Chinese solar energy resources has obvious geographical characteristics. Except for the two autonomous regions of Tibet and Xinjiang, the western region is higher than the eastern region, and the southern region is substantially lower than the northern region. The distribution characteristic reflects the restriction of solar energy resources by conditions such as climate and geography. The Chinese Academy of Meteorological Science divides China into four bands of solar resources, as shown in Figure 3c [15]. The climate is partitioned according to the requirements of the thermal engineering design (Figure 3b). The climate factors of the air temperature, the average temperature of the coldest month (January), and the hottest month (July) are taken as the main indexes. The annual average temperature between 5 °C and 25 °C is taken as the auxiliary index. Thus, China is divided into five zones: severe cold; cold; hot in summer and cold in winter; hot in summer and warm in winter; and temperate regions [16]. Compared with Lanzhou and Shenyang (Figure 3d), Shouguang has better temperature and illumination resources, which ensure the basic photothermal environment requirements in the CSG.

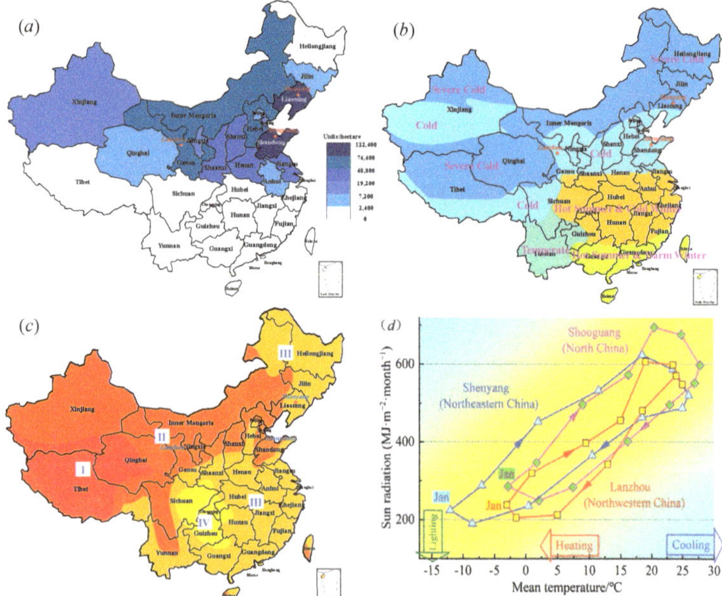

Figure 3. Distribution of solar greenhouse and various environmental resources in China: (**a**) solar greenhouse area distribution in Chinese provinces [8]; (**b**) climatic subdivision of the Chinese thermal engineering; (**c**) solar energy resource distribution in China, I (high areas), II (higher areas), III (general areas), and IV (low areas) [15]; (**d**) comparison of the average temperature of the three areas and the cumulative solar radiation per month in Shouguang (north), Lanzhou (northwest), and Shenyang (northeast) [17–19].

2.2. Solar Technology for Chinese Solar Greenhouses

The poor heat-preserving performance of the current Chinese solar greenhouses and the shadowing effect in PV greenhouses have directed a series of studies in academia on utilizing advanced solar technology for PT (photothermic) and PV utilization technology, which are the main subjects studied by Chinese scholars, as shown in Figure 4.

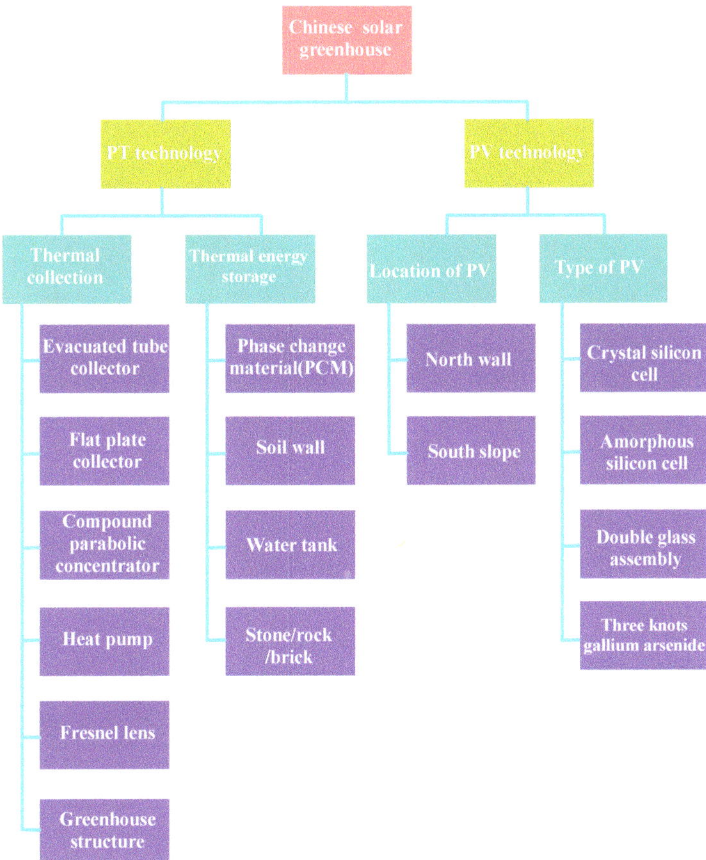

Figure 4. Classification of photothermal and photovoltaic technology for a Chinese solar greenhouse.

3. Application of Photovoltaic Technology in CSGs

3.1. Development of Photovoltaic Technology in CSGs

In photovoltaic applications combined with buildings, the photovoltaic system can be combined with various buildings, structures, and envelope structures that can receive solar illumination. Photovoltaic technologies are widely used in production in conjunction with various agricultural infrastructures, such as photovoltaic systems, integrated glasshouses, and tunnel greenhouses [20]. In view of the special conditions of strong light during the day and low temperatures at night in most areas in northwest China, the solar greenhouse is suitable for photovoltaic applications. Currently, photovoltaic systems are still in the startup and exploration phases for CSGs [21]. The CSG is usually located in a suburban county or a remote area away from the town. Generally, the cost of power transmission and supply is high, and some remote areas even have no power supply. Power utilization facilities in the advanced CSG are numerous, so a stable power supply is needed if the economic and efficient operation of the CSG is to be satisfied [22]. Related technologies are applied to traditional CSGs, such as illumination, temperature and humidity regulation, ventilation, CO_2 regulation, and irrigation spraying systems. Photovoltaic cell panels covered on the south slope surface of the CSG are usually thick and heavy, and the greenhouse is required to be supported by better stand columns and beam frames. The southern slope angle of the existing CSG is gentle, and the generating efficiency of the photovoltaic cell panel is influenced to a certain degree.

The sunlight incident on the surface of the CSG film comprises three types: ultraviolet light, visible light, and infrared light. Visible light is the most suitable for plant growth, wherein the blue light within 400~520 nm and the red orange light within 610~720 nm are effective energy regions for photosynthesis [23]. Meanwhile, the greenhouse structure absorbs NIR more than the plants. As a result, the temperature of the greenhouse air increases, and this leads to a reduction in crop production. Thus, NIR is better utilized by semi-transparent photovoltaic modules to produce electricity (Figure 5).

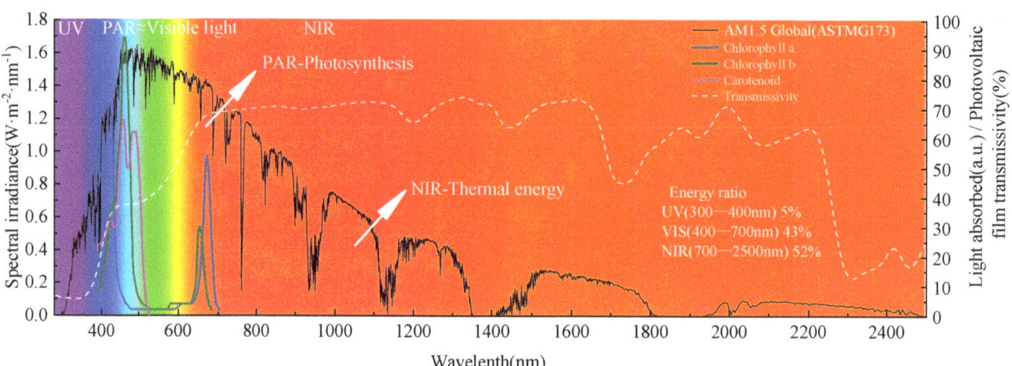

Figure 5. Solar energy distribution at the Earth's sea level (solid black line) showing photovoltaic film transmissivity (dashed white line) and photosynthetic absorption spectra (solid color line) [24]. Different regional colors represent spectral colors of different bands.

In common solar photovoltaic cells (Table 1), the light transmittance of the thin-film cell can be customized to any percentage, such as 10%, 20%, and 30%. The light-transmitting principle is realized by laser scribing and the small-hole imaging principle. The power of the film assembly decreases with increasing light transmission, and the size of the assembly can be selected or customized based on the original structural features and modulus [25]. Most photovoltaic components used in the current Chinese photovoltaic agricultural greenhouse are thin-film components, with the advantages of good weak light properties, low cost, long power generation time, and a light transmission spectrum beneficial to plant growth. However, the thin-film components have low photoelectric conversion efficiency and a large annual attenuation rate, so the thin-film components have a short service life, generally 10~15 years. Compared with other related component materials, polysilicon has higher economy and practicability due to its advantages of maintenance-free properties, long service life, and low cost.

The dual-glass assembly is composed of three raw materials: glass, ethylene vinyl acetate (EVA) film, and polysilicon (Figure 6). EVA is a kind of thermosetting adhesive film with the advantages of adhesive force, durability, and optical characteristics. So, the EVA adhesive film is increasingly being applied to novel photovoltaic assemblies. The photovoltaic cell glass adopts toughened glass with the advantages of high strength, impact resistance, and bending resistance. Its bending strength is 3~5 times that of common glass. It has good thermal stability and light transmission. One of the features of the dual-glass assembly is variable optical transmission. Therefore, different light transmittances can be customized according to the illumination requirements of different plants in the greenhouse. It can prevent ultraviolet rays from damaging plants in the daytime and excessive heat from entering a greenhouse, while at night it can also prevent indoor infrared heat from radiating outwards for heat preservation and insulation [26].

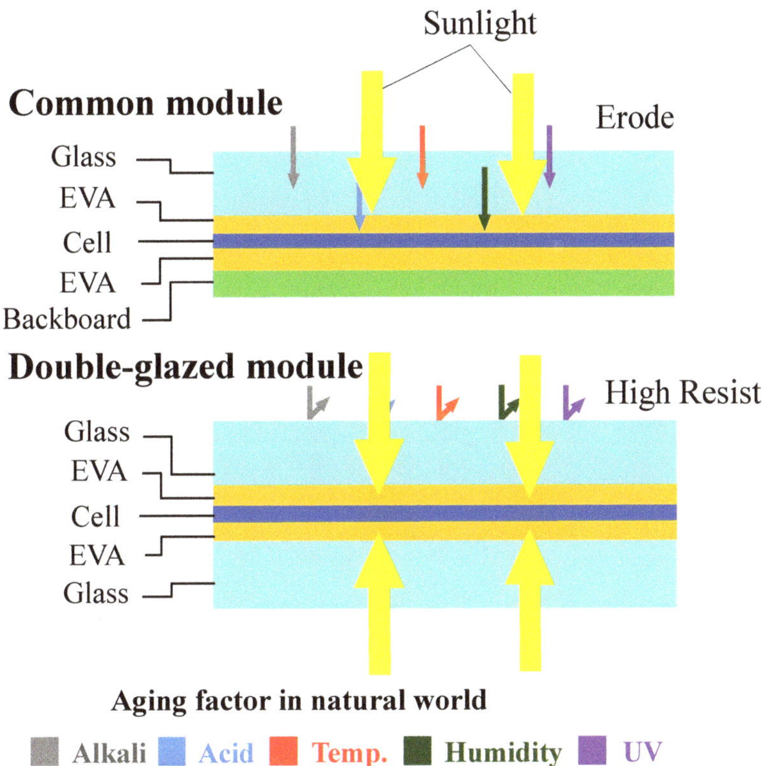

Figure 6. Difference between a dual-glass assembly photovoltaic cell and a conventional photovoltaic cell.

In order to assess the impact of transparent solar cells on plant growth, the average action spectrum of 27 herbaceous plants (including common greenhouse crops such as tomatoes, lettuce, and cucumbers) was used. Figure 7 shows the photon flux density, $b_s(\lambda)$, and the photon flux density weighted by the plant action spectrum, $b_s(\lambda)a(\lambda)$. This modified spectrum can be understood as the photon flux spectrum required for optimal plant growth.

The impact of the absorption by the organic photovoltaic (OPV) device on crop growth was determined by calculating a crop growth factor, $G(x)$, as a function of active layer thickness, x, which can be defined as

$$G(x) = \frac{\int T(x, \lambda) b_s(\lambda) a(\lambda) \mathrm{d}\lambda}{\int b_s(\lambda) a(\lambda) \mathrm{d}\lambda} \tag{1}$$

where $T(x, \lambda)$ is the total transmission of the complete simulated solar cell stack. The rate of photosynthesis in a crop is governed by the integral of the solar spectrum and action spectra. $G(x)$ therefore represents the ratio of the rate of photosynthesis under a clear sky to the rate under a greenhouse material (such as an OPV covering) with spectrally dependent transparency, $T(x, \lambda).b_s(\lambda)$ is AM1.5 spectrum irradiance. $a(\lambda)$ represents the average action spectrum of 27 herbaceous plants (including common greenhouse crops such as tomatoes, lettuce, and cucumbers). The growth of crops in the greenhouse is subsequently assumed to be reduced by 1% for every 1% reduction in the rate of photosynthesis [27].

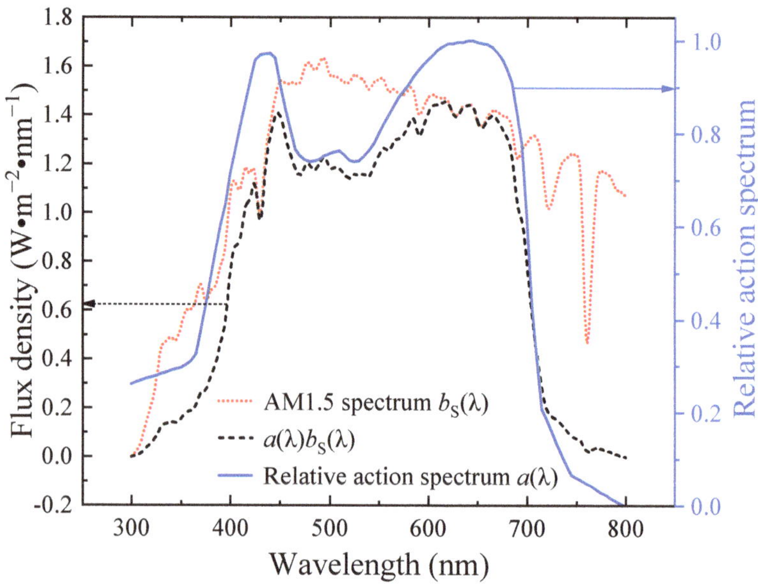

Figure 7. Relative action spectrum for plants (solid blue line), AM1.5 photon flux density (dotted red line), and action spectrum modified photon flux density (dashed black line) [28].

In order to achieve a certain light transmittance, photovoltaic assemblies of solar greenhouses have the following two options [25]. First, by adjusting the interval between the opaque solar cells, which can be realized by silica-based thin-film solar cells and crystalline silicon solar cells. The shortage lies in the fact that the solar cell part is sheltered from the sunlight of the full wave band, which has a certain influence on vegetation. In addition, the silicon-based thin-film solar cells can achieve certain light transmittance through laser scribing or a transparent back electrode to satisfy the lighting requirements of the photovoltaic greenhouse. Moreover, plants can uniformly receive light through this application mode. In addition, the silicon-based thin-film solar cells can reduce ultraviolet rays, which are unfavorable for plant growth. If the transparent back electrode assembly is adopted, unijunction silicon-based thin-film solar cells can be selected. In this way, the defect that the silicon-based thin-film solar cells have strong absorptivity for light with a wavelength of 440 nm (Figure 5) can be overcome, which is active in photosynthesis.

Combining the above two modes, the best solution is to install the silicon-based film light-transmitting assembly according to the former, so that the photovoltaic greenhouse can generate electricity without influencing the growth of crops. So, the income of farmers is increased [29]. Solar power generation is combined with shade-tolerant crop production, and the economic value exceeds 30% compared with that of traditional agriculture [30]. The investment recovery period of the photovoltaic greenhouse is estimated using the electric energy generated by the photovoltaic assembly, so that the sustainable production of greenhouse crops can be better promoted. The calculated recovery period for the dynamic photovoltaic greenhouse under clear air conditions in Italy is 6 years [31]. The annual return rate of photovoltaic CSGs varies from 9% to 20%, and the investment return period is less than 9 years [32]. With the development of photovoltaic technology, the combination of photovoltaic assembly and CSGs will be a mature choice in the near future.

Table 1. Common CSG photovoltaic component materials [33,34].

Comparison Item	Single-Crystal Silicon	Polycrystalline Silicon	Amorphous Silicon Thin Film	Double-Glass Assembly
Technical maturity	Maturity reached	Currently, the technology of ingot-casting polycrystalline silicon is commonly used	Becoming mature	Becoming mature
Photoelectric efficiency conversion rate	13~18%	12~16%	5~9%	Improved by about 4% compared with common assembly
Price	Materials and manufacturing procedures are complicated; cost is high	Simple material manufacture, power consumption savings, and a lower total production cost than that of monocrystalline silicon	Production process is relatively simple, and total production cost is low	Cost is higher than that of crystalline silicon solar cells
Light transmitting or not	No	No	Yes	Yes
Illumination adaptability	Output power is directly proportional to illumination intensity	Output power is directly proportional to illumination intensity	Good weak light response and high charging efficiency	Output power is directly proportional to illumination intensity
Temperature adaptability	Insufficient exertion of efficiency under high-temperature conditions	Insufficient exertion of efficiency under high-temperature conditions	Requirement on the ambient temperature is lower	Heat dissipation performance is better than that of solar cells with the back plate
Operation maintenance	Extremely low failure rate of the assemblies	Extremely low failure rate of the assemblies	Easy to deposit dust and difficult to clean	Good surface aging resistance and easy maintenance
Service life	Guaranteed 25 years of service	Guaranteed 25 years of service	Decay is rapid, service life is only 10~15 years	Guaranteed 30 years of service life
Appearance	Black, atroceruleous	Irregular dark blue	Dark blue	Various colors

3.2. Photovoltaic Cell on the Top of the North Wall

Taking the combined benefits of the shading of the adjacent greenhouse on the southern side with the solar panels into account, solar panels are usually placed on top of the broad north wall of CSG. To ensure the lighting interval between the north and south adjacent CSGs, the photovoltaic assembly bracket and the CSG structure are completely independent. Photovoltaic power generation and greenhouse production can be simultaneously performed [34]. However, due to the height of the photovoltaic panels (Figure 8a), the interval between adjacent CSGs is correspondingly increased. This reduces the effective utilization of the land by agricultural production compared to conventional CSGs. Therefore, in order to improve the land utilization rate, the greenhouse is designed in the form of a synergy CSG, with vegetables planted in the south and mushrooms planted in the north. The shorter photovoltaic panels above the north wall (Figure 8b) can avoid shading the plants inside the north greenhouse. The installation can make full use of the north wall passageway or skeleton. The mode has a north wall-enclosing channel, so the cleaning and the appearance inspection are convenient and fast, with low maintenance costs [35].

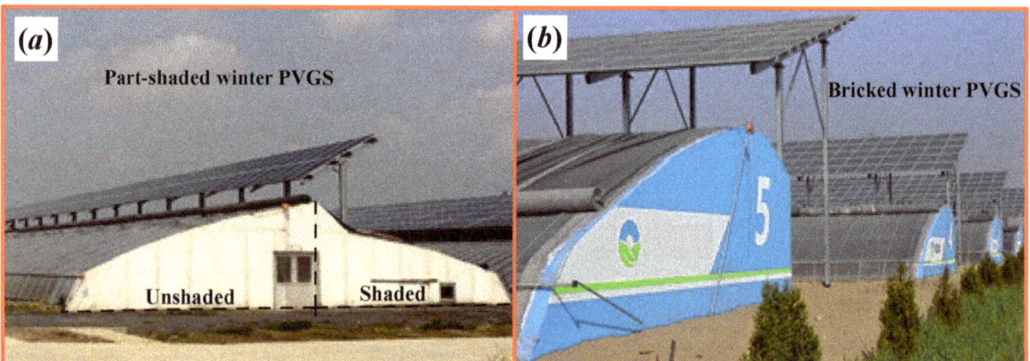

Figure 8. Photovoltaic cells arranged on the north wall of CSG: (**a**) part-shaded winter PVGs; (**b**) bricked winter PVGs.

3.3. Photovoltaic Cells on the South Slope of CSGs

3.3.1. Coverage Area of the Photovoltaic Assembly

The design mode of combining the photovoltaic panel with the front roof framework of the CSG is adopted, with the greenhouse structure fully utilized and structural materials reduced [36]. The greenhouse roof correspondingly increases the area of the photovoltaic panel and the photovoltaic power generation amount. However, because the photovoltaic panel blocks the lighting of crops in the greenhouse while generating power [37], the light becomes a competitive resource required by both photovoltaic power generation and agricultural production, besides planting shade-loving crops (such as mushrooms), which do not need light or need weak light in the greenhouse.

When the solar radiation is strong in the summer, the photovoltaic assembly can play a good role in shading sunlight. In order to ensure that crop growth is not affected as much as possible, the Italian government stipulates that the installation ratio of photovoltaic assemblies on the roof of the greenhouse must not exceed 50% for different coverage rates of crystalline silicon solar cells on the glass greenhouse [38,39]. Compared to foreign countries, Chinese scholars have less research on the combination of crystalline silicon photovoltaics and CSGs. Zan et al. use ECOTECT software to establish models of monocrystalline silicon photovoltaic greenhouses with roof coverage rates of 7.61%, 15.22%, 22.83%, and 30.44%, respectively, based on Kunming climate characteristics. Compared with CSGs without photovoltaic assembly coverage, the lighting coefficient is reduced by about 16%, 38%, 49%, and 58%, respectively [40].

In terms of the effect of the amorphous silicon photovoltaic assembly coverage area on the CSG, Zhao et al. monitored the light environment and tomato growth conditions in a thin-film photovoltaic CSG and a plastic-film CSG (Figure 9) in Yanan city (36° N, 108° E). Under the layout with a ratio of 1:1 between the area of the film photovoltaic and the PC board, the total radiation transmittance within 2 h before and after noon on a clear day in the summer is 38.7%, and the photosynthetic effective light quantum flux density transmittance is 38.9%. They are, respectively, 30.3% and 17.6% lower than that of a plastic-film solar greenhouse. And during this period, the thin-film photovoltaic solar greenhouse can block 3949.8 kJ/m^2 more heat into the room than the plastic thin-film solar greenhouse on a clear day [25]. Zhao et al. also experimented on the southern slope of a CSG in Yangling, Shaanxi (34°16′ N, 108°04′ E). Every three rows of film photovoltaic assemblies and PC sunlight plates are laid from west to east at intervals of 1:2 and 1:3. It is found that the total solar radiation with the 1:3 ratio in winter is increased by 50.3 W/m^2, and the average transmittance of the total solar radiation is increased by 9.1%, but the transmittance of the average photosynthetic active quantum flux density is almost unchanged. The average light transmittance of the inclined plane formed by the amorphous silicon cell assembly and the PC board is changed within a range of 34.7~41.7% (Figure 10c) [41].

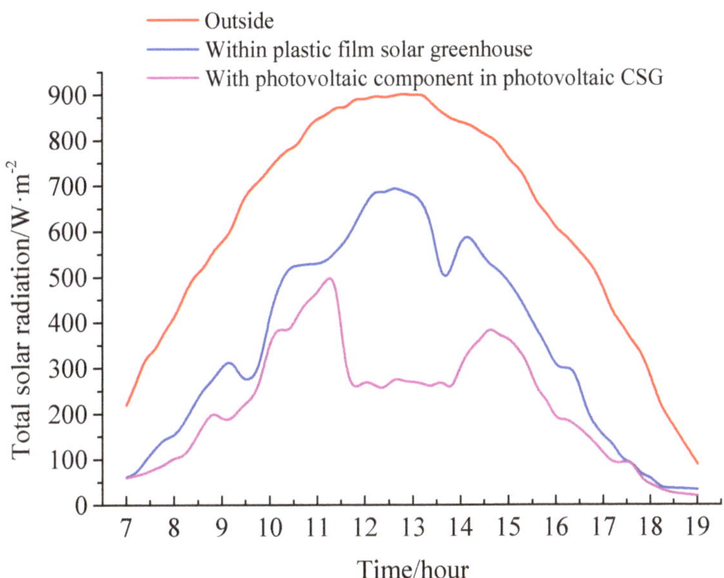

Figure 9. Light environment inside and outside the photovoltaic CSG and the plastic-film CSG on a sunny day (5 July 2012).

Figure 10. Schematic diagram of photovoltaic cells arranged on the front slope of the CSG: (**a**) east–west direction of the crystalline silicon solar cell, with a coverage rate of 81%; (**b**) checkerboard-arranged crystalline silicon solar cell; (**c**) north–south direction of the amorphous silicon cell; (**d**) distribution of double glass assemblies [42].

The research shows that the film photovoltaic assembly can significantly block solar radiation and reduce indoor temperatures in the summer. This has little influence on the photosynthesis of plants and light transmittance compared with a PC plate. However, the coverage rate of the photovoltaic assembly of less than 50% is determined according to specific planting objects. When the coverage rate of the crystalline silicon photovoltaic greenhouse is 30%, the crystalline silicon photovoltaic greenhouse has an obvious inhibition effect on photosynthetic effective radiation, but the adverse effect of a thin-film photovoltaic CSG with the same coverage rate is much smaller [38]. The coverage area of the photovoltaic assembly can directly influence the power generation amount, so the photovoltaic coverage areas can be designed in a photovoltaic CSG that does not take power generation as its main purpose. For example, if the electricity generated is only used by the system itself and the coverage rate is low, the quality and yield of agricultural production can be ensured with reductions in the investment of grid-connected equipment and the energy loss in the grid-connected process.

3.3.2. Arrangement of the Photovoltaic Assembly

On the premise of the same coverage area, the influence of the arrangement mode on the environment in the greenhouse is very important. Zan et al. establish a monocrystalline silicon photovoltaic CSG model with one row of compact, two rows of compact, two rows of chessboard, and four rows of chessboard arrangements and compare the one row of compact and two rows of chessboard type greenhouses, respectively, under the condition that the roof solar cell coverage rate is 7.61% [40]. The results show that during the winter daytime, the temperature in one row of a compact greenhouse is higher than that in two rows of a chessboard-type greenhouse, with a maximum temperature difference of 2.5 °C. The results are close to those of the two rows of compact and four rows of chessboard-type greenhouses, with a coverage rate of 15.22% for roof photovoltaic cells. The lighting coefficients of the bottoms of the two rows of compact and four rows of chessboard-type greenhouse are, respectively, superior to those of one row of compact and two rows of compact type (Figure 10 a,b).

The comparisons between the compact type and the chessboard type show that the sealed heat preservation effect of the compact greenhouse is better than that of the chessboard-type arrangement. However, chessboard-type arrangement lighting is better than the compact type of arrangement. This is more favorable for the uniform irradiation of illumination because the photosynthesis and quality of crops are less influenced [43].

In the aspect of research on the arrangement of the amorphous silicon thin-film photovoltaic assemblies, Zhao et al. find that the average light transmittance of the thin-film photovoltaic CSG is 33.0% in January, which is 11.6% lower than that of the plastic thin-film CSG. By counting the days of lowest temperature during the test period, it was found that the number of days below 8°C in the photovoltaic greenhouse (2 days) was significantly less than that in the plastic-film greenhouse (10 days, 33%) (Figure 10c) [25]. The film photovoltaic assembly covering 9.8% of the roof area of the CSG cannot influence the yield and quality of tomatoes. However, in the crystalline silicon photovoltaic CSG, the light transmittance of the checkerboard-arranged film solar sun is obviously better than that of a straight-line arrangement to improve the quality of crops. So, the checkerboard arrangement of the photovoltaic assembly can enable the illumination in the CSG to be more uniform and more suitable for the growth of crops. In addition, the low-temperature days in winter show that the winter temperature environment of the photovoltaic CSG is obviously better than that of a plastic-film CSG, and the photovoltaic CSG is more favorable for fruit and vegetable cultivation in winter.

The whole greenhouse roof area is often covered with opaque conventional PV panels to maximize energy production [44]. However, this scenario is unsuitable for green plant cultivation (Figure 11a). In fact, on average, among the most common PV greenhouse types, the annual global radiation decreases by 0.8% for each additional 1.0% of PV coverage on the roof. The greenhouse internal light environment varies greatly according to whether the PV modules are concentrated as a single array (Figure 11b,c) or dispersed over the roof (Figure 11d–f) when the roof is partially covered with PV modules. PV panel shadows are dissected and scattered on the wider area of greenhouse interior space when the PV modules are distributed in a dispersed manner on the roof (Figure 11d–f). Stripe (Figure 11d,e) or checkerboard (Figure 11f) arrangements of PV arrays using conventional opaque PV panels have been studied. An important implication exists related to the stripe direction. Straight lines of PV modules aligned east–west, mounted on the south roof of an east–west oriented greenhouse, are suitable for electricity production (Figure 11d) [37]. The inclination of south-facing roofs of east–west-oriented greenhouses in the northern hemisphere usually contributes to PV electricity production. By contrast, all plants in the greenhouse receive direct sunlight frequently during a sunny day when the strings of PV modules are aligned north–south, irrespective of the greenhouse orientation (Figure 11e) or if the PV modules are placed in a checkerboard arrangement (Figure 11f) [45,46].

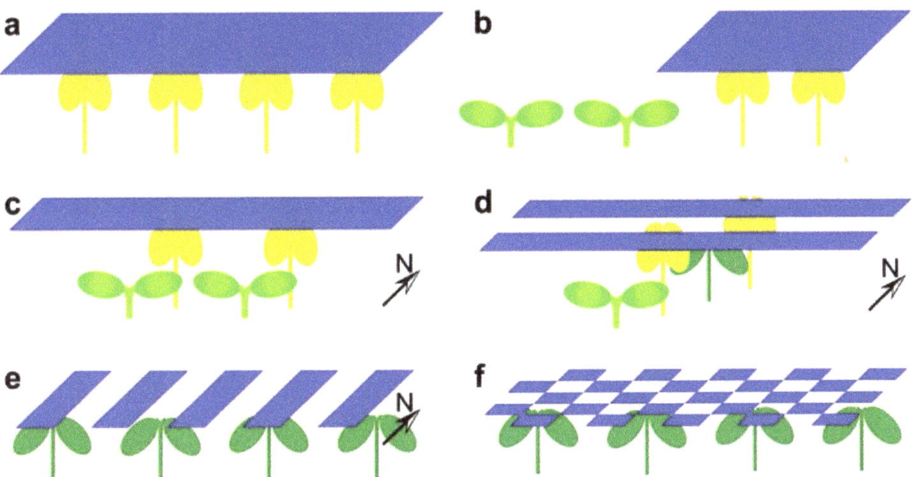

Figure 11. Some examples of possible opaque PV module arrangements above greenhouse crops: (**a**) 100% coverage; (**b**) concentrated partial coverage; (**c**) east–west straight; (**d**) stripe along east–west; (**e**) stripe along north–south; (**f**) checkerboard. Non-shaded, intermediately shaded, and heavily shaded plants are presented, respectively, as bright green, dark green, and yellow plants.

4. Application of Solar Thermal Technology in CSGs

4.1. Profile of Solar Thermal Energy Utilization in CSGs

Solar energy is a renewable, clean energy source, and it is also a major source of light and heat for plant growth in CSGs. The solar heat collector is applied to the greenhouse active heating technology, so the requirement for fossil energy can be reduced. Moreover, the requirements for energy conservation and emission reduction are satisfied in an environmentally friendly way. The application principle of the active heating technology in CSG is that the solar heat collectors in different forms are used for storing solar radiation energy in the daytime, such as a heat-preserving water tank, a wall body, and soil. The stored heat is released into the greenhouse to increase the indoor temperature at night or in low-temperature weather, such as rainy days. When studying the heating effect of the solar heat collector in greenhouses, Chinese scholars mostly adopt the mode of installing the solar hot water collector outdoors to collect more solar heat (Figure 12a).

To overcome the influence of the cold weather on agricultural production, different modern heating modes are researched on the influence of indoor temperature, relative humidity, air flow rate, soil temperature, and indoor air quality on plants, as shown in Table 2.

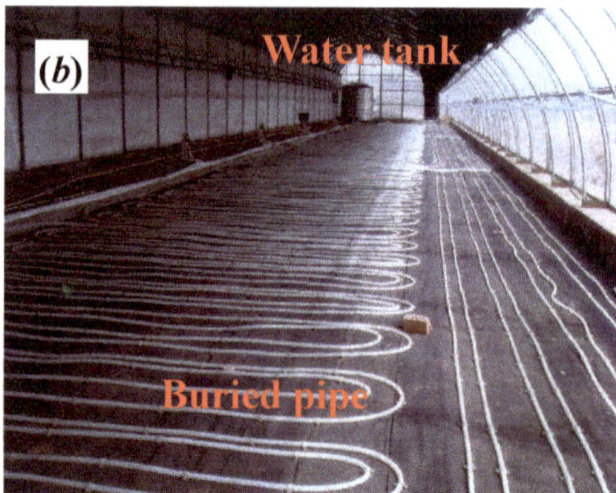

Figure 12. Application of solar heat collection in CSGs: (**a**) solar heat collectors arranged on the north wall of the greenhouse along the ridge direction of the greenhouse; (**b**) heat dissipation pipes and heat storage water tanks arranged in the greenhouse.

Table 2. Modern CSG heating mode.

Heating Mode	Measures	Disadvantages	Advantages	References
Boiler heating	Electric boiler, coal-fired boiler, or gas-fired boiler.	(1) Combustion products, such as smoke dust, sulfide, and nitrogen oxide, are easy to generate and have inevitable harm to the environment; and (2) energy consumption is too large.	Mature technology, high efficiency, and higher freight costs.	[47]
Solar heating	Direct utilization of solar energy; solar radiation panels.	The solar energy resource shows periodic variation, which causes the energy source of the greenhouse to be unstable.	Clean energy, both heat and light sources, and abundant resources.	[48,49]
PCM heating	Applied to greenhouse building materials to realize heat storage–release processes.	(1) Utilization efficiency is low; (2) investment is large; and (3) phase-change materials applied to the internal parts of the greenhouse are not mature at present.	Realize that short-term heat storage occupies less space.	[50,51]
Cross-seasonal heat storage	Solar cross-seasonal heat storage and solar cross-seasonal phase-change heat storage technology.	(1) Investment is large; (2) requirement for the heat preservation of the greenhouse is higher; (3) revenue is not directly proportional to investment; and (4) heat storage technology has a higher requirement.	Clean energy and large-scale greenhouse requirements.	[52]

The low ground temperature is an important environmental factor that restricts the normal production of vegetables in the CSG in the winter. Particularly, the difficulty of increasing the ground temperature of the greenhouse is higher after continuously cloudy days. Due to the fact that the soil has thermal inertia, the heat required to increase the ground temperature is relatively large. The heat is directly obtained from the air, so the heat transfer efficiency for storing the heat in the soil is low. Therefore, in practical production, the method of heating indoor air is rarely actively adopted to increase the ground temperature. Some researchers directly lay the radiator on the ground or in the soil so that the heating efficiency is higher (Figure 12b).

There are many methods to increase the ground temperature of greenhouses, such as heating with electric heating wires, biological reactors, underground heat exchange, and hot water. The heating of the soil with electric heating wires is the most direct and effective method, but the method has a large power consumption, so the method is essentially not adopted in large-scale vegetable production except for the use of greenhouse seedling culture or local heating in the greenhouse [53]. The biological reactor technology is a soil temperature-increasing form that has been widely popularized and applied in CSGs in recent years. It can improve the CO_2 concentration in the greenhouse besides the ground temperature. However, the time difference in contradiction with the requirements of the greenhouse in the early and later periods of planting is larger [54]. Underground heat exchange technology has been transferred from Japan to China since the 1980s. It is mainly suitable for areas with better sunlight and stronger soil heat storage capacity [55]. The solar energy underground heat exchange device fully utilizes solar energy, and the relatively surplus energy in the daytime is transported to the underground through fans to heat the soil in the greenhouse and store the energy. When the temperature is low at night, fans are used to carry the heat stored in the soil to the air, so that the temperature in the facility is increased. Low-temperature hot water (the water temperature is mostly below 50 °C) is easy to obtain, simple to control, and uniform in soil and indoor temperature distribution. So, it is one of the most common methods for soil heating. The solar panel collector is used for collecting solar radiant heat. Water is used as a heating medium and directly supplied to the ground or soil of the greenhouse in an environmentally friendly way.

4.2. North Wall Materials to Improve the CSG Thermal Environment

The north wall body of the CSG plays an important role in forming an indoor thermal environment. Tests on the temperature distribution and change rule of the wall body along the thickness direction in the CSG show that in a general solid wall body structure, the inner surface of the wall body is greatly influenced by solar radiation and the indoor thermal environment. The diurnal temperature can clearly fluctuate, but the fluctuation is also attenuated quickly [56]. For example, for a brick composite wall with a thickness of about 500 mm (Figure 13), the wave motion is obviously reduced to 200 mm from the inner surface of the wall. At 300 mm, the temperature of the wall body is close to stable, which indicates that the heat storage and release positions of the wall body are mainly concentrated in the thickness range within 200 mm of the inner side of the wall body, and other wall body materials are little or not directly involved in the heat storage and release process [57]. The limited heat storage area, insufficient heat storage capacity, and too low temperatures at night occur in the wall body of the CSG. On the other hand, the phenomenon that the temperature of the greenhouse is too high on sunny days and the waste heat is often removed by ventilation is undoubtedly a contradiction. Therefore, the materials in the deep part of the wall body are transferred to participate in the heat storage and release process by changing the structure of the wall body. In this way, the total heat storage and release amount of the wall body is increased, the solar energy utilization efficiency of the CSG is further improved, and the greenhouse environment is improved.

In different regions of China, due to different climates, the construction forms of greenhouse walls are different. And due to different solar radiation distribution conditions, the construction forms of the greenhouse wall are more diversified. The CSG wall is developed into a composite heterogeneous wall from a simple substance wall. The influence of a combination form on the thermal engineering performance of the wall is gradually considered. The CSG wall is further developed into various walls constructed with different combinations of materials and in different positions. Table 3 shows several typical construction forms of the solar greenhouse walls in China.

Table 3 shows four typical applications of the greenhouse wall: Shouguang type, northern Jiangsu type, Chinese standard type, and novel phase-change heat storage type. It develops from a simple substance wall body to a composite material wall body. There is a breakthrough in the construction mode and the phase-change material for the greenhouse wall body. The wall body of Shouguang type is made of the soil wall, which is a typical means for increasing the thickness of the wall body to increase the indoor temperature at night. However, the land utilization rate of the CSG is low in this way. The novel phase-change heat storage type of CSG is constructed according to an ideal wall structure [59]. The application of phase-change latent heat storage in the CSG is realized. At present, four typical greenhouse walls are verified, applied, and popularized to build a thermal environment suitable for plant growth.

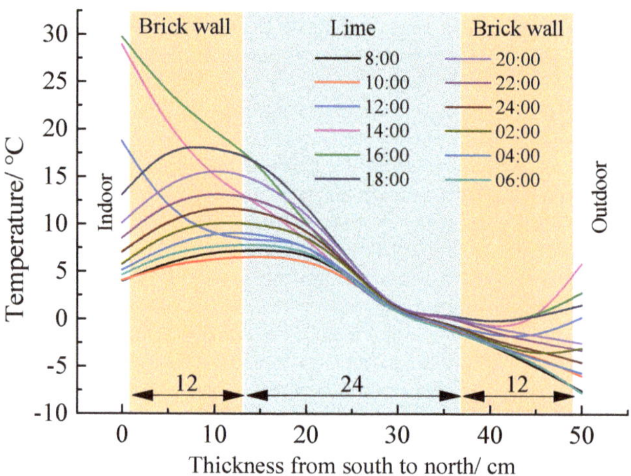

Figure 13. Time course of temperature at different depths of the north wall (Lime wall as an example), Note: Data in the plot were averaged for 20–21 December 2006.

Table 3. Construction form of typical CSG wall [58].

Applications	Wall Structure Diagram (Left Is South, Right Is North)	Construction Parameters	Remarks
Shouguang	Soil wall	The wall body is in an isosceles trapezoid shape, which is made of a soil wall. The length of the upper bottom edge is 2 m, and the length of the lower bottom edge is 7 m.	In order to satisfy the growing environment of plants in the greenhouse, people in cold regions increase the temperature of the greenhouse by increasing the thickness of the wall body.
North of Jiangsu province	Straw brick Color plate	Straw brick (0.49 m) and color plate (2 mm).	The heat preservation effect is good.
International standard	Brick wall Polystyrene board Brick wall	Brick wall (0.37 m), brick wall (0.24 m), and polystyrene board (0.1 m).	Has the functions of heat preservation, heat accumulation, and heat insulation.
Passive phase-change heat storage	PCM board Brick wall Insulation board	PCM board (5 cm), brick wall (0.8 m). and insulation board (5 cm).	According to an ideal wall structure, a novel phase-change heat storage type of greenhouse wall is constructed.

According to the theory of thermal engineering, it is better to place a latent heat energy storage material (such as a phase-change material) with strong heat storage capacity on the inner surface layer of the wall body in the solar greenhouse to ensure that the projected solar radiation energy can be efficiently received and stored. In the CSGs, the material with large thermal resistance and good heat preservation performance (such as light polystyrene board) is placed on the outer surface layer of the wall body, so that the heat loss from the outer surface of the wall body to the outdoor environment is reduced to the maximum extent. Sensible heat energy storage materials (such as heavy building block bricks) with strong heat storage capacity are placed in the middle layer of the wall body of the CSG to further improve the structural stress and the overall heat storage capacity of the wall body [60].

To accumulate more surplus heat energy in the daytime for heating at night without consuming energy sources such as electric power, Zhao et al. adopt the structure of a hollow wall body (Figure 14). The natural force is formed by utilizing the temperature differences and gravity differences of the air in the greenhouse and the wall body to introduce the greenhouse air into the hollow part of the wall body. The surplus heat energy from the Sun is stored in the deep part of the wall body, so that the large-area contact heat exchange between the air and the wall body is utilized. The heat accumulation and release effects are comprehensively exerted through the inner wall body material, the outer wall body material, and even the roof material at the top part. The purposes of storing solar heat energy more effectively, increasing the heat release capacity at night, and increasing the temperature at night have been achieved [61]. The upper and lower parts of the inner side of the hollow part of the wall body are uniformly provided with a plurality of vent holes along the horizontal direction. The vent holes communicate with the interior of the wall body and the indoor space. In this way, a new path is provided by forming air circulation flow around the inner side wall body. The height difference between the upper vent hole and lower vent hole is set at 2 m to ensure sufficient natural thermal pressure difference for air flow.

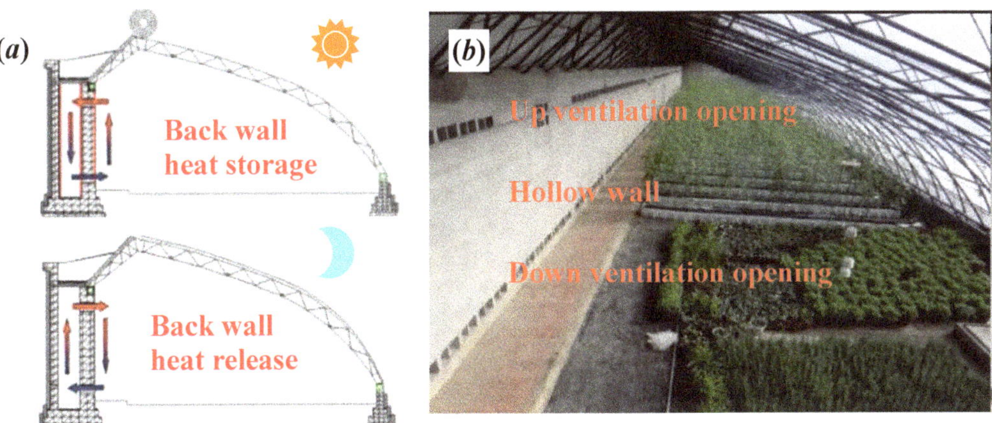

Figure 14. Schematic diagram of a natural convective hollow north wall on thermal storage/release in a CSG: (**a**) airflow direction during daytime and night time; (**b**) structure of hollow north wall in a solar greenhouse.

In Figure 15a, in the process of heat storage in the wall body by natural convection circulation of air, the temperatures of the indoor surface and the two wall surfaces of the hollow layer are higher in the daytime. Meanwhile, the air temperature of the hollow layer is higher than that of the two wall surfaces. This shows that the area of the wall body participating in heat exchange is not only the indoor surface but also the wall surfaces on both sides of the hollow layer. That is why the heat exchange area of the wall body has increased. The temperature of the convection air is lower than that of the north and south wall surfaces of the hollow layer at night. This indicates that the convection air is heated in the circulation process at night. On 2–4 February 2017, it was cloudy, and on 5–6 February, it was sunny. In Figure 15b, under the action of natural convection of air, the heat storage amount and the heat release amount of the hollow layer in the heat storage wall body account for 31% and 35% of the total heat storage and release amount of the greenhouse, respectively. Compared with the control wall body, the wall body structure has the advantages that the daytime heat storage capacity is improved by 15.1%, and the nighttime heat release capacity is improved by 14.7%. These advantages show that the wall body structure can improve the heat storage and release capacity of the wall body [62].

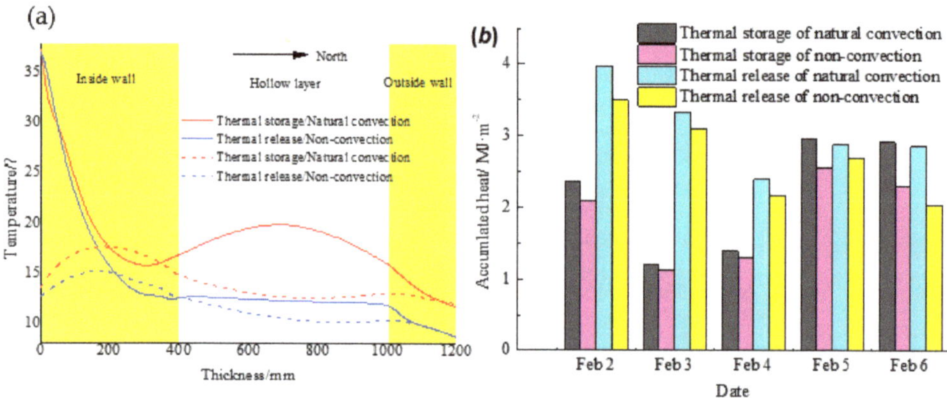

Figure 15. (**a**) Temperature distribution in the wall at a typical thermal storage time and release time on a sunny day. Typical thermal storage time (1 February 2017 14:00); Typical thermal release time (2 February 2017 04:00); Note: Wall thickness is calculated along the indoor-to-outdoor direction. (**b**) Accumulated heat of the wall.

Pebbles are good sensible heat energy storage materials. Broad pebble beds are generally buried in underground soil in the middle of greenhouses. As heat storage materials for CGSs in China, pebbles are mainly used in two forms: one is directly used as a wall material, and the other is used as a heat storage bed. The heat transfer rate of the pebbles is relatively high. The pores among the pebbles are also favorable for convection heat transfer. However, the temperature in the first half night is high, and the temperature in the second half night is low due to the fact that the pebbles release heat too fast in winter. Therefore, it is necessary to study how to slow down the release of the heat accumulated by the pebbles or how to transfer the heat accumulated by the pebbles to other materials for storage [63].

Many studies have assessed the selection and preparation of phase-change materials (PCMs) [64–66], their packaging [66], and their applications in greenhouses [67]. The main screening materials used included paraffin [68], $Na_2SO_4 \cdot 10H_2O$ [69], $CaCl_2 \cdot 6H_2O$ [64], $Na_2HPO_4 \cdot 12H_2O$ [66], and fatty acids [70]. The packaging methods used included blend soaking [69], block packaging [69], rice husk adsorption [65], and graphite adsorption [70]. The method used to combine the materials with the greenhouse was mainly to place the packed phase-change material on the north wall of the CSG [64] or use the plate or block directly built on the inner side of the CSG north wall [67] (Figure 16).

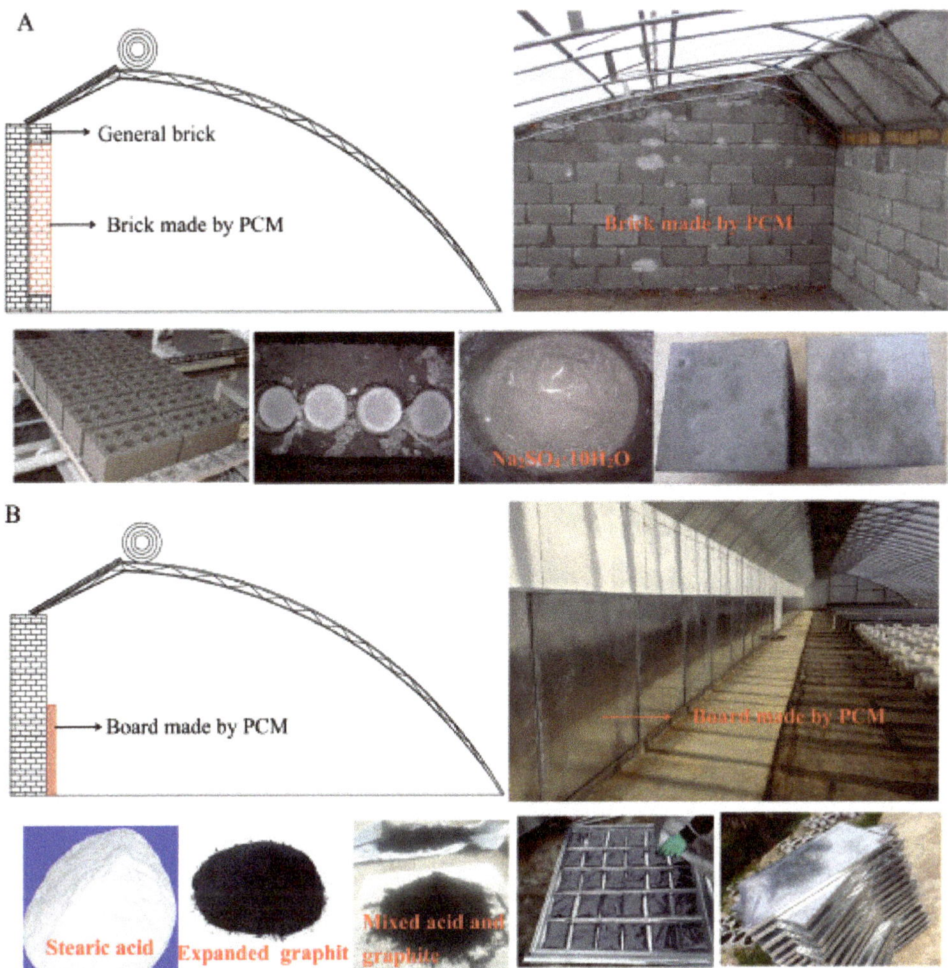

Figure 16. Schematic diagram of phase-change materials (PCM) north wall CSG (CSG): (**A**) brick made by a PCM in a CSG; (**B**) board made by a PCM in a CSG [71].

The phase-change heat storage material for CSG not only needs to overcome the problems of the material itself, but also needs to satisfy the conditions required by plant growth [72]. The following requirements should also be satisfied: (1) The phase-change temperature should be maintained within a temperature range suitable for crop growth. In general, most crops grow at an optimum temperature of about 25 °C, so the melting point of the phase-change material is preferably around this temperature range. The phase-change process of the phase-change material requires a certain amount of time. However, the adjustment to the environment is somewhat delayed. So, the selection of the phase-change temperature of the material between 20 and 25 °C is more practical and can better play the role of the phase-change material. (2) Large latent heat and small volume change in the phase-change process. (3) No toxicity, no corrosiveness, and no flammability. (4) Low cost and wide source.

The phase-change material has good heat storage and release characteristics. For example, the change in the temperature of the environment in the greenhouse can be slowed down, improving solar energy utilization efficiency. The phase-change heat storage material is also suitable to be applied to the CSG. However, there is still much work to be perfected on the aspect of popularization and application of phase-change technology in CSGs, such as the establishment of optimal phase-change temperature intervals of phase-change materials in different regions, the determination of optimal thicknesses of phase-change heat storage material layers of different types of greenhouses, and the determination of optimal laying areas [73]. The work of reducing plant shading to achieve the optimal input-output ratio is still under further study.

4.3. Equipment to Improve the CSG Thermal Environment

As shown in Figure 17, the active solar heat storage–release (AHS) system of the CSG is composed of heat collection/release, heat storage, and control devices. In the daytime, the solar radiant heat is absorbed and stored, while at night, the heat is released to warm the greenhouse. Zhang et al. adopted an aluminum alloy finned tube as a heat collector to improve the stability and reliability of the CSG initiative heat accumulation and release system operation. Results show that under three different solar radiation intensities, the temperature difference between the south and the north of the test area is large, with a maximum average temperature difference between the south and the north inside a plant population of 2.8 °C, 2.6 °C, and 2.4 °C, respectively [74]. By calculating the heat transfer process model of the CSG and the active heat storage and release system, for a polystyrene board solar greenhouse of 640 m^2 in the Beijing area, if the average temperature is 12 °C at night, at least 118 m^2 of a heat collection plate is required to be configured with a water supply volume of 10.4 m^3. When the design value of the air temperature increases by 1 °C at night, a 5 m^2 larger area of heat-collecting plates and 0.2 m^3 more volume of water are required [75]. Xu instead uses black hollow plates as heat collectors (Figure 17b). In January in the Beijing area, the daily heat collection quantity of the heat collector with an area of 84.4 m^2 is usually 0.40~0.88 MJ/m^2 (calculated according to the ground area of the greenhouse). The requirement that the night temperature of the CSG with the 400m^2 polystyrene board wall be increased by 3~5 °C can be satisfied. In addition, the water storage volume configuration of 0.02~0.03 m^3/m^2 (calculated according to the ground area of the greenhouse) can satisfy the heating heat requirements on 2~3 continuously cloudy days. The system can raise the temperature of the tomato canopy obviously, to ensure the overwintering production of tomatoes and promote their growth. In this way, the yield is increased, and the fruit can mature in advance and come into the market [76].

The main function of the solar greenhouse frame is to support the pressure of the solar greenhouse covering and to bear the external pressure (wind load and snow load) caused by nature. It is an innovation for the solar greenhouse to use galvanized steel pipe roof trusses as a heat collector and a radiator [77]. However, because the steel pipes are close to the outside, when the night system operates, a large amount of heat is dissipated to the outside, as the heat preservation measures are not implemented well when the system radiates to the inside (Figure 18).

The heating capacity of a single air source heat pump is rapidly reduced along with the reduction of the environmental temperature, particularly when the environmental temperature is reduced to be below 0 °C. The heating efficiency cannot satisfy the temperature adjustment requirement, even when the machine is shut down due to a serious frosting phenomenon. The single solar heat pump temperature regulating system completely depends on sunlight irradiation, and solar energy is used as intermittent energy and greatly changes along with day and night and weather conditions. So, the single solar heat pump temperature-regulating system cannot satisfy the temperature-increasing requirements of the greenhouse. At present, composite heat pump temperature-regulating systems, such as solar energy combined ground source heat pumps and water source heat pumps, have the problems of complex equipment and large construction investment costs [78].

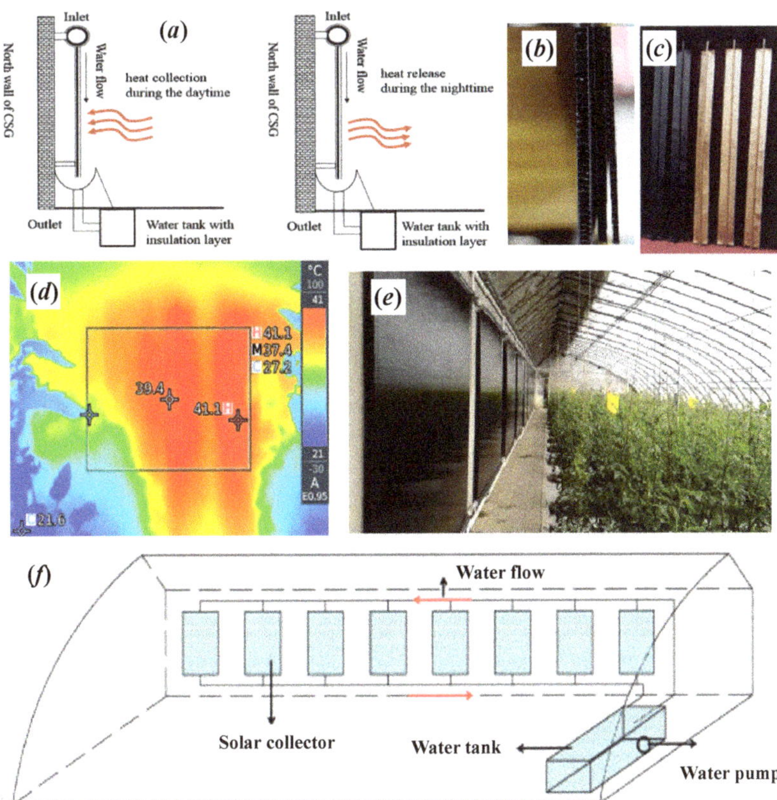

Figure 17. Schematic diagram of the water circulation curtain thermal storage system in a Chinese solar greenhouse: (**a**) active solar heat storage–release (AHS) system and its two operating modes, heat collection and release; (**b**) black polypropylene hollow sheet; (**c**) aluminum alloy finned tube; (**d**) uneven temperature distribution on the hollow polycarbonate sheet surface during the daytime due to the tall plant shading (note: the image was made with infrared light when there was no water flowing in the collector); (**e**) test greenhouse experimental setup with the hollow polycarbonate sheet-constructed solar collector; (**f**) schematic diagram of the water-circulating solar heat collection and release system and its location.

The multi-surface solar air collector with double receiver tubes (MSC-DRT) developed by Chen et al. for the CSG has the characteristics of simple operation and maintenance, small occupied space, and high heat collection efficiency (Figure 19a) [79]. The light inlet of the multi-curved-surface heat collector is narrow, which has little influence on the shading of the rear greenhouse. It is arranged on the north wall of the CSG with a high integration degree with the greenhouse (Figure 19b). Under the conditions that the daily average temperature of the outdoor environment is lower than −1.2 °C and the average inlet air temperature is lower than 3 °C, the daily accumulated heat collection amount of the heat collector per unit area is about 4.84 MJ/(m^2·d), and the daily average outlet temperature can reach 56 °C, which can provide favorable conditions for the active heat storage of the wall body of the CSG (Figure 19c). The length of the heat collector can be considered within the range of 16 m~18 m. In practical applications, the outer curved surface of the heat collector and the air supply pipeline have a certain amount of heat loss. Especially, the heat loss of the air supply pipe along the outdoor part is larger.

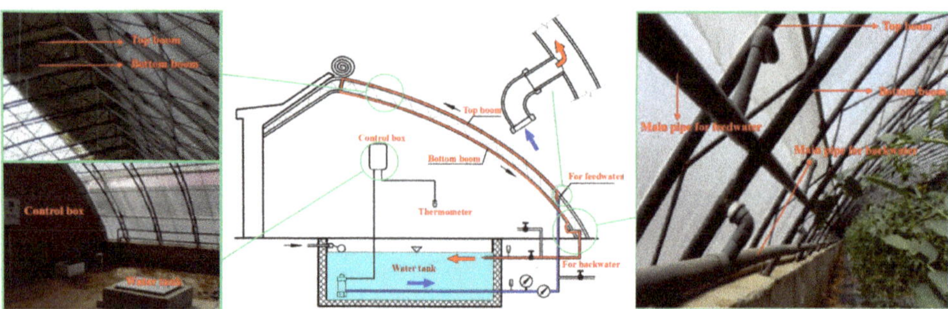

Figure 18. Water-circulating solar heat collection and release system with the indoor collector constructed from the pipe network formed by the greenhouse steel-roof truss.

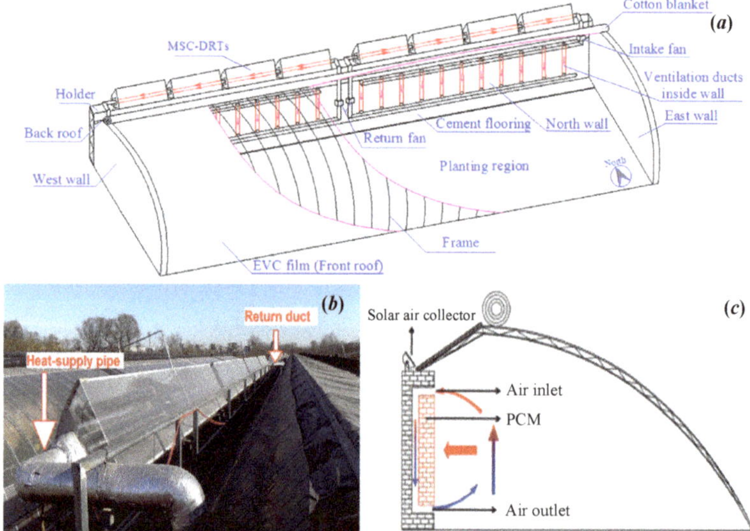

Figure 19. Schematic diagram of a solar air collector with dual collector tubes and phase-change materials in Chinese solar greenhouses: (**a**) thermal performance experimental system of air collector with double collector tubes; (**b**) physical picture; (**c**) phase-change material wall with vertical air channels integrating solar concentrators [80].

Figure 20a shows the three-dimensional schematic of the developed novel MSC-DRT, and Figure 20b presents its cross-section with an indication of its dimensions. The main components of the novel MSC-DRT include a glass cover with high transmittance, a multi-surface reflector with side and bottom surfaces, and two receiving tubes, as shown in Figure 20a. The concentrating ratio C is the ratio of the collector aperture to the heat absorption areas. For the described dimensions of the concentrating collector, C is approximately 1.4.

In order to improve the concentration and sunbeam convergence ratio of the MSC-DRT, the multi-surface reflector was designed with the off-axis location of the receiver tubes in mind. Two tubes were used to split the airflow and increase the heating surface/volume ratio. In addition, this design allows the enhancement of the convergence rate of the direct and reflected sunbeams over a wide range of incidence angles. The three parabolic curves

AB, *CD*, and *EOF*, which make up the reflecting surfaces of the collector, can be described in the coordinate system shown in Figure 20b as follows:

$$AB : y = \frac{1}{0.92}(x + 0.23)^2 \tag{2}$$

$$CD : y = \frac{1}{0.92}(x - 0.23)^2 \tag{3}$$

$$EOF : y = \frac{x^2}{0.16} \tag{4}$$

where x and y are coordinates in units of length (meters).

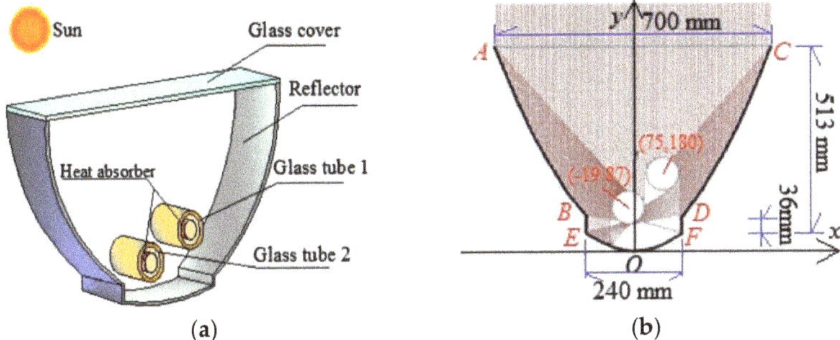

Figure 20. Novel MSC-DRT and its cross-section with an indication of its dimensions: (**a**) three-dimensional schematic; (**b**) cross-section of the MSC-DRT with an indication of its dimensions.

Most current studies of the application of solar energy heating in greenhouses involve short-cycle heat storage. Despite the fact that short-cycle heat storage requires low investment, there are often more than a week of continuous cloudy days in the winter when the solar system cannot work. Li et al. carried out experimental and numerical investigations on underground heat storage using conducting oil as heat transfer fluid for growing plants in cold climates [52]. A curved Fresnel lens was used to maintain a temperature of 8 °C in a volume of 716 m³. The pipe depth varied from 0.5 to 3 m. The corresponding time required for the soil surface to reach a stable temperature was 7~100 days.

In order to save the land utilization rate of the CSG and effectively utilize the non-planting area on the top of the CSG, Wu et al. adopted a Monte Carlo ray tracing method to research the performance of the CPV/T system of the cylindrical surface Fresnel lens heat collector (Figure 21) [81]. To achieve a maximum total monthly average direct radiation daily amount, a round surface-paraboloid combined south roof structure is selected with a space utilization rate of 18.2% for the non-planting area. The experimental study is carried out by using a gallium arsenide high-concentration solar cell as a receiver. The test result under the sunny weather condition shows that the maximum generating efficiency of the system at noon is about 18%, and the maximum heat collection efficiency of the cooling water is about 45%. During noon (11:00~13:00), the total thermoelectric efficiency can reach 55%. Through the test, scattered light not utilized by the lens system is found, and the whole-day illuminance of the plant canopy in the test area is reduced by about 10~40% compared with that in the non-test area. In addition, the proportion of scattered light in plant canopies on cloudy days can be increased, and the heat collection efficiency of the corresponding device can be reduced.

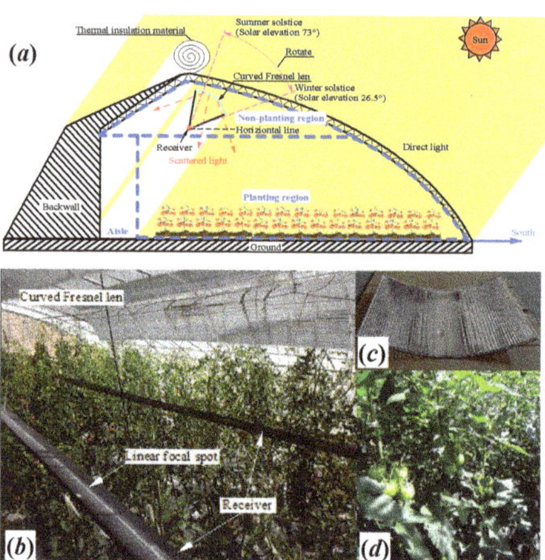

Figure 21. Energy and optical analysis of photovoltaic thermal integrated with a rotary linear curved Fresnel lens inside Chinese solar greenhouses: (**a**) schematic diagram of the experimental principle; (**b**) surface Fresnel mirror position map in the Chinese solar greenhouses; (**c**) curved Fresnel mirror object; (**d**) light distribution of the plant canopy under the lens.

5. Solar Energy Supply and Energy Demand in CSGs

Solar energy utilization technology is an important component of the sustainable development of the CSG. It can reduce the consumption of the conventional energy sources of the CSG in cold areas during the winter [82]. Solar energy technology is increasingly applied to greenhouse production, particularly two types of technologies such as those shown in Figure 22. One type is that solar energy is directly combined with photothermal heat storage technology, and heat is transferred to greenhouse soil and air through a heat exchanger [83]. The other type is that solar energy is converted into direct current to supply power to a storage battery by utilizing photovoltaic power generation technology, and the direct current can also supply power to greenhouse electric equipment through an inverter. The solar heat storage system mainly obtains low-grade energy (heat energy), which is mainly used to improve the heat preservation and heat storage capacity of the CSG in winter, while the solar power generation technology can obtain high-grade energy (electric energy), which can be directly used for illuminating lamps, light supplement lamps, rolling blinds machines, irrigation equipment, and plant care equipment in the greenhouse and provide energy for greenhouse environment regulation equipment [84].

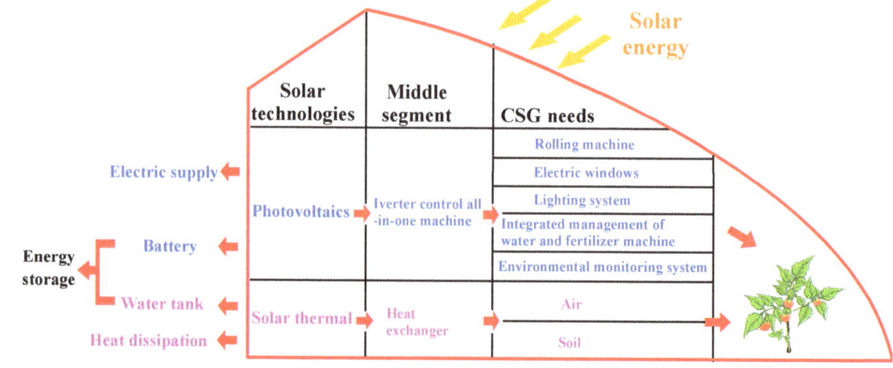

Figure 22. Different solar technologies covering different CSG energy needs.

However, the optical mechanism between photovoltaic power generation and the normal growth of plants in the CSG is still an urgent problem to be solved in the solar greenhouses. Solar energy is not only an indispensable energy source for the photosynthesis of green plants, but also a main heat source for the CSG. Therefore, when designing a solar energy system for the CSG, the problem of lighting the greenhouse should be solved first, and the photosynthetically active radiation (PAR) should be transmitted into the greenhouse to the maximum extent.

6. Summary

An important aspect of greenhouse sustainability and efficiency is energy-efficient operation. The north wall body of the Chinese solar greenhouse plays an important role in the formation of an indoor thermal environment. The heat storage and release positions of the wall body are mainly concentrated in the thickness range within 200 mm of the inner side of the wall body. In the vast northern areas of China, due to the cold climate in late autumn, winter, and early spring, the temperature difference between day and night is large. Without heating, the requirement for crop growth is hardly satisfied by the temperature of a CSG at night. As for heating, the traditional heating method needs to consume a large amount of non-renewable fossil fuel, generating a large amount of harmful gas to pollute the environment simultaneously [85]. Solar energy is a clean energy, large in quantity and wide in distribution, that can be applied to greenhouse warming. It can replace non-renewable fossil fuels for heating and is one of the most promising energy sources [86]. In the last two decades, China has performed more relevant research [87]. The heat collection mode of external solar heat collectors can collect solar heat outside the CSG, but it occupies outdoor land. Moreover, the heat loss in the conveyance of hot water is large and requires the installation of costly indoor heat release devices. The heat-collection mode of internal solar heat collectors can collect surplus heat energy indoors, but it has the defect of competing for light and heat with plants. Therefore, the existing solar heat collector cannot be completely used for collecting heat energy inside and outside the CSG. A new heat-collection mode and a new system are designed by combining the change rule of the CSG structure and the internal environment.

Photovoltaic solar greenhouses combine greenhouses with photovoltaic power generation systems to achieve integration of solar power generation with facility production, attracting increasing attention in the facility horticulture industry [88,89]. The annual rate of return of photovoltaic solar greenhouses varies from 9% to 20%, with a return-on-investment period of less than 9 years. Exploratory construction and operation of the photovoltaic power generation in CSG have also been carried out in China in the last decade, but due to the lack of research on the facility's luminous environment aimed at the crop mechanism, the influence of the photovoltaic CSGs on crop production is uncertain. Meanwhile, in China, most PV agricultural companies specialize in PV businesses and lack experience in crop planting.

Because of the characteristics of the severe cold in winter in the north of China and the national conditions of more people, less land, an energy shortage, and emission reduction pressure, the CSG will be the most practical and widely applied mainstream horticulture facility in China for a period of time. Unfortunately, due to cultural and linguistic differences, CSG has long been the focus of only Chinese researchers. One of the advantages of CSGs is low-cost production, so the principle must be followed by the energy equipment invested. From the perspective of improving energy utilization rate, a sensor that is simple to use and low in cost and a low-cost intelligent control scheme should be developed to control heat preservation quilts, roller shutters, natural ventilation dehumidification, and temperature fluctuation. Moreover, the existing control equipment and energy equipment should be gradually upgraded. To make this feasible, researchers and designers should continue to investigate the unique characteristics of greenhouses and conduct interdisciplinary research that aims to maximize renewable energy.

Author Contributions: G.W.: conceptualization, writing of the photovoltaic utilization section, reviewing, and editing. H.F.: conceptualization, original idea, supervision, writing of definitions, and designing sections. Y.Z.: writing of the photothermal utilization section. K.L.: writing the north wall materials to improve the CSG thermal environment section. D.X.: polish, reviewing, and editing. All authors have read and agreed to the published version of the manuscript.

Funding: This work was supported by the Beijing Municipal Science and Technology Project (Z211100004621002), the Central Public-Interest Scientific Institution Basal Research Fund (BSRF202112, Y2021PT04), and the Agricultural Science and Technology Innovation Program (ASTIP).

Institutional Review Board Statement: Not applicable.

Informed Consent Statement: Consent to participate is not applicable. All authors consented to the publication of this manuscript.

Data Availability Statement: All data generated or analyzed during this study are included in this published article. Code availability Not applicable. Declarations.

Conflicts of Interest: The authors declare no competing interest.

Nomenclature

Abbreviations

AHS	Active solar heat storage–release
CSG	Chinese solar greenhouse
CPV/T	Concentrating photovoltaic/thermal
EVA	Ethylene vinyl acetate
OPV	Organic photovoltaic
PAR	Photosynthetically active radiation
PCM	Phase-change material
PVG	Photovoltaic greenhouse
MSC-DRT	Multi-surface solar air collector with double receiver tubes
NIR	Near-infrared spectrum

Symbols

$T(\lambda)$	Transmittance spectrum
$b_s(\lambda)$	AM1.5 spectrum irradiance
$a(\lambda)$	Average action spectrum

Greek Letters

λ	Wavelength

References

1. Evans, A. Focus on energy and water. *FloraCult. Int.* **2007**, *9*, 15–17.
2. Farfan, J.; Lohrmann, A.; Breyer, C. Integration of greenhouse agriculture to the energy infrastructure as an alimentary solution. *Renew. Sustain. Energy Rev.* **2019**, *110*, 368–377. [CrossRef]
3. Qi, F.; Zhou, X.Q.; Zhang, Y.F.; Li, Z. Development of world greenhouse equipment and technology and some implications to China. *Trans. CSAE* **2008**, *24*, 279–285. (In Chinese with English Abstract)
4. Stefani, L.; Zano, M.; Modesti, M.; Ugel, E.; Vox, G.; Scheltini, E. Super durable plastic film tested for greenhouses. *FlowerTECH* **2008**, *11*, 22–24.
5. Zhai, J. The first lagrest country of facility horticulture. *Science and Technology Daily*, 22 August 2017; 001. (In Chinese)
6. Dai, J.F.; Luo, W.H.; Li, Y.X.; Qiao, X.J.; Wang, C. A Microclumate model-based energy consumption prediction system for greenhouse heating. *Sci. Agric. Sin.* **2006**, *39*, 2313–2318. (In Chinese with English Abstract)
7. Chen, J.T.; Ma, Y.W.; Pang, Z.Z. A mathematical model of global solar radiation to select the optimal shape and orientation of the greenhouses in southern China. *Sol. Energy* **2020**, *205*, 380–389. [CrossRef]
8. Department of Agricultural Mechanization, Ministry of Agriculture and Rural Affairs of China. *National Agricultural Mechanization Statistical Yearbook*; China Machine Press: Beijing, China, 2016; p. 65. (In Chinese)
9. Li, Z.H.; Zhang, X.J.; Wu, Z.W.; Ding, X.M.; Du, L.Y.; Chen, S.Y. Thinking about development of the facility agriculture based on the integration of Jing-Jin-Ji. *J. Chin. Agric. Mech.* **2016**, *37*, 241–245. (In Chinese with English Abstract)
10. Tong, G.H.; Christopher, D.M.; Li, T.L.; Wang, T.L. Passive solar energy utilization: A review of cross-section building parameter selection for Chinese solar greenhouses. *Renew. Sustain. Energy Rev.* **2013**, *26*, 540–548. [CrossRef]
11. Liu, S.; Zhang, J.; Zhang, B.; Wang, J.; Yuan, Z.; Yang, Q.; Li, G. Experimental study of solar thermal storage for increasing the earth temperature of greenhouse. *Acta Enegriae Solaris Sin.* **2003**, *24*, 461–465. (In Chinese with English Abstract)

12. Wang, C.; Cheng, X.; Shuai, C.; Huang, F.; Zhang, P.; Zhou, M.; Li, R. Evaluation of energy and environmental performances of Solar Photovoltaic-based Targeted Poverty Alleviation Plants in China. *Energy Sustain. Dev.* **2020**, *56*, 73–87. [CrossRef]
13. Wu, J.; Ge, Z.; Han, S.; Xing, L.; Zhu, M.; Zhang, J.; Liu, J. Impacts of agricultural industrial agglomeration on China's agricultural energy efficiency: A spatial econometrics analysis. *J. Clean. Prod.* **2020**, *260*, 121011. [CrossRef]
14. Chen, D.S. Advance of the research on the architecture and environment of the Chinese energe-saving sunlight greenhouse. *Trans. CSAE* **1994**, *10*, 123–129. (In Chinese with English Abstract)
15. Wang, B.Z. Solar energy resource division in China. *Acta Energiae Solaris Sin.* **1983**, *4*, 221–228. (In Chinese with English Abstract)
16. *GB 50176-2016*; Code for Thermal Design of Civil Building. China Architecture and Building Press: Beijing, China, 2016. (In Chinese)
17. Wu, L.R.; Wang, J.M.; Liu, H.J.; Sun, X. Spatiotemporal variation of solar radiation and sunshine hours in Shaanxi province. *Bull. Soil Water Conserv.* **2010**, *30*, 212–214. (In Chinese with English Abstract)
18. Data Cloud Portal of Chinese Academy of Sciences. Resource Discipline Innovation Platform. Available online: http://www.data.ac.cn/table/tbc24 (accessed on 12 January 2023). (In Chinese).
19. Liu, X.A.; Fan, L.S.; Wang, Y.H.; Wang, Q.F.; Ren, C.Y.; Li, Z.Q. The calculation methods and distributive character of solar radiation in Liaoning province. *Resour. Sci.* **2002**, *24*, 82–87. (In Chinese with English Abstract)
20. Cossu, M.; Cossu, A.; Deligios, P.A.; Ledda, L.; Li, Z.; Fatnassi, H.; Poncet, C.; Yano, A. Assessment and comparison of the solar radiation distribution inside the main commercial photovoltaic greenhouse types in Europe. *Renew. Sustain. Energy Rev.* **2018**, *94*, 822–834. [CrossRef]
21. Fu, X.Q.; Zhou, Y.Z.; Sun, H.B.; Wang, Y. Park-level agricultural energy internet: Concept, characteristic and application Value. *Trans. CSAE* **2020**, *36*, 152–161. (In Chinese with English Abstract)
22. Zhang, Y.; Ni, X.Y.; Zhang, K.X.; Xu, Y.J. Cooling performance for tomato root zone with intelligent ecological planting matrix temperature control system driven by photovoltaic in greenhouse. *Trans. CSAE* **2020**, *36*, 212–219. (In Chinese with English Abstract)
23. Lin, K.H.; Huang, M.Y.; Huang, W.D.; Hsu, M.H.; Yang, Z.W. The effects of red, blue, and white light-emitting diodes on the growth, development, and edible quality of hydroponically grown lettuce(*Lactucasativa* L. var. capitata). *Sci. Hortic.* **2013**, *150*, 86–91. [CrossRef]
24. Marucci, A.; Monarca, D.; Cecchini, M.; Colantoni, A.; Manzo, A.; Cappuccini, A. The semitransparent photovoltaic films for mediterranean greenhouse: A new sustainable technology. *Math. Probl. Eng.* **2012**, *2012*, 451934. [CrossRef]
25. Zhao, X.; Zou, Z.R.; Xu, H.J.; Zhao, J.T.; Li, J. Effects of summer light environment on tomato growth in photovoltaic solar greenhouse. *J. Northwest A F Univ. (Nat. Sci. Ed.)* **2013**, *41*, 93–99. (In Chinese with English Abstract)
26. Ge, Z.W.; He, T.T.; Xiao, Y.Y.; Dai, J.B.; Jiang, X.F. Study on High energy efficiency photovoltaic facility agricultural system in tropical area of China. *MATEC Web Conf.* **2018**, *153*, 08003. [CrossRef]
27. Wang, D.; Liu, H.R.; Li, Y.H.; Zhou, G.Q.; Zhan, L.L.; Zhu, H.M.; Lu, X.H.; Chen, H.Z.; Li, C.Z. High-performance and eco-friendly semitransparent organic solar cells for greenhouse applications. *Joule* **2021**, *5*, 945–957. [CrossRef]
28. Emmott, C.J.M.; Rohr, J.A.; Campoy-Quiles, M.; Kirchartz, T.; Urbina, A.; Ekins-Daukes, N.J.; Nelson, J. Organic Photovoltaic Greenhouses: A Unique Application for Semitransparent PV? *Energy Environ. Sci.* **2015**, *4*, 1317–1328. [CrossRef]
29. Hosseini-Fashami, F.; Motevali, A.; Nabavi-Pelesaraei, A.; Hashemi, S.J.; Chau, K. Energy-Life cycle assessment on applying solar technologies for greenhouse strawberry production. *Renew. Sustain. Energy Rev.* **2019**, *116*, 109411. [CrossRef]
30. Dinesh, H.; Pearce, J.M. The potential of agrivoltaic systems. *Renew. Sustain. Energy Rev.* **2016**, *54*, 299–308. [CrossRef]
31. Marucci, A.; Cappuccini, A. Dynamic photovoltaic greenhouse: Energy efficiency in clear sky conditions. *Appl. Energy* **2016**, *170*, 362–376. [CrossRef]
32. Wang, T.; Wu, G.; Chen, J.; Cui, P.; Chen, Z.; Yan, Y.; Zhang, Y.; Li, M.; Niu, D.; Li, B.; et al. Integration of solar technology to modern greenhouse in China: Current status, challenges and prospect. *Renew. Sustain. Energy Rev.* **2017**, *70*, 1178–1188. [CrossRef]
33. Perez-Alonso, J.; Perez-Garcia, M.; Pasamontes-Romera, M.; Callejon-Ferre, A.J. Performance analysis and neural modelling of a greenhouse integrated photovoltaic system. *Renew. Sustain. Energy Rev.* **2012**, *16*, 4675–4685. [CrossRef]
34. Xue, J.L. Photovoltaic agriculture-New opportunity for photovoltaic applications in China. *Renew. Sustain. Energy Rev.* **2017**, *73*, 1–9. [CrossRef]
35. Li, C.S.; Wang, H.Y.; Miao, H.; Ye, B. The economic and social performance of integrated photovoltaic and agricultural greenhouses systems: Case study in China. *Appl. Energy* **2017**, *190*, 204–212. [CrossRef]
36. Gupta, R.; Tiwari, G.N.; Kumar, A.; Gupta, Y. Calculation of total solar fraction for different orientation of greenhouse using 3D-shadow analysis in Auto-CAD. *Energy Build.* **2012**, *47*, 27–34. [CrossRef]
37. Cossu, M.; Ledda, L.; Urracci, G.; Sirigu, A.; Cossu, A.; Murgia, L.; Pazzona, A.; Yano, A. An algorithm for the calculation of the light distribution in photovoltaic greenhouses. *Sol. Energy* **2017**, *141*, 38–48. [CrossRef]
38. Cossu, M.; Murgia, L.; Ledda, L.; Deligios, P.A.; Sirigu, A.; Chessa, F.; Pazzona, A. Solar radiation distribution inside a greenhouse with south-oriented photovoltaic roofs and effects on crop productivity. *Appl. Energy* **2014**, *133*, 89–100. [CrossRef]
39. Marucci, A.; Zambon, I.; Colantoni, A.; Monarca, D. A combination of agricultural and energy purposes: Evaluation of a phototype of photovoltaic greenhouse tunnel. *Renew. Sustain. Energy Rev.* **2018**, *82*, 1178–1186. [CrossRef]
40. Zan, J.Y.; Liu, Z.M.; Liao, H.; Zhang, J.G.; Li, S.Z.; Liu, X.P. Simulation research on photovoltaic greenhouse temperature. *J. Yunnan Norm. Univ.* **2014**, *34*, 42–47. (In Chinese with English Abstract)

41. Zhao, X.; Zou, Z.R. Preliminary study on power generation using photovoltaic solar greenhouse in winter. *J. Northwest AF Univ. (Nat. Sci. Ed.)* **2014**, *42*, 177–182. (In Chinese with English Abstract)

42. Liu, H.; Sheng, G.; Fu, Q.; Zhou, Y. The current applications, challenges and the future prospects for thin-film solar photovoltaic greenhouse in Hangzhou city. *J. Zhejiang Agr. Sci.* **2012**, *6*, 782–787. (In Chinese with English Abstract)

43. Fatnassi, H.; Poncet, C.; Bazzano, M.M.; Brun, R.; Bertin, N. A numerical simulation of the photovoltaic greenhouse microclimate. *Sol. Energy* **2015**, *120*, 575–584. [CrossRef]

44. Akira, Y.; Marco, C. Energy sustainable greenhouse crop cultivation using photovoltaic technologies. *Renew. Sustain. Energy Rev.* **2019**, *109*, 116–137.

45. Yano, A.; Kadowaki, M.; Furue, A.; Tamaki, N.; Tanaka, T.; Hiraki, E. Shading and electrical features of a photovoltaic array mounted inside the roof of an east–west oriented greenhouse. *Biosyst. Eng.* **2010**, *106*, 367–377. [CrossRef]

46. Kadowaki, M.; Yano, A.; Ishizu, F.; Tanaka, T.; Noda, S. Effects of greenhouse photovoltaic array shading on Welsh onion growth. *Biosyst. Eng.* **2012**, *111*, 290–297. [CrossRef]

47. Xu, G.Y.; Liu, M.C.; Li, W.; Xu, C.; Wu, L. Effect of electric boiler heating system for soil warming in solar greenhouse. *Chin. Agric. Sci. Bull.* **2011**, *27*, 171–174. (In Chinese with English Abstract)

48. Zhou, S.; Zhang, Y.; Yang, Q.C.; Cheng, R.; Zhou, B. Performance of active heat storage-release unit assisted with a heat pump in a new type of Chinese solar greenhouse. *Appl. Eng. Agric.* **2016**, *32*, 641–650.

49. Fang, H.; Yang, Q.C.; Zhang, Y.; Sun, W.T.; Liang, H. Performance of a solar heat collection and release system for improving night temperature in a Chinese solar greenhouse. *Appl. Eng. Agric.* **2015**, *31*, 283–289.

50. Guan, Y.; Chen, C.; Han, Y.Q.; Ling, H.S.; Yan, Q.Y. Experimental and modelling analysis of a three-layer wall with phase-change thermal storage in a Chinese solar greenhouse. *J. Build. Phys.* **2015**, *38*, 548–559. [CrossRef]

51. Li, X.L.; Tie, S.N.; Zhang, F.J. Thermal improvement of solar greenhouse on Qinghai-Tibet plateau by modified mirabilite phase change energy storage materials. *Bull. Chin. Ceram. Soc.* **2018**, *37*, 3646–3651. (In Chinese with English Abstract)

52. Li, Z.Y.; Ma, X.L.; Zhao, Y.S.; Zheng, H.F. Study on the performance of a curved Fresnel solar concentrated system with seasonal underground heat storage for the greenhouse application. *J. Sol. Energy Eng.* **2019**, *141*, 011004. [CrossRef]

53. Zhao, Y.L.; Yu, X.C.; Li, Y.S.; He, C.X.; Yan, Y. Application of electric carbon crystal soil-warming system for tomato production in greenhouse. *Trans. CSAE* **2013**, *29*, 131–138. (In Chinese with English Abstract)

54. Yuan, D.Z.; Liao, Y.C.; Wang, Y.F. Effects of straw bioreactor on growth environment and yield of cucumber in greenhouse. *J. Northwest AF Univ.* **2014**, *42*, 186–192. (In Chinese with English Abstract)

55. Bai, Y.K.; Chi, D.C.; Wang, T.L.; Zhao, D.; Ren, B.J. Experimental research of heating by fire-pit and underground heating exchange system in a solar greenhouse. *Trans. CSAE* **2006**, *22*, 178–181. (In Chinese with English Abstract)

56. Wang, W.; Zhang, J.S.; Wang, Y.B. The research progress on the structure and the properties of solar greenhouse walls in China. *J. Shanxi Agric. Sci.* **2015**, *43*, 496–498. (In Chinese with English Abstract)

57. Wen, X.Z.; Li, Y.L. Analysis of temperature within north composite wall of solar greenhouse. *J. Shanxi Agric. Univ. (Nat. Sci. Ed.)* **2009**, *6*, 525–528. (In Chinese with English Abstract)

58. Liu, P.P.; Yang, H.J.; Guan, Y.; Chen, C.; Hu, W.J. Verification and analysis of the solar greenhouse thermal environment simulation by Energyplus. *Build. Energy Effic.* **2016**, *44*, 60–64. (In Chinese with English Abstract)

59. Chen, D.S.; Zhang, H.S.; Liu, B.Z. Comprehensive Study on the Meteorological Environment of the Sunlight Greenhouse—I. Preliminary study on the thermal effect of the wall body and covering materials. *Trans. CSAE* **1990**, *6*, 1577–1583. (In Chinese with English Abstract)

60. Bai, Q.; Zhang, Y.H.; Sun, L.X. Analysis on heat storage layer and thickness of soil wall in solar greenhouse based on theory of temperature-wave transfer. *Trans. CSAE* **2016**, *32*, 207–213. (In Chinese with English Abstract)

61. Ren, X.; Cheng, J.; Xia, N.; Zhao, S.; Cui, W.; Ma, C.; Wang, Q. Study on the effect of natural convective hollow wall on thermal storage/release in solar greenhouse. *J. China Agric. Univ.* **2017**, *22*, 115–122. (In Chinese with English Abstract)

62. Zhao, S.; Zhuang, Y.; Zheng, K.; Ma, C.; Cheng, J.; Ma, C.; Chen, X.; Zhang, T. Thermal performance experiment on air convection heat storage wall with cavity in Chinese solar greenhouse. *Trans. CSAE* **2018**, *34*, 223–231. (In Chinese with English Abstract)

63. Zhang, Y.; Zou, Z.R. The innovative structure of the solar greenhouse on the back wall of the pebbles in non-cultivated land. *Agric. Eng. Technol. (Greenh. Hortic.)* **2015**, *35*, 28–30. (In Chinese with English Abstract)

64. Berroug, F.; Lakhal, E.K.; El Omari, M.; Faraji, M.; El Qarnia, H. Thermal performance of a greenhouse with a phase change material north wall. *Energy Build* **2011**, *43*, 3027–3035. [CrossRef]

65. Wang, Y.X.; Liu, S.; Wang, P.Z.; Shi, G.Y. Preparation and characterization of microencapsulated phase change materials for greenhouse application. *Trans. CSAM* **2016**, *47*, 348–358. (In Chinese with English Abstract)

66. Chen, S.Q.; Zhu, Y.P.; Chen, Y.; Liu, W. Usage strategy of phase change materials in plastic greenhouses, in hot summer and cold winter climate. *Appl. Energy* **2020**, *277*, 115416. [CrossRef]

67. Guan, Y.; Chen, C.; Ling, H.S.; Han, Y.Q.; Yan, Q.Y. Analysis of heat transfer properties of three-layer wall with phase-change heat storage in solar greenhouse. *Trans. CSAE* **2013**, *29*, 166–173. (In Chinese with English Abstract)

68. Shi, W.; Cheng, S.X. Temperature control effect of paraffin graphite composite phase change materials in greenhouse. *Bull. Chin. Ceram. Soc.* **2017**, *36*, 4112–4116. (In Chinese with English Abstract)

69. Zhang, L.M.; Zou, Z.R.; Lu, G.D.; Qiao, Z.W. The preparation of compound phase change material of greenhouse wall and analyzed by ANSYS. *Agric. Mech. Res.* **2008**, *4*, 158–160. (In Chinese with English Abstract)

70. Zhang, Q.; Wang, H.L.; Mi, X. Preparation and characterization of lauric-myristic-capric acid/expanded graphite form-shaped composite phase change material. *New Chem. Mater.* **2015**, *43*, 46–48. (In Chinese with English Abstract)
71. Cao, K.; Xu, H.J.; Zhang, R.; Xu, D.W.; Yan, L.L.; Sun, Y.C. Renewable and sustainable strategies for improving the thermal environment of Chinese solar greenhouses. *Energy Build.* **2019**, *202*, 109414. [CrossRef]
72. Chen, C.; Li, Z.; Guan, Y.; Han, Y.Q.; Ling, H.S. Effects of building methods on thermal properties of phase change heat storage composite for solar greenhouse. *Trans. CSAE* **2012**, *28* Suppl. S1, 186–191. (In Chinese with English Abstract)
73. Zhang, Y.; Zou, Z.R.; Li, J.M.; Hu, X.H. Preparation of the small concrete hollow block with PCM and its efficacy in greenhouses. *Trans. CSAE* **2010**, *26*, 263–267. (In Chinese with English Abstract)
74. Ma, Q.L.; Yang, Q.C.; Ke, X.L.; Zhang, Y. Performance of an active heat storage-release system for canopy warming in solar greenhouse. *J. Northwest AF Univ. (Nat. Sci. Ed.)* **2020**, *48*, 57–64. (In Chinese with English Abstract)
75. Lu, W.; Zhang, Y.; Fang, H.; Ke, X.L.; Yang, Q.C. Modelling and experimental verification of the thermal performance of an active solar heat storage-release system in a Chinese solar greenhouse. *Biosyst. Eng.* **2017**, *160*, 12–24. [CrossRef]
76. Xu, W.W.; Song, W.T.; Ma, C.W. Performance of a water-circulating solar heat collection and release system for greenhouse heating using an indoor collector constructed of hollow polycarbonate sheets. *J. Clean. Prod.* **2020**, *253*, 119918. [CrossRef]
77. Ma, C.; Jiang, Y.; Cheng, J.; Zhao, S.; Xia, N.; Wang, P.; Yang, P. Analysis and experiment of performance on water circulation system of steel pipe network formed by roof truss for heat collection and release in Chinese solar greenhouse. *Trans. CSAE* **2016**, *32*, 209–216. (In Chinese with English Abstract)
78. Onder, O. Use of solar assisted geothermal heat pump and small wind turbine systems for heating agricultural and residential buildings. *Energy* **2010**, *35*, 262–268.
79. Ling, H.S.; Chen, C.; Guan, Y.; Wei, S.; Chen, Z.G.; Li, N. Active heat storage characteristics of active-passive triple wall with phase change material. *Sol. Energy* **2014**, *110*, 276–285. [CrossRef]
80. Chen, C.; Han, F.; Mahkamov, K.; Wei, S.; Ma, X.; Ling, H.; Zhao, C. Numerical and experimental study of laboratory and full-scale prototypes of the novel solar multi-surface air collector with double-receiver tubes integrated into a greenhouse heating system. *Sol. Energy* **2020**, *202*, 86–103. [CrossRef]
81. Wu, G.; Yang, Q.C.; Zhang, Y.; Fang, H.; Feng, C.Q.; Zheng, H.F. Energy and optical analysis of photovoltaic thermal integrated with rotary linear curved Fresnel lens inside a Chinese solar greenhouse. *Energy* **2020**, *197*, 117215. [CrossRef]
82. Zhang, S.H.; Guo, Y.; Zhao, H.J.; Wang, Y.; Chow, D.; Fang, Y. Methodologies of control strategies for improving energy efficiency in agricultural greenhouses. *J. Clean. Prod.* **2020**, *274*, 122695. [CrossRef]
83. Gorjian, S.; Calise, F.; Kant, K.; Ahamed, M.S.; Copertaro, B.; Najafi, G.; Zhang, X.; Aghaei, M.; Shamshiri, R.R. A review on opportunities for implementation of solar energy technologies in agricultural greenhouses. *J. Clean. Prod.* **2020**, *285*, 124807. [CrossRef]
84. La Notte, L.; Giordano, L.; Calabrò, E.; Bedini, R.; Colla, G.; Puglisi, G.; Reale, A. Hybrid and organic photovoltaics for greenhouse applications. *Appl. Energy* **2020**, *278*, 115582. [CrossRef]
85. Huang, J.P.; Li, R.; He, P.; Dai, Y.J. Status and prospect of solar heat for industrial processes in China. *Renew. Sustain. Energy Rev.* **2018**, *90*, 475–489.
86. Golzar, F.; Heeren, N.; Hellweg, S.; Roshandel, R. A novel integrated framework to evaluate greenhouse energy demand and crop yield production. *Renew. Sustain. Energy Rev.* **2018**, *96*, 487–501. [CrossRef]
87. Huang, J.P.; Fan, J.H.; Furbo, S. Feasibility study on solar district heating in China. *Renew. Sustain. Energy Rev.* **2019**, *108*, 53–64. [CrossRef]
88. Sujata, N.; Tiwari, G.N. Energy and exergy analysis of photovoltaic/thermal integrated with a solar greenhouse. *Energy Build.* **2008**, *40*, 2015–2021.
89. Yano, A.; Furue, A.; Kadowaki, M.; Tanaka, T.; Hiraki, E.; Miyamoto, M.; Ishizu, F.; Noda, S. Electrical energy generated by photovoltaic modules mounted inside the roof of a north-south oriented greenhouse. *Biosyst. Eng.* **2009**, *103*, 228–238. [CrossRef]

MDPI
St. Alban-Anlage 66
4052 Basel
Switzerland
www.mdpi.com

Energies Editorial Office
E-mail: energies@mdpi.com
www.mdpi.com/journal/energies